CHEMISTRY PROBLEMS

the text of this book is printed on 100% recycled paper

The Author

Paul R. Frey received his doctor's degree in physical chemistry from Oregon State University in 1936. Since that time he has been associated with the chemistry department at Colorado State University. Dr. Frey also has held teaching positions at York College, Colorado University, and Oregon State University.

The author has written numerous articles based upon original research, principally in the fields of dipole moment values of organic compounds and the utilization of carotene in the animal body. His articles have been published in the *Journal of the American Chemical Society*, *American Journal of Physiology*, *Journal of Nutrition*, *Science*, *Journal of the American Veterinary Medical Association*, and the *Journal of the Colorado-Wyoming Academy of Science*. Professor Frey is the author of a standard textbook, *College Chemistry*. In addition, he is a member of the American Chemical Society, the Society for Experimental Biology and Medicine, the American Association for the Advancement of Science, the Colorado-Wyoming Academy of Science, Sigma Xi, Phi Kappa Phi, Phi Lambda Upsilon, and Sigma Pi Sigma.

CHEMISTRY PROBLEMS

AND HOW TO SOLVE THEM

PAUL R. FREY

Professor Emeritus of Chemistry
Colorado State University

Seventh Edition

BARNES & NOBLE BOOKS
A DIVISION OF HARPER & ROW, PUBLISHERS
New York, Hagerstown, San Francisco, London

L. C. catalogue card number: 68-55293

ISBN: 0-06-460046-7

Manufactured in the United States of America

80 12 11 10

Preface

Because of space limitations, most texts written for the first year course in college chemistry contain a minimal number of practice problems and worked examples. This book was written to make available to the student supplementary problems and a more detailed analysis of the solution of problems. The thorough understanding of the physical and chemical principles involved in a problem is essential in order to apply intelligently the mathematics used in the solution of the problem. Consistent use is made of dimensions and the rules applying to significant digits. Each type of problem is presented as an independent unit, which makes the order of study of the problems immaterial. Answers are given for all problems.

New material has been added to be consistent with present course contents in the first year course in chemistry. Problems of stoichiometry are solved by means of mole ratios obtained from equations. Dimensional analysis is used consistently in the solution of problems. Sections have been added on algebraic equations, proportionality, the trigonometric functions, molecular geometry, the principal thermodynamic functions, Gibbs free energy, bond energies, the theory of acid-base titration, complex ion equilibria, galvanic cells, standard electrode potentials, and the Nernst equation. More sophisticated problems have been added to most of the chapters. Three new chapters have been added. Chapter 3 deals with the graphical treatment of experimental data, Chapter 4 with the statistical treatment of experimental data, and Chapter 5 with crystal formation, crystal geometry, and diffraction in crystals.

I wish to express my appreciation to Dr. Roy A. Keller for his suggestions and critical review of the sixth edition of *Chemistry Problems*.

The author is greatly indebted to the many coworkers who submitted criticisms of the earlier editions of the book, and to Dr. Gladys Walterhouse and Mr. H. John Fisher for their many constructive suggestions as editors of this and previous editions.

Colorado State University PAUL R. FREY

Table of Contents

1

Review of Mathematics

Chemical and physical changes obey definite natural laws, most of which can be expressed as a mathematical formula. Problems involving mathematics thus become an essential part of a course in chemistry. Chemistry problems consist of two essential parts: (1) the information given in the problem, and (2) what you are to find. Your ability to work chemistry problems will depend upon the care with which you study the information given in the problem and the steps you outline for the solution of the problem.

A brief review of mathematics follows.

Fractions

1.1. Simple Fractions. A *simple fraction* consists of a ratio of two integers, as $\frac{3}{5}$. In a *proper fraction* the numerator is less than the denominator; in an *improper fraction* the numerator is greater than the denominator, as $\frac{7}{3}$.

When one is *adding* or *subtracting* simple fractions the quantities involved must be reduced to a *common denominator.*

Example 1.1. Find the value of x in the expression:

$$x = \tfrac{3}{4} - 3\tfrac{1}{6} + 8\tfrac{5}{12}.$$

Solution. The common denominator is 12. That is, 12 is divisible by 4, 6, and 12. Therefore:

$$x = \frac{3}{4} - 3\tfrac{1}{6} + 8\tfrac{5}{12} = \frac{3}{4} - \frac{19}{6} + \frac{101}{12} = \frac{9 - 38 + 101}{12} = 6.$$

When *multiplying* simple fractions, multiply the numerators for a new numerator, and the denominators for a new denominator; then reduce the resulting fraction to its *lowest terms.*

Example 1.2. Find the value of x in the expression:

$$x = 16 \times \tfrac{3}{4} \times \tfrac{5}{6}.$$

Solution.

$$x = \frac{16}{1} \times \frac{3}{4} \times \frac{5}{6} = \frac{16 \times 3 \times 5}{4 \times 6} = 10.$$

When *dividing* one fraction by another, *invert* the divisor and multiply.

Example 1.3. Find the value of x in the expression:

$$x = 6\tfrac{1}{2} \div \tfrac{3}{4}.$$

Solution.

$$x = 6\tfrac{1}{2} \div \frac{3}{4} = 6\tfrac{1}{2} \times \frac{4}{3} = \frac{13 \times 4}{2 \times 3} = 8\tfrac{2}{3}.$$

Only *common factors* may be canceled when they appear in both the numerator and denominator of a fraction.

Example 1.4. Find the value of x in the expression:

$$x = 36 \times \frac{18}{5 \times 18} \times \frac{24 + 6}{6}.$$

Solution.

$$x = 36 \times \frac{18}{5 \times 18} \times \frac{24 + 6}{6} = \frac{36 \times \cancel{18} \times (24 + 6)}{5 \times \cancel{18} \times 6} = \frac{36 \times \cancel{30}}{\cancel{30}} = 36.$$

Problems

1.1. Find the value of x.

a. $x = 3\tfrac{3}{4} + 5\tfrac{7}{8}$. *Ans.* $9\tfrac{5}{8}$.

b. $x = 2\tfrac{1}{3} \times 6\tfrac{1}{4}$. *Ans.* $14\tfrac{7}{12}$.

c. $x = 6\tfrac{5}{8} - 3\tfrac{3}{4}$. *Ans.* $2\tfrac{7}{8}$.

d. $x = 5\tfrac{1}{4} \div 1\tfrac{2}{3}$. *Ans.* $3\tfrac{3}{20}$.

e. $x = \left(\frac{231}{6} \times \frac{6}{11}\right) + \left(\frac{18 + 3}{3}\right)$. *Ans.* 28.

f. $x = \frac{760}{124} \times \frac{(272 + 18)}{580} \times \frac{744}{760}$. *Ans.* 3.

g. $x = \frac{3\frac{1}{3} \times 12\frac{3}{4}}{3} \div \tfrac{1}{6}$. *Ans.* 85.

1.2. Decimal Fractions. A *decimal fraction* is a *proper fraction* in which the denominator is some power of 10, usually signified by a decimal point, as $0.3 = \frac{3}{10}$ and $0.03 = \frac{3}{100}$.

When one is *adding* or *subtracting* decimal fractions, the decimal points must be in a vertical column.

Example 1.5. Find the value of x.

(a) $x = 16.34 + 176.00 + 3.41$, and (b) $x = 1.436 - 0.471$.

Solution.

(a)
$$\begin{array}{r} 16.34 \\ 176.00 \\ \underline{3.41} \\ x = 195.75 \end{array}$$

(b)
$$\begin{array}{r} 1.436 \\ \underline{0.471} \\ x = 0.965 \end{array}$$

The *product* of two decimal fractions contains as many decimal places as the *sum* of the decimal places in the two quantities multiplied.

Example 1.6. Solve for x: $x = 7.33 \times 4.7$.

Solution. $x = 7.\underline{33} \times 4.\underline{7} = 34.\underline{451}$.

The *quotient* involving two decimal fractions contains as many decimal places as there are places in the *dividend* minus the number of places in the *divisor*. Note: zeros added to the dividend must be counted as decimal places.

Example 1.7. Solve for x: $x = 258.98 \div 28.4$.

Solution. 28.4 | 258.980 | 9.12 That is, $x = 9.12$.

$$\begin{array}{r} \underline{255\ 6} \\ 3\ 38 \\ \underline{2\ 84} \\ 540 \\ \underline{568} \end{array}$$

To change a simple fraction to a decimal fraction, divide the numerator by the denominator. For example, $\frac{4}{5} = 0.8$.

Problems

1.2. Find the value of x.

a. $x = 1.634 + 0.880 + 15.223$. *Ans.* 17.737.

b. $x = 221.68 - 96.73$. *Ans.* 124.95.

c. $x = 3.48 \times 0.775$. *Ans.* 2.697.

d. $x = 3654 \div 311.1$. *Ans.* 11.75.

e. $x = 12.36 \div 54.7$. *Ans.* 0.226.

1.3. Change each of the following to the corresponding decimal fraction.

a. $\frac{13}{15}$ *Ans.* 0.87.

b. $3\frac{7}{8}$ *Ans.* 3.9.

1.4. Reduce $\dfrac{2.125}{0.27}$ to the simplest decimal fraction. *Ans.* 7.9.

1.5. Reduce 0.16 to the simplest common fraction. *Ans.* $\frac{4}{25}$.

1.6. Mental exercises.

a. 25×10

b. 33×100

c. 2.75×10

d. 125×1000

e. $32 \div 10$

f. $764 \div 100$

g. $61.7 \div 100$

h. $2174 \div 1000$

i. 0.089×100

j. $1.42 \div 1000$

Exponential Numbers

1.3. Introduction. In scientific work it is often necessary to use quite large numbers such as 602,000,000,000,000,000,000, or extremely small numbers such as 0.00000000048. A method whereby such numbers can be presented in a more condensed form will be given. For example, 10,000 may be written 10^4 where 4 is the *exponent* to the *base* 10. Table 1.1 gives a sequence of powers of 10. The table could be extended indefinitely.

Numbers other than 10 may be used as a base. For example, $6^2 = 36$ and $12^2 = 144$. The base 10 is the most convenient to use.

In Table 1.1 it will be observed that $10^0 = 1$. Any expression (other than zero) to the exponent zero is equal to one. That is, $x^0 = 1$ if $x \neq 0$. It is important to remember this fact since zero frequently occurs as an exponent in mathematical expressions. Also, reference to Table 1.1 will show that a quantity expressed in the negative exponential form may be written as the reciprocal of the number with the sign of the exponent changed. That is:

TABLE 1.1

SOME POWERS OF TEN

10^3	=	1000.
10^2	=	100.
10^1	=	10.
10^0	=	1.
10^{-1}	=	0.1
10^{-2}	=	0.01
10^{-3}	=	0.001

$$10^{-1} = \frac{1}{10}, \quad 10^{-2} = \frac{1}{10^2}, \quad 10^{-3} = \frac{1}{10^3}, \quad \text{or} \quad x^{-a} = \frac{1}{x^a}.$$

Any factor may be interchanged between the numerator and denominator of a fraction by changing the sign of the exponent.

1.4. Numbers Expressed as a Power of Ten. The question now arises as to how a number such as 600 could be expressed as a power of ten. Since $10^2 = 100$ and $10^3 = 1000$, it is evident that a fractional exponent having a value between 2 and 3 must be used to express 600 as a power of ten. That is, in the expression $10^x = 600$, x has a value lying between 2 and 3. Methods of finding such fractional exponents will be discussed in Sec. 1.8.

It is possible to express a number as the product of two numbers, one of which is an integer power of 10.

Example 1.8. Express each of the following as a product of two numbers, one of which is an integer power of ten.

(a) 600, (b) 0.006, and (c) 674,000.

Solution.

(a) $600 = 6 \times 100 = 6 \times 10^2$.

(b) $0.006 = \frac{6}{1000} = 6 \times \frac{1}{1000} = 6 \times \frac{1}{10^3} = 6 \times 10^{-3}$.

(c) $674{,}000 = 674 \times 1000 = 674 \times 10^3$

$= 67.4 \times 10{,}000 = 67.4 \times 10^4$

$= 6.74 \times 100{,}000 = 6.74 \times 10^5$

$= 0.674 \times 1{,}000{,}000 = 0.674 \times 10^6$.

In Example 1.8c the quantity 674,000 is expressed as a power of ten in four different ways, all of which are correct. It is purely a matter of choice as to which form of answer shall be used. The form most commonly used is the one containing one integer value to the left of the decimal place, which in the above example would be 6.74×10^5.

Problems

1.7. Express each of the following as the product of two numbers, one of which is an integer power of ten.

a. 22,400 *Ans.* 2.24×10^4.
b. 364 *Ans.* 3.64×10^2.
c. 0.364 *Ans.* 3.64×10^{-1}.
d. 0.000364 *Ans.* 3.64×10^{-4}.
e. 2004 *Ans.* 2.004×10^3.

1.8. Change each of the following to the nonexponential form.

a. 2.3×10^3 *Ans.* 2300.
b. 2.3×10^{-3} *Ans.* 0.0023.
c. 0.0076×10^4 *Ans.* 76.
d. 0.0076×10^{-2}. *Ans.* 0.000076.
e. $\frac{23.4}{10^3}$ *Ans.* 0.0234.

1.9. Mental exercises. Determine the value of x.

a. $10^3 = x$.
b. $10^x = 10{,}000$.
c. $10^x = 0.0001$.
d. $10^{-4} = x$.

1.5. Addition and Subtraction of Exponential Quantities. In order to add or subtract exponential numbers, each number must be expressed to the same *power* and referred to the same *base*. For example, $10^3 + 10^3 = 2 \times 10^3$.

The above limitations greatly restrict the use of exponents in the operations of addition and subtraction. In fact, exponential expressions are seldom used in such mathematical operations.

1.6. Multiplication and Division of Exponential Quantities. In order to multiply or divide exponential quantities, each number needs only to be referred to the same base. Exponential quantities will therefore be found quite useful in simplifying the operations of multiplication and division. In multiplication the exponents are *added;* in division the exponent of the divisor is *subtracted* from that of the dividend.

Example 1.9. Find the value of x in each of the following:

(a) $x = 10^2 \times 10^3$.

(b) $x = 10^7 \div 10^3$.

(c) $x = 3200 \times 75{,}000$.

(d) $x = 0.0035 \div 0.00078$.

(e) $x = \dfrac{0.088 \times 760 \times 220{,}000}{4300}$.

Solution.

(a) $x = 10^2 \times 10^3 = 10^{2+3} = 10^5$.

(b) $x = \dfrac{10^7}{10^3} = 10^{7-3} = 10^4$.

(c)
$$\begin{aligned} x &= 3200 \times 75{,}000 \\ &= (3.2 \times 10^3)(7.5 \times 10^4) \\ &= (3.2 \times 7.5)(10^3 \times 10^4) \\ &= 24 \times 10^7. \end{aligned}$$

Observe that rearranging the order in which a series of quantities are multiplied does not affect the answer.

(d) $x = \dfrac{0.0035}{0.00078} = \dfrac{3.5 \times 10^{-3}}{7.8 \times 10^{-4}} = 0.45 \times 10^{-3-(-4)} = 0.45 \times 10 = 4.5.$

(e)
$$\begin{aligned} x &= \frac{0.088 \times 760 \times 220{,}000}{4300} \\ &= \frac{(8.8 \times 10^{-2})(7.6 \times 10^2)(2.2 \times 10^5)}{(4.3 \times 10^3)} \\ &= \left(\frac{8.8 \times 7.6 \times 2.2}{4.3}\right)\left(\frac{10^{-2} \times 10^2 \times 10^5}{10^3}\right) \\ &= \frac{147}{4.3} \times 10^{-2+2+5-3} \\ &= 34 \times 10^2. \end{aligned}$$

By means of the following rule it is possible to determine the *numerical value* and *sign* of an exponent when expressing a number as a power of ten. **Rule:** A shift of the decimal point to the *right* requires the use of a *negative* exponent; to the *left*, a *positive* exponent. In either case the exponent is equal numerically to the number of places the decimal point has been moved. For example:

$$\underset{\rightarrow}{0.088} = 8.8 \times 10^{-2}, \quad \text{and} \quad \underset{\leftarrow}{760.} = 7.6 \times 10^2.$$

Problems

1.10. Evaluate x in each of the following as a power of ten.

a. $x = 22{,}400 \times 250$. *Ans.* 5.6×10^6.

b. $x = 1.48 \times 0.0095$. *Ans.* 1.41×10^{-2}.

c. $x = \dfrac{73{,}000 \times 0.00075}{170}$. *Ans.* 3.22×10^{-1}.

d. $x = 380 \times \frac{810}{760} \times \frac{300}{270}$. *Ans.* 4.5×10^2.
e. $x = 72{,}400 \div 340$. *Ans.* 2.13×10^2.
f. $x = 340 \div 72{,}400$. *Ans.* 4.70×10^{-3}.
g. $x = 0.027 \div 0.009$. *Ans.* 3.
h. $x = 2.68 \div 0.044$. *Ans.* 6.09×10.
i. $x = (10^3)^2$. *Ans.* 10^6.
j. $x = (10^{-3})^2$. *Ans.* 10^{-6}.

1.7. Approximation of Answers. In problems involving the successive multiplication and division of numbers, there is always the possibility of misplacing the decimal point in the answer. In such problems one may determine by inspection whether or not the answer is of the proper order of magnitude. This method involves what is termed "approximation of numbers," in which each number other than the power of ten is changed to the nearest integer value. For example, a number such as 7.65×10 becomes 8×10 as an approximation, and 1.25×10^2 becomes 1×10^2.

Example 1.10. Determine by the method of approximation the location of the decimal point in the following answer.

$$76.5 \times \frac{125}{2000} \times \frac{1600}{275} = 278.$$

Solution.

$$\frac{(7.65 \times 10)(1.25 \times 10^2)(1.6 \times 10^3)}{(2 \times 10^3)(2.75 \times 10^2)} = \frac{(8 \times 1 \times 2)(10^6)}{(2 \times 3)(10^5)} = \frac{16}{6} \times 10 = 30.$$

That is, the answer is of the order of magnitude of 30. The correct answer is therefore 27.8. With practice one can readily acquire the ability to carry out most of the above operations mentally.

Problems

1.11. By the method of approximation determine the location of the decimal point in each of the following.

a. $\dfrac{2.76 \times 1300}{230} = 1560.$ *Ans.* 15.60.

b. $\dfrac{1600 \times 4.6}{486 \times 76.5} = 1980.$ *Ans.* 0.1980.

c. $38 \times \frac{735}{760} \times \frac{300}{273} = 4038.$ *Ans.* 40.38.

d. $\dfrac{341 \times 0.00782}{1742 \times 3.07} = 4985.$ *Ans.* 4.985×10^{-4}.

1.8. Fractional Exponents. Reference was made to fractional exponents in Sec. 1.4.

The expression $x^{\frac{a}{b}}$ is the general form for a quantity containing a fractional exponent. Fractional exponents may be written using the root sign, $\sqrt{\ }$. That is, $x^{\frac{a}{b}} = \sqrt[b]{x^a}$. Specific examples are, $100^{\frac{1}{2}} = \sqrt{100} = 10$; $1000^{\frac{1}{3}} = \sqrt[3]{1000} = 10$; and $10^{\frac{2}{3}} = \sqrt[3]{10^2} = 4.64$.

The square root of a quantity is obtained by dividing the exponent by two; the cube root is obtained by dividing the exponent by three.

Example 1.11. Evaluate x in each of the following:

(a) $x = \sqrt{10^4}$.
(b) $x = \sqrt[3]{10^{-9}}$.
(c) $x = \sqrt{900}$.
(d) $x = \sqrt{80{,}000}$.
(e) $x = \sqrt{0.0001}$.
(f) $x = \sqrt[3]{0.000089}$.

Solution.

(a) $x = \sqrt{10^4} = 10^{\frac{4}{2}} = 10^2 = 100$.
(b) $x = \sqrt[3]{10^{-9}} = 10^{-\frac{9}{3}} = 10^{-3} = 0.001$.
(c) $x = \sqrt{900} = \sqrt{9 \times 100} = \sqrt{9 \times 10^2} = \sqrt{9} \times \sqrt{10^2} = 3 \times 10 = 30$.
(d) $x = \sqrt{80{,}000} = \sqrt{8 \times 10^4} = \sqrt{8} \times 10^2 = 2.83 \times 10^2 = 283$.
(e) $x = \sqrt{0.0001} = \sqrt{10^{-4}} = 10^{-\frac{4}{2}} = 10^{-2} = 0.01$.
(f) $x = \sqrt[3]{0.000089} = \sqrt[3]{89 \times 10^{-6}} = \sqrt[3]{89} \times 10^{-2} = 4.5 \times 10^{-2}$.

In the above examples note that the quantity under the root sign has been expressed as a power of ten such that the exponent is divisible by the root. Logarithms would be required to obtain the value of a quantity such as $\sqrt[3]{89}$ with any degree of accuracy.

Problems

1.12. Evaluate x in each of the following:

a. $x = \sqrt{640{,}000}$. *Ans.* 8×10^2.
b. $x = 0.000064^{\frac{1}{2}}$. *Ans.* 8×10^{-3}.
c. $x = \sqrt{4 \times 10^8}$. *Ans.* 2×10^4.
d. $x = (64 \times 10^{-9})^{\frac{1}{3}}$. *Ans.* 4×10^{-3}.
e. $x = \sqrt{(4 \times 10^3)(16 \times 10^5)}$. *Ans.* 8×10^4.
f. $x = \sqrt{16 \times 10^{-6}} \times \sqrt{9 \times 10^4}$. *Ans.* 1.2.
g. $x = \sqrt{22{,}400}$. *Ans.* 1.50×10^2.
h. $x = (22{,}400)^{\frac{1}{3}}$. *Ans.* 2.82×10.
i. $x = \sqrt[3]{0.0000042}$. *Ans.* 1.61×10^{-2}.
j. $x = (602{,}000{,}000)^{\frac{1}{4}}$. *Ans.* 1.57×10^2.
k. $x = \sqrt{(248)^2}$. *Ans.* 248.
l. $x = \sqrt{(10{,}000)^{\frac{1}{2}}}$. *Ans.* 10.

1.9. Logarithms. In Example 1.11, and in the accompanying problems, note that the quantity under the root sign has been expressed as a power of 10 such that the exponent is divisible by the root. In solving exponential quantities that are not integer powers of 10 we must resort to the use of *logarithms*.

In the expression $a^x = y$, x is the logarithm of y to the base a, where a must be a positive number other than one. A *logarithm* is therefore an exponent and, as such, follows the rules applying to exponents (Secs. 1.3 to 1.7). The discussion which follows will be limited to the use of the base 10.

A logarithm is divided into two parts, the integer part called the *characteristic*, and the decimal fraction called the *mantissa.* Remember: the mantissa of a logarithm is always positive, whereas the characteristic may be either positive or negative. The following table shows some important principles relating to logarithms.

$$\begin{aligned}
\log 635 &= \log(6.35 \times 10^2) = 2.8028\\
\log 63.5 &= \log(6.35 \times 10^1) = 1.8028\\
\log 6.35 &= \log(6.35 \times 10^0) = 0.8028\\
\log 0.635 &= \log(6.35 \times 10^{-1}) = \bar{1}.8028\\
\log 0.0635 &= \log(6.35 \times 10^{-2}) = \bar{2}.8028
\end{aligned}$$

The table shows: (1) that the location of the decimal point determines the value of the characteristic, but does not affect the mantissa; (2) that to find the characteristic, the number must be expressed as the product of two other numbers, one of which is an integer power of 10, and the other a number containing but one digit to the left of the decimal point (Sec. 1.4); then the power of 10 is the value of the characteristic.

From the above we see that the characteristic of the logarithm may be determined by inspection. The mantissa values must be obtained from tables, called *logarithm tables* (Appendix IV).

Logarithms follow the rules of exponents. That is:

$$\begin{aligned}
\log AB &= \log A + \log B.\\
\log \frac{A}{B} &= \log A - \log B.\\
\log A^n &= n \log A.\\
\log \sqrt[n]{B} &= \frac{\log B}{n}.
\end{aligned}$$

Example 1.12. Multiply 235 by 86.

Solution. First find the characteristic of each number.

$$235 = 2.35 \times 10^2. \quad \text{Characteristic} = 2.$$
$$86 = 8.6 \times 10^1. \quad \text{Characteristic} = 1.$$

Next find the mantissa value for each. These values will be found in the table in Appendix IV. Add the logarithms. Then:

$$\begin{aligned} \log 235 &= 2.3711 \\ \log 86 &= 1.9345 \\ \text{add} \quad & 4.3056 \end{aligned}$$

Now find the number in the logarithm table in Appendix IV corresponding to the mantissa value 3056. The number is 2021. That is:

$$\text{antilogarithm } 4.3056 = 2.021 \times 10^4 = 20{,}210.$$

Example 1.13. Divide 4.86 by 0.096.

Solution. The logarithm of 0.096 must be subtracted from the logarithm of 4.86. Therefore:

$$\begin{aligned} \log 4.86 &= \log(4.86 \times 10^0) = 0.6866 \\ \log 0.096 &= \log(9.6 \times 10^{-2}) = \bar{2}.9823 \end{aligned}$$

When the characteristic is negative it is advisable to add and subtract 10 to and from the characteristic. Remember, the mantissa is always positive. Then:

$$\begin{aligned} \log 4.86 &= 10.6866 - 10 \\ \log 0.096 &= 8.9823 - 10 \\ \text{subtract} \quad & 1.7043 \end{aligned}$$

$$\text{antilogarithm } 1.7043 = 5.061 \times 10^1 = 50.61.$$

Example 1.14. Simplify the expression $\frac{273 \times 0.00783}{486}$

Solution. Subtract the logarithm of 486 from the sum of the logarithms of 273 and 0.00783. Then:

$$\begin{aligned} \log 273 &= 2.4362 \\ \log 0.00783 &= 7.8938 - 10 \\ \text{add} \quad & 10.3300 - 10 \\ \log 486 &= 2.6866 \\ \text{subtract} \quad & 7.6434 - 10 = \bar{3}.6434 \end{aligned}$$

$$\text{antilogarithm } \bar{3}.6434 = 4.399 \times 10^{-3} = 0.004399$$

Example 1.15. Evaluate 27.5^3.

Solution. Since $27.5^3 = 27.5 \times 27.5 \times 27.5$, then:

$$\log 27.5^3 = 3 \times \log 27.5 = 3(1.4393) = 4.3179$$
$$\text{antilogarithm } 4.3179 = 2.078 \times 10^4 = 20{,}780$$

The mantissa table in Appendix IV gives values for numbers containing three places only. By means of interpolation it is possible to calculate the mantissa of a number containing four places. For example, find the mantissa of 3264 from the table in Appendix IV. The mantissa of 3260 = 5132, and of 3270 = 5145. Evidently the mantissa of 3264 lies between the above values. Considering that there are ten proportional parts between 3260 and 3270, then 3264 is four proportional parts greater than 3260. The difference between the above mantissa values is 5145 − 5132 = 13. In the column marked proportional parts, for a difference of 13 in the mantissa values, four proportional parts require a correction of 5 to be added to the lower value 5132. That is:

$$\log 3264 = 4.5132 + 0.0005 = 4.5137$$

Similarly, it is possible to determine the antilogarithm of a number to four places.

Problems

1.13. Perform the indicated operations using logarithms.

a. 782×74. *Ans.* 57870.

b. $7340 \div 29$. *Ans.* 253.

c. 0.0756×0.34. *Ans.* 0.0257.

d. $0.314 \div 2.14$. *Ans.* 0.1467.

1.14. Simplify the expression:

$326 \times \frac{273}{296} \times \frac{723}{760}$. *Ans.* 286.

1.15. Evaluate:

32614×78546. *Ans.* 2.562×10^9.

1.16. Evaluate:

a. $\sqrt{7645}$. *Ans.* 87.44.

b. $\sqrt[3]{8743}$. *Ans.* 20.60.

c. $\sqrt{0.000765}$. *Ans.* 2.770×10^{-2}.

d. $\sqrt[3]{0.00174}$. *Ans.* 0.1203.

1.17. Evaluate:

a. 76.3^2. *Ans.* 5822.

b. 15.2^3. *Ans.* 3512.

c. 0.176^2. *Ans.* 0.03098

d. 0.034^3. *Ans.* 3.93×10^{-5}.

Elementary Algebra

1.10. Algebraic Equations. Only coefficients of like factors may be added or subtracted. For example:

$$4x + 3x = 7x,$$

$$6x^2 - 2x^2 + x = 4x^2 + x,$$

and

$$3x^2 - 2x - 5x = 3x^2 - 7x.$$

The rules relating to exponential quantities, as presented previously, apply to algebraic equations. Exponents of like factors are *added* when the exponential quantities are *multiplied*, and *subtracted* when the quantities are *divided*. For example:

$$3x \times 2x^2 = 6x^3,$$

$$12x^2 \div 3x = 4x,$$

and

$$\frac{12x^2 \times x^2}{4x^3} = 3x.$$

The discussion which follows will be limited to the solution of problems involving only one unknown, usually represented by x.

Example 1.16. Solve for x.

$$36x = 756 + 8x.$$

Solution.

$36x = 756 + 8x$. Collecting like terms gives

$$36x - 8x = 756$$

or

$$28x = 756$$

and

$$x = 27.$$

Example 1.17. Solve for x.

$$9x^2 + 5x - 8 = 0.$$

Solution. This is a *quadratic* equation of the general form

$$ax^2 + bx + c = 0,$$

and may be solved for x by substituting the values of the constants a, b, and c in the formula:

$$x = \frac{-b \pm \sqrt{b^2 - 4ac}}{2a}.$$

Then

$$x = \frac{-5 \pm \sqrt{5^2 - (4)(9)(-8)}}{(2)(9)}$$

$$= 0.7 \quad \text{or} \quad -1.3.$$

Problems

1.18. Determine the value of x in each of the following expressions.

a. $3x - 9 = 0$. *Ans.* $x = 3$.

b. $\frac{26 - x}{x} = 12.$ *Ans.* $x = 2.$

c. $(560)(43 - 21x) = (150)(42.9).$ *Ans.* $x = 1.5.$

d. $2x^2 - 3x - 50 = 0.$ *Ans.* $x = 5.8$ or $-4.3.$

1.11. Proportionality. Proportionality involves the relationships between variables. For example, the distance, d, a car will travel in a given period of time, t, depends upon its velocity, v. In this example d, v, and t are variables, since they may assume a succession of values. With related variables it is possible to derive a mathematical formula showing their relationship.

Two types of proportionality will be discussed. (1) direct proportionality, and (2) inverse proportionality.

(1) *Direct proportionality.* In the above example we see that the distance the car travels at a constant velocity varies *directly* with the time the car is in motion. That is:

$$\mathrm{d} \propto \mathrm{t} \ (\mathrm{v\ constant}) \quad \text{or} \quad \mathrm{d} = k\mathrm{t}$$

where k is called the *proportionality constant*, which, in this case, is velocity. That is:

$$\mathrm{d} = \mathrm{vt} \quad \text{or} \quad \mathrm{v} = \frac{\mathrm{d}}{\mathrm{t}}.$$

The quantity $\frac{\mathrm{d}}{\mathrm{t}}$ is a *ratio.* An equality between two such related ratios is called a *proportion.*

There are many natural proportionality constants. One such constant is π, which represents the ratio between the circumference, C, of a circle, and its diameter, D. That is:

$$\pi = \frac{\mathrm{C}}{\mathrm{D}} = 3.1416\ldots\ldots$$

Direct proportionality between variables may be as a power, root, log (etc.) of one or more of the variables. The area of a circle, A, varies directly as the square of the radius, r. That is:

$$\mathrm{A} \propto \mathrm{r}^2, \quad \text{or} \quad \mathrm{A} = k\mathrm{r}^2,$$

where k is again equal to 3.1416. . . . . .

(2) *Inverse proportionality.* The word *inverse* means opposite or inverted. An inverse proportionality between two variables indicates that as the numerical value of one increases, the numerical value of the other decreases, or visa versa. This inverse relationship may be as the power, log, root (etc.) of one or more of the variables. For

example, a car is required to cover a stated distance, d, at a velocity, v, in time, t. With d constant, how are v and t related? Evidently the greater the velocity, v, the shorter the time, t. That is:

$$t \propto \frac{1}{v} \quad \text{or} \quad t = k\frac{1}{v}.$$

The proportionality constant in this case is the distance, d. Therefore, $t = \frac{d}{v}$.

Problems

1.19. A sample of brass weighing 4.55 lb contained 3.18 lb of copper and 1.37 lb of zinc. How much copper would there be in 500 lb of the brass? *Ans.* 349 lb.

1.20. A 12 ft rod of iron weighed 33.4 lb. What would be the weight of a section of the rod measuring 4 ft 7 in, assuming the rod to be of uniform cross section? *Ans.* 12.8 lb.

1.21. If seven-tenths of a ton of coal costs $11.25, what will 3.5 tons cost? *Ans.* $56.25.

1.22. What is the area of a circle whose radius is 6.0 in.? *Ans.* 113 in.^{2}

1.23. What is the volume of a sphere whose radius is 6.0 in.? *Note:* The volume of a sphere varies as the cube of the radius, and $k = \frac{4}{3}\pi$. *Ans.* 905 in.^{3}

1.24. The surface area of a sphere varies as the square of the radius, and $k = 4\pi$. What is the surface area of a sphere 12.0 in. in diameter? *Ans.* 453 in.^{2}

1.25. Given the formula K.E. = $\frac{1}{2}mv^2$, where K.E. represents the kinetic energy in ergs of a mass of m grams moving with a velocity of v cm/sec, express the proportionality relationship between the variables: (a) K.E. and v, (b) K.E. and m, and (c) m and v.

1.12. Percentage. Percentage is a *ratio* indicating the number of parts out of 100. That is, 5% means five parts out of 100, or $\frac{5}{100}$. A common fraction represents a ratio, and a decimal fraction a given portion of a unit value.

Example 1.18. Express each of the following as percentage: (a) $\frac{11}{36}$, and (b) 0.24.

Solution.

(a) Since percentage is parts per 100, then:

$$\frac{11}{36} \times 100 = 31\%.$$

(b) Since a decimal fraction represents parts out of one, then:

$$0.24 \times 100 = 24\%.$$

From the above examples we see that both common fractions and decimal fractions may be expressed as percentages by multiplying the fractions by 100.

Problems

1.26. Convert each of the following to percentage.

a. 0.74 *Ans.* 74%.
b. 0.01 *Ans.* 1%.
c. $\frac{6}{100}$ *Ans.* 6%.
d. $\frac{23}{25}$ *Ans.* 92%.
e. $1\frac{3}{4}$ *Ans.* 175%

1.27. Convert each of the following to the corresponding decimal fraction

a. 16% *Ans.* $\frac{4}{25}$.
b. 1.5% *Ans.* $\frac{3}{200}$.
c. 0.3% *Ans.* $\frac{3}{1000}$.
d. $6\frac{3}{4}\%$ *Ans.* $\frac{27}{400}$.

1.28. What is 15% of $1500? *Ans.* $225.

1.29. What is 2.75% of $1600? *Ans.* $44.

1.30. What is 0.25% of $2000? *Ans.* $5.

1.31. A sample of brass contains 65.0 per cent copper and 35.0 per cent zinc How much copper and how much zinc would there be in 140 pounds of the brass? *Ans.* 91 lb copper, 49 lb zinc.

1.32. An average sample of sea water contains $5 \times 10^{-6}\%$ bromine. How many tons of sea water would be required in order to obtain one ton of bromine? *Ans.* 2×10^7 tons.

1.33. An average sample of sea water contains 3.5% dissolved solids. How many pounds of sea water would be required to obtain one pound of dissolved solids by evaporation of the water? *Ans.* 28.6 lb.

1.34. Stainless steel consists of approximately 74% iron and 26% chromium. How much iron and how much chromium would be required to prepare 250 tons of stainless steel? *Ans.* 185 tons iron, 65 tons chromium.

Trigonometry

1.13. The Trigonometric Functions. Trigonometry is that branch of mathematics which deals with the magnitudes of the sides and angles of right angle triangles. In Fig. 1.1 a circle is inscribed such that its center coincides with the intersection of the x- and y-axes at the origin, O. This divides the circle into the four quadrants, I, II, III, and IV. The right angle triangle is in quadrant I, forming the angle Θ

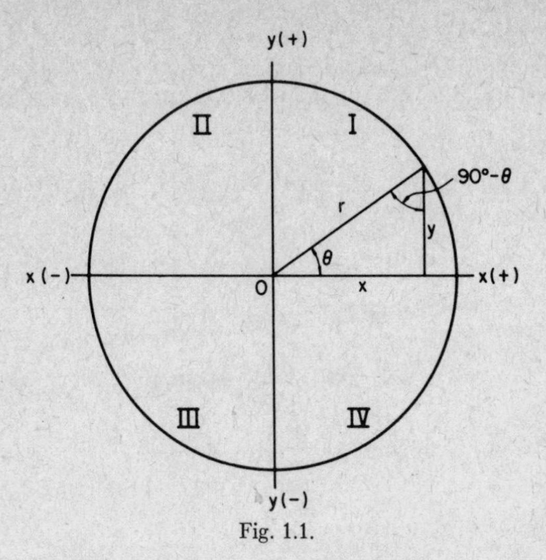

Fig. 1.1.

with the x-axis. The three sides of the triangle (x, y, and r) may be related by six ratios called the *trigonometric functions*.

From Fig. 1.1 we see that x and y may have positive or negative values. The signs, + or −, of the trigonometric functions will therefore be determined by the quadrant in which Θ lies. In quadrant I the six functions are all positive (Table 1.2).

TABLE 1.2
THE SIX TRIGONOMETRIC FUNCTIONS

Name	*Abbreviation*	*Functions in Terms of x, y, and r*
Sine	sin	$\sin = \frac{y}{r}$
Cosine	cos	$\cos = \frac{x}{r}$
Tangent	tan	$\tan = \frac{y}{x}$
Cotangent	cot	$\cot = \frac{x}{y}$
Secant	sec	$\sec = \frac{r}{x}$
Cosecant	csc	$\csc = \frac{r}{y}$

It is evident that for each value of Θ there will be a numerical value for each of the six trigonometric functions. These values have been calculated and put into tables called *Natural Trigonometric Functions* (Appendix I).

Example 1.19. What is the numerical value of each of the six trigonometric functions when $\Theta = 45°$?

Solution. When $\Theta = 45°$, then $x = 1$, $y = 1$, and $r = \sqrt{1^2 + 1^2} = \sqrt{2}$

Then:

$$\sin 45° = \frac{1}{\sqrt{2}} = 0.7071$$

$$\cos 45° = \frac{1}{\sqrt{2}} = 0.7071$$

$$\tan 45° = \frac{1}{1} = 1.0000$$

$$\cot 45° = \frac{1}{1} = 1.0000$$

$$\sec 45° = \frac{\sqrt{2}}{1} = 1.4142$$

$$\csc 45° = \frac{\sqrt{2}}{1} = 1.4142.$$

Example 1.20. Calculate (a) the values of the sine, tangent, and secant, and (b) the value of Θ, for the triangle in Fig. 1.1, when $x = 36.0$ cm and $y = 9.00$ cm.

Solution. First find the value of r.

(a) $r = \sqrt{36.0^2 + 9.00^2} = 37.1$ cm.

And:

$$\sin \Theta = \frac{9.00}{37.1} = 0.243$$

$$\tan \Theta = \frac{9.00}{36.0} = 0.250$$

$$\sec \Theta = \frac{37.1}{36.0} = 1.031$$

(b) Reference to Appendix I shows that Θ must have a value between 14° and 15° in order that $\tan \Theta = 0.250$. By interpolation we find that Θ is equal to approximately 14°3′.

Appendix I gives values for 0° to 45°. The following relationships hold when Θ is greater than 45°.

$$\sin \Theta = \cos (90° - \Theta)$$
$$\tan \Theta = \cot (90° - \Theta)$$
$$\sin \Theta = \sin (180° - \Theta)$$
$$\tan \Theta = -\tan (180° - \Theta)$$

Other relationships may be found by reference to a standard book on trigonometry.

1.14. Molecular Geometry. In dealing with molecules it is very seldom that spatial relationships involve right angles. However, any triangle may be resolved into two right triangles by drawing a line from the appropriate angle perpendicular to the opposite side.

Example 1.21. The three atoms in a molecule of water, H_2O, exist at the corners of an isosceles triangle. The distance between each hydrogen atom and the oxygen atom has been found to be 0.96 A, while the distance between the two hydrogen atoms is 1.52 A. What is the bond angle at the oxygen atom in a molecule of water?

Solution. By drawing a line from the oxygen atom perpendicular to the line joining the hydrogen atoms, two right triangles of equal size will have been formed. The base of each of these triangles will be:

$$\frac{1.52}{2} = 0.76 \text{ A}.$$

Let A be the angle at the oxygen atom. Then:

$$\sin A = \frac{0.76}{0.96} = 0.79.$$

From Appendix I we see that $A = 52°$ (approximately). The bond angle at the oxygen atom would be $2 \times 52° = 104°$.

Many molecules are unsymmetrical and, furthermore, do not lie in one plane. In such cases we must resort to other relationships among the trigonometric functions. There are many such relationships, two of which are shown:

$$a^2 = b^2 + c^2 - 2bc \cos A, \text{ and } \frac{a}{\sin A} = \frac{b}{\sin B} = \frac{c}{\sin C},$$

where A, B, and C are the angles of the triangle, and a, b, and c the respective sides opposite the angles.

Example 1.22. Three atoms, A, B, and C, are in contact forming a solid triangular molecule. By connecting the centers of the atoms we form the triangle ABC. Given the following radii of the atoms, $A = 1.75$ A, $B = 1.40$ A, and $C = 2.35$ A, calculate the bond angles at points A, B, and C.

Solution. Draw a diagram of the triangle formed by connecting the centers of the three atoms. The lengths of the sides of the triangle may then be calculated.

$$a = 2.35 \text{ A} + 1.40 \text{ A} = 3.75 \text{ A}.$$
$$b = 2.35 \text{ A} + 1.75 \text{ A} = 4.10 \text{ A}.$$
$$c = 1.75 \text{ A} + 1.40 \text{ A} = 3.15 \text{ A}.$$

Since $a^2 = b^2 + c^2 - 2bc \cos A$, then:

$$\cos A = \frac{b^2 + c^2 - a^2}{2bc}$$

$$= \frac{(4.10)^2 + (3.15)^2 - (3.75)^2}{(2)(4.10)(3.15)} = 0.486,$$

and the angle $A = 61°$.

To obtain angle B we use the relationship that:

$$\frac{a}{\sin A} = \frac{b}{\sin B} \quad \text{or} \quad \frac{3.75}{0.875} = \frac{4.10}{\sin B}$$

and $\sin B = 0.957$. The angle $B = 73°$.

Since the three angles of a triangle are equal to 180°, then:

$$\text{Angle } C = 180° - (61° + 73°) = 46°.$$

Problems

1.35. From Appendix I determine the sine and tangent values for each of the following angles. (a) 35°, (b) 130°, (c) 225°, and (d) 300°.

1.36. The H-S distance in an H_2S molecule is 1.35 A, and the H-H distance is 1.96 A. What is the bond angle at the sulfur atom?

Ans. Approximately 93°.

1.37. A certain molecule consists of three atoms in contact. When the centers of the atoms are connected, an unsymmetrical triangle ABC is formed. Given that the angle at A is 140°, side b 1.25 A, and side c is 2.25 A, determine the length of side a. *Ans.* 3.31 A.

1.38. The N_2O_4 molecule is symmetrical, with the two nitrogen atoms bonded to each other and two oxygen atoms bonded to each nitrogen atom. The N-N distance is 1.65 A, and the angle NNO is 126°. What is the distance between the two oxygen atoms bonded to a nitrogen atom? *Ans.* 1.90 A.

1.39. Assume the benzene molecule, C_6H_6, to be symmetrical and in one plane, with the six carbon atoms forming a regular hexagon. The C-C bond distance is 1.40 A. What is the distance between (a) any 1 and 3 carbon atoms, and (b) any 1 and 4 carbon atoms?

Ans. (a) 1.21 A, and (b) 2.80 A.

2

Units of Measurement

The metric system of measure and the centigrade scale of temperature are used almost exclusively in chemistry. Every student therefore must be familiar with these units of measure. It is for this reason that a survey is given at this time of the common metric units involving length, weight, and volume, and a comparison of the Fahrenheit, centigrade, and absolute scales of temperature. Table 2.4, p. 24, which gives the relationship between corresponding English and metric units, will be found helpful in visualizing the relative magnitude of the more commonly used metric units. The unit of time used universally is the second.

Most quantities consist of both a numerical value and a dimensional value. The statement that a beaker contains 215 ml of a liquid conveys two important ideas — the numerical value 215 and the dimensional value milliliters, both of which are necessary in order to express correctly the volume of liquid. In the solution of a problem both the numerical and the dimensional values may be treated mathematically. When this is done, the answer represents the correct numerical and dimensional values. The process of treating dimensions mathematically is called *dimensional analysis*.

Metric Units of Length, Mass, and Volume

2.1. Introduction. In any system of measurement there must be arbitrarily established units of *length*, *mass*, and *volume*. Examination of the apparatus used in laboratories will show that length is commonly measured in *meters*, mass in *grams*, and volume in *cubic centimeters* or *milliliters*. These are the fundamental units of length, mass, and volume in the metric system.

Metric units are based on a decimal system in which the prefix designates the multiple or submultiple value of the quantity in terms of the fundamental unit. See Table 2.1, p. 22.

The principal advantage of the metric system is that values of length, mass, and volume repeat in multiples of ten. English units are inconsistent — 12 in. equals one foot, 3 ft. equals one yard, and $5\frac{1}{2}$ yd. equals one rod. The metric system also represents a more international standard than does the English system. The meter, gram, and liter as used in the United States are identical to the corresponding units used in any other country. On the other hand, in England a quart is larger than the quart used in the United States, and in France an inch is longer than the inch used in England.

The *centimeter-gram-second* (cgs) system, called the metric system, and the *foot-pound-second* (fps) system, called the English system, are the two principal surviving standards in the world today. The unit of time, the second, is the same in each of the two systems.

TABLE 2.1

PREFIXES USED IN THE METRIC SYSTEM AND THEIR VALUES

Prefix	Value	
pico =	1.0×10^{-12}	part of fundamental unit
nano =	1.0×10^{-9}	
micro =	1.0×10^{-6}	
milli =	0.001	
centi =	0.01	
deci =	0.1	
	1.0	fundamental unit
deka =	10	times fundamental unit
hecto =	100	
kilo =	1000	
mega =	1.0×10^{6}	
giga =	1.0×10^{9}	
tera =	1.0×10^{12}	

2.2. The More Commonly Used Metric Units of Length. The fundamental unit of length in the metric system is the *meter* (m). Originally the meter was intended to represent one ten-millionth of the earth's quadrant. Recent measurements with more precise measuring instruments than used originally have shown the meter to be slightly in error. Rather than make the slight corrections necessary, the meter is defined as the distance between two parallel lines on a platinum-iridium bar kept in the International Bureau of Weights and Measures at Sèvres, France. Replicas of the bar are available to any country desiring them. The United States keeps all such standards in the National Bureau of Standards, Washington, D.C.

Recently scientists have established the length of the meter to be 1,650,763.73 times the wave length of the orange-red spectral line of krypton-86. This gives an absolute reference standard for the length of the meter that may be duplicated at any time.

In Tables 2.2 and 2.3, the more commonly used units are given in **boldface** type.

TABLE 2.2

METRIC UNITS OF LENGTH USING THE METER AS THE FUNDAMENTAL UNIT

1 **millimeter** (mm)	=	0.001 m.
1 **centimeter** (cm)	=	0.01 m.
1 decimeter (dm)	=	0.1 m.
1 **meter** (m)	=	1 m.
1 **kilometer** (km)	=	1000 m.

Three other units of length are used to express the size of extremely small particles such as atoms, and the magnitudes relating to extremely short waves such as light and X-rays. The units are the *micron* (μ), the *millimicron* (mμ), and the *angstrom* (A or Å).

$$1.00 \text{ A.} = 10^{-8} \text{ cm.}$$
$$1.00 \text{ m}\mu = 10^{-7} \text{ cm} = 10 \text{ A.}$$
$$1.00 \ \mu = 10^{-4} \text{ cm} = 10^{3} \text{ m}\mu = 10^{4} \text{ A.}$$

2.3. The More Commonly Used Metric Units of Mass. The fundamental unit of mass in the metric system is the *gram* (g). The gram was originally intended to represent the volume occupied by 1 cm^3 of water at 3.98° C, the temperature at which water possesses its greatest density. As with the meter, later measurements have shown that one gram of water at 3.98° C occupies a volume of 1.000027 $cm.^3$ It is only in extremely accurate work that this discrepancy in volume must be taken into consideration.

A direct relationship thus exists between the centimeter as a unit of length and the gram as a unit of mass. Table 2.3 gives metric units of mass in terms of decimal fractions of a gram because of the common practice of recording weights in this manner.

TABLE 2.3

METRIC UNITS OF MASS USING THE GRAM AS THE FUNDAMENTAL UNIT

1 milligram (mg.)	=	0.001 g
1 centigram (cg.)	=	0.01 g
1 decigram (dg.)	=	0.1 g
1 gram (g.)	=	1 g
1 kilogram (kg.)	=	1000 g.

2.4. The More Commonly Used Metric Units of Volume. Units of area and volume in the metric system are obtained in the same manner as are corresponding units in the English system.

The *cubic centimeter* (cc. or cm^3) is the metric unit of volume most commonly used in laboratory work. It is, however, a relatively small unit of volume. A larger unit of volume, the *liter* (*l*.), is defined as the

TABLE 2.4

CONSTANTS RELATING ENGLISH AND METRIC UNITS

A[1]		C		B
centimeters	×	10^8	=	angstroms
centimeters	×	10^7	=	millimicrons
centimeters	×	10^4	=	microns
cubic centimeters	×	2.64×10^{-4}	=	gallons
cubic inches	×	16.4	=	cubic centimeters
feet	×	30.5	=	centimeters
inches	×	2.54	=	centimeters
kilograms	×	2.20	=	pounds
kilometers	×	0.621	=	miles
liters	×	0.264	=	gallons
liters	×	61.0	=	cubic inches
liters	×	1.06	=	quarts
meters	×	1.09	=	yards
meters	×	3.28	=	feet
meters	×	39.4	=	inches
ounces	×	28.3	=	grams
pounds	×	454	=	grams
quarts	×	946	=	cubic centimeters
square inches	×	6.45	=	square centimeters

[1] $A = \frac{B}{C}$

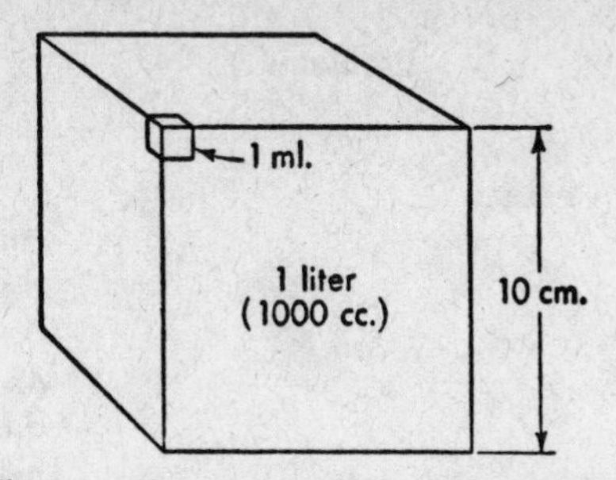

Fig. 2.1. The metric unit of volume, the liter.

volume occupied by one kilogram of water at 3.98° C. The liter was also designed to correct for the error involved when the cubic centimeter was adopted. One liter is therefore equal to 1000.027 cm.[3] Ordinarily in the laboratory it is assumed that one milliliter (ml) is equal in volume to one cubic centimeter. That is, cc, cm^3, and ml may be used interchangeably. See Fig. 2.1.

2.5. The Relationship between English and Metric Units of Length, Mass, and Volume. By means of constants such as given in Table 2.4, one may change values in one system of units into the corresponding values in the other system of units. See Fig. 2.2.

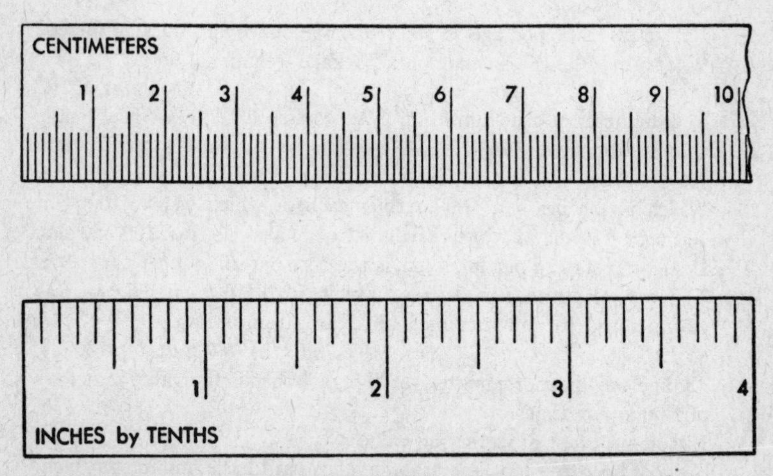

Fig. 2.2. A comparison of the English and metric units of length.

Problems

Part I

2.1. Linear units. Change:

a. 125 mm to cm. *Ans.* 12.5 cm.
b. 1.25 m to cm. *Ans.* 125 cm.
c. 2600 m to km. *Ans.* 2.6 km.
d. 1.65 cm to A. *Ans.* 1.65×10^8 A.
e. 2800μ to mm. *Ans.* 2.8 mm.
f. 25 in to cm. *Ans.* 63.5 cm.
g. 18 ft to m. *Ans.* 5.49 m.
h. 2.0 in. to A. *Ans.* 5.08×10^8 A.

2.2. Volume units. Change:

a. 1500 ml to *l*. *Ans.* 1.5 *l*.
b. 0.025 *l* to ml. *Ans.* 25 ml.
c. 15 qt to *l*. *Ans.* 14.2 *l*.
d. 15 *l* to gal. *Ans.* 3.96 gal.
e. 2.35 qt to ml. *Ans.* 2223 ml.

2.3. Weight units. Change:

a. 2.10 g to mg. *Ans.* 2100 mg.
b. 1.65 kg to g. *Ans.* 1650 g.
c. 3500 mg to g. *Ans.* 3.5 g.
d. 35 lb to kg. *Ans.* 15.9 kg.
e. 2.36 kg to lb. *Ans.* 5.19 lb.

2.4. How many cm ^{3} are there in a rod of uniform cross section measuring 10 cm by 2 cm by 90 cm ? *Ans.* 1800 cm.^{3}

2.5. Calculate the volume in liters of a box measuring 12 cm by 120 mm by 1.2 m. *Ans.* 17.3 *l*.

2.6. First-class U.S. postage is six cents per ounce or fraction thereof. What would be the postage on a package weighing 1.0 kg.? *Ans.* $2.16.

2.7. The diameter of a helium atom is approximately 2.0 A. How many helium atoms could be laid side by side the length of a meter stick? *Ans.* 5×10^9.

2.8. Which is the heavier, 100 marbles each of which weighs 10 g , or 200 marbles each of which weighs 0.1 oz ? *Ans.* 100 marbles.

2.9. If sugar costs 12¢ per lb , what is the cost per kilogram? *Ans.* 26¢.

2.10. The wave length of an infrared ray is 7.8×10^{-5} cm. Express as (a) microns, (b) millimicrons, and (c) angstrom units. *Ans.* (a) 0.78μ; (b) 780 mμ; (c) 7800 A.

2.11. A dime weighs approximately 2500 mg. What is the value in dollars of 1.00 kg of dimes? *Ans.* $40.

2.12. List in the order of increasing value:

a. Units of length — foot, meter, kilometer, yard, mile, millimeter, inch, and centimeter.

b. Units of mass — kilogram, ounce, milligram, pound, and gram.
c. Units of volume — milliliter, quart, cubic centimeter, cubic inch, cubic decimeter, and liter.

Part II

2.13. Linear units. Change:
a. 0.235 m to mm. *Ans.* 235 mm.
b. 0.015 km to cm. *Ans.* 1500 cm.
c. 255 A to mμ. *Ans.* 25.5 mμ.
d. 0.75 mm to μ. *Ans.* 750 μ.
e. 2800 mμ to cm. *Ans.* 2.8×10^{-4} cm.
f. 25 cm to in. *Ans.* 9.85 in.
g. $\frac{5}{16}$ in. to mm. *Ans.* 7.9 mm.
h. 5×10^5 mμ to in. *Ans.* 0.02 in.

2.14. Volume units. Change:
a. 1500 cm^3 to ml. *Ans.* 1500 ml.
b. 2.25 m^3 to *l.* *Ans.* 2250 *l.*
c. 1.27 *l.* to cc. *Ans.* 1270 cc.
d. 25 $in.^3$ to *l.* *Ans.* 0.41 *l.*
e. 1.75 yd^3 to kl. *Ans.* 1.338 kl.

2.15. Weight units. Change:
a. 0.0025 kg to mg. *Ans.* 2500 mg.
b. 125 g to kg. *Ans.* 0.125 kg.
c. 1000 g to lb. *Ans.* 2.2 lb.
d. 1.25 tons to kg. *Ans.* 1136 kg.
e. 1000 g to oz. *Ans.* 35.3 oz.

2.16. How many cm^3 are there in one cubic meter? *Ans.* 10^6 $cm.^3$

2.17. How many liters are there in one cubic meter? *Ans.* 10^3 *l.*

2.18. What is the weight of one gallon of water in kilograms? *Ans.* 3.78 kg.

2.19. How many cubic inches are there in one liter? *Ans.* 61 $in.^3$

2.20. Calculate the volume in liters of a box measuring 24 in. by 6.5 in. by 10 in. *Ans.* 25.6 *l.*

2.21. A speed of 60 $\frac{\text{mi}}{\text{hr}}$ corresponds to how many kilometers per hour?
Ans. 97 $\frac{\text{km}}{\text{hr}}$

2.22. The wave lengths of the visible rays in sunlight vary from 4000 A. for violet to 7000 A for red. Express as centimeters.
Ans. 4×10^{-5} cm.; 7×10^{-5} cm.

2.23. The wave lengths of X-rays vary from 10^{-9} cm to 10^{-6} cm. Express as angstrom units. *Ans.* 0.10 A.; 100 A.

2.24. A uniform iron bar 15 in. long weighs 2 lb 4 oz. Calculate the weight of the bar in grams per centimeter of length. *Ans.* 26.8 g.

2.25. At room temperature the linear velocity of an oxygen molecule is about 4×10^4 cm per sec. What would the velocity of the oxygen molecule be in mi per hour? *Ans.* 894 $\frac{\text{mi}}{\text{hr}}$

2.26. If gasoline costs 30¢ per gallon, what is the cost per liter?
Ans. 7.9¢.

2.27. When placed on water, 1.00 ml of oil spreads uniformly over an area of 500 $cm.^2$ Express the thickness of the film in (a) microns, and (b) millimicrons. *Ans.* 20μ; (b) 2×10^4 $m\mu$.

2.28. Soap bubble films average about 60 A in thickness. Express as (a) millimeters, and (b) centimeters.
Ans. (a) 6×10^{-6} mm ; 6×10^{-7} cm.

2.29. How much area in m^2 would 1.00 cm^3 of oil cover on water if the oil film were 4 A. in thickness? *Ans.* 2.5×10^3 m^2.

2.30. Parcels shipped to areas in which the humidity is high are sometimes coated with a gel, one gram of which will cover approximately 5×10^6 cm^2 of surface. Express this as micrograms of gel per square meter. *Ans.* 2×10^3 μg.

2.31. A nickel weighs approximately 5 g. What is the weight in kilograms of $100 worth of nickels? *Ans.* 10 kg.

2.32. Calculate your height in meters and in centimeters.

2.33. Calculate your weight in grams and in kilograms.

2.34. How many nanograms are there in one microgram? *Ans.* 1000 ng.

2.35. How many pounds are there in one gigagram?
Ans. 2.20×10^6 lb.

Temperature Scales

2.6. Introduction. In the laboratory, temperatures are usually recorded as *centigrade*, or *Celsius*, (°C) or *absolute* (°K). The symbol K for absolute temperatures is used in honor of Lord Kelvin, the first person to postulate the existence of such a scale and its possible use. The *Fahrenheit* (°F) scale is used domestically and, to some extent, in industry in the United States and England. The mercury expansion type of thermometer is the most commonly used instrument to measure temperatures.

2.7. Interconversion of Centigrade and Absolute Temperatures. In the solution of mathematical problems it is common practice to represent absolute temperatures by T and centigrade temperatures by t in formulas involving the quantities.

From Fig. 2.3, p. 29, it will be observed that for any given temperature the values in °C and °K differ by the constant value 273. This constant difference is due to the choice of the zero point on each of the two scales. That is:

$$T = 273 + t$$

or

$$t = T - 273.$$

Example 2.1. Normal body temperature is about 37° C. What would this be on the absolute scale?

Solution.

$$T = 273 + t$$

or

$$T = 273 + 37$$
$$= 310° \text{ K.}$$

2.8. Interconversion of Fahrenheit and Centigrade Temperatures. The interconversion of Fahrenheit and centigrade temperatures is based upon the relationship shown in Fig. 2.3. That is:

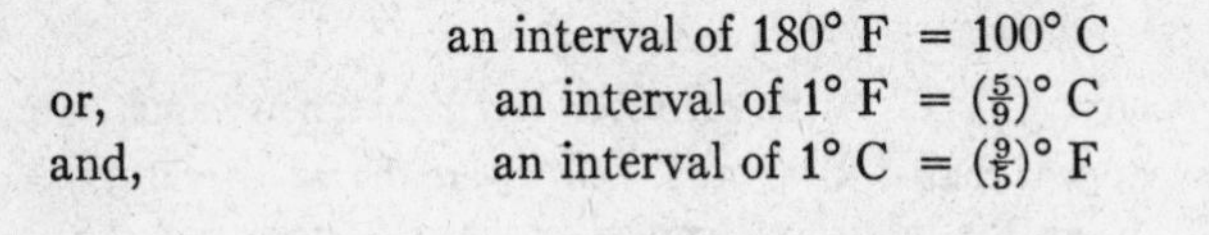

an interval of 180° F = 100° C

or, an interval of 1° F = $(\frac{5}{9})$° C

and, an interval of 1° C = $(\frac{9}{5})$° F

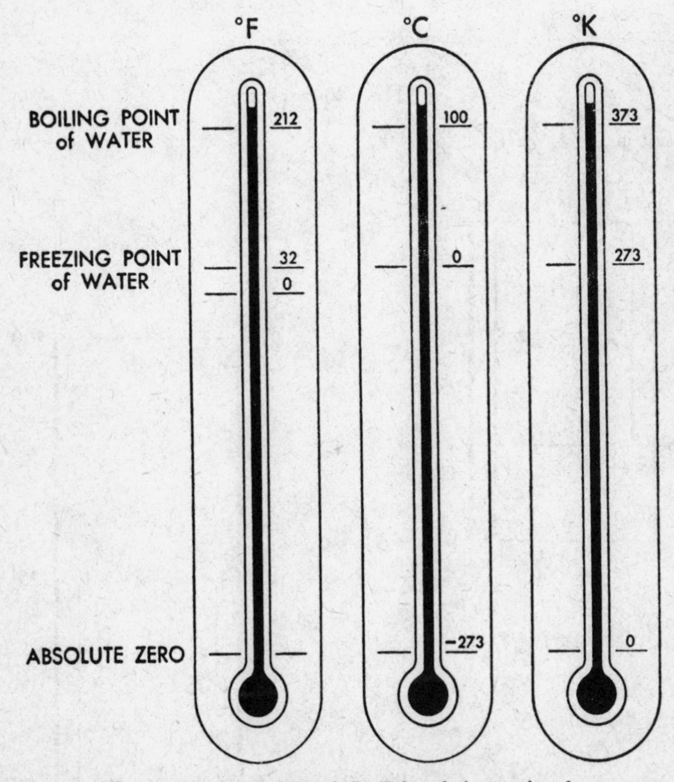

Fig. 2.3. A comparison of the Fahrenheit, centigrade, and absolute scales of temperature.

Example 2.2. Convert 86° F to the corresponding centigrade value.

Solution. Let us see how we may reason out the answer to the problem. First, choose a reference temperature such as the boiling point of water as shown in Fig. 2.4. The solution to the problem now lies in converting the Fahrenheit scale interval "a" to the corresponding centigrade scale interval "b." The temperature to be found is "c," which is $100 - b$.

$$a = 212 - 86 = 126,$$
$$b = 126 \times \tfrac{5}{9} = 70,$$
and
$$c = 100 - 70 = 30° \text{ C}.$$

That is, 86° F is equal to 30° C.

Example 2.3. Convert 25° C to the corresponding Fahrenheit value.

Solution. It so happens that −40° C and −40° F are the same temperature. Using −40° C as the reference temperature we see, from Fig. 2.5, that:

$$a = 40 + 25 = 65,$$
$$b = 65 \times \tfrac{9}{5} = 117,$$
and
$$c = 117 - 40 = 77° \text{ F}.$$

That is, 25° C is equal to 77° F.

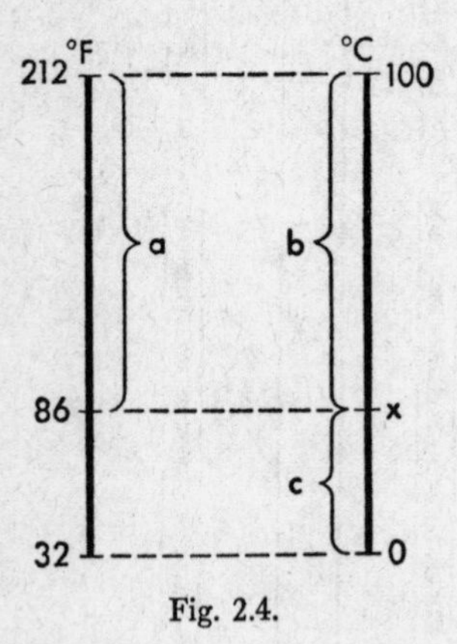

Fig. 2.4.

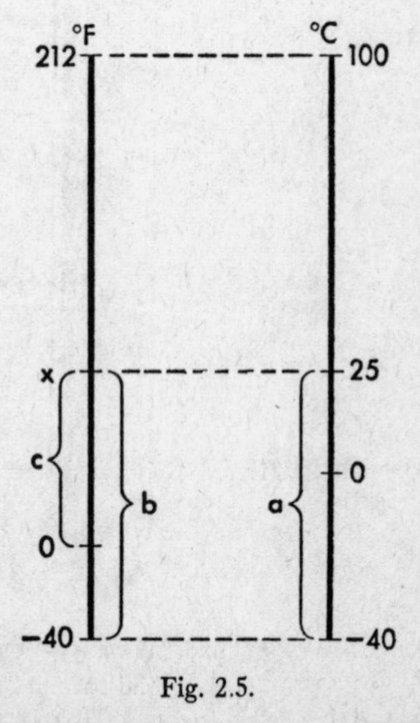

Fig. 2.5.

Problems

Part I

2.36. Convert:

a. 50° C to °F. *Ans.* 122° F.

b. 50° F to °C. *Ans.* 10° C.

c. 50° F to °K. *Ans.* 283° K.

d. 50° K to °F. *Ans.* −369° F.

2.37. A solution of salt water was found to freeze at 14° F. What is the freezing point of the solution in °C ? *Ans.* −10° C.

2.38. Three thermometers are side by side. One is calibrated in °F, another in °C, and the third in °K. The Fahrenheit thermometer registers −22°. What is the temperature on each of the other two thermometers? *Ans.* −30° C., 243° K.

2.39. Silver melts at 960.8° C. What is the melting point in °F ? *Ans.* 1761° F.

2.40. The sublimation temperature of dry ice is −109° F. What is the temperature in °C ? *Ans.* −78.3° C.

Part II

2.41. Mercury boils at 630° K. Calculate the boiling point of mercury in (a) °C, and (b) °F. *Ans.* (a) 357° C.; (b) 675° F.

2.42. Assuming the surface of the sun to be 10,832° F calculate the temperature in °K. *Ans.* 6273° K.

2.43. By means of Fig. 2.6 convert: (a) 50° C to °F, (b) 50° F to °C, and (c) 50° F to °K. *Ans.* (a) 122° F; (b) 10° C; (c) 283° K.

2.44. Show that Fahrenheit and centigrade temperatures are numerically the same at −40° C.

2.45. Mercury is used in some thermometers, and alcohol containing a colored pigment in others. Mercury freezes at −38.9° C and boils at 356.6° C; alcohol freezes at −117.3° C and boils at 78.5° C. Over what limits of temperature range in °F could each type of thermometer be used?
Ans. Mercury, −38° F. and 674° F.; alcohol, −179° F and 173° F.

2.46. If the freezing point of water had been used on the Fahrenheit scale as the zero of temperature, what would be the boiling point of water? *Ans.* 180°.

2.47. Zero absolute corresponds to what temperature on the Fahrenheit scale? *Ans.* −459° F.

2.48. A 10 degree interval on the centigrade scale corresponds to how many degrees interval on (a) the Fahrenheit scale, and (b) the absolute scale? *Ans.* (a) 18°; (b) 10°.

2.49. List the following in order of increasing temperature: 100° K, 100° C., and 100° F.

Dimensional Analysis

2.9. The Treatment of Dimensions in Addition and Subtraction. Numbers which are added or subtracted must have the same dimensions. Obviously there would be no significance to the sum obtained by adding quantities such as 15 in. and 27 cm without first converting to like units.

2.10. The Treatment of Dimensions in Multiplication and Division. Quantities expressed either in like or unlike units may be multiplied or divided. The dimensions of the numbers may be treated mathematically in the same manner as the numbers.

Example 2.4. What is the area of a rectangular board 3.0 ft long and 2.0 ft wide?

Solution.

$$3.0 \text{ ft} \times 2.0 \text{ ft} = (3.0 \times 2.0)(\text{ft} \times \text{ft}) = 6.0 \text{ ft.}^2$$

Observe that both the numerical and the dimensional values have been multiplied.

Example 2.5. A car traveled 260 miles in 5.0 hours. Calculate the speed of the car in miles per hour $\left(\frac{\text{mi}}{\text{hr}}\right)$.

Solution.

$$\frac{260 \text{ mi}}{5.0 \text{ hr}} = \frac{260}{5.0}\,\frac{\text{mi}}{\text{hr}} = 52\,\frac{\text{mi}}{\text{hr}}.$$

In this case the dimensional value could not be simplified.

Example 2.6. How many centimeters are there in 15.0 in.?

Solution. There are $2.54 \frac{\text{cm}}{\text{in.}}$ Therefore:

$$2.54 \frac{\text{cm}}{\cancel{\text{in.}}} \times 15.0 \,\cancel{\text{in.}} = 38.1 \text{ cm.}$$

Cancellation of like units gives the desired dimensions, centimeters.

Example 2.7. How many feet are there in 200 cm ?

Solution. There are $30.5 \frac{\text{cm}}{\text{ft}}$ Therefore:

$$\frac{200 \text{ cm}}{30.5 \frac{\text{cm}}{\text{ft}}} = \frac{200}{30\ 5}\,\cancel{\text{cm}} \times \frac{\text{ft}}{\cancel{\text{cm}}} = 6.56 \text{ ft.}$$

Example 2.8. Given that there are 2.54 $\frac{\text{cm}}{\text{in}}$. Determine the number of inches per centimeter $\left(\frac{\text{in}}{\text{cm}}\right)$.

Solution. Dimensionally $\frac{\text{cm}}{\text{in.}}$ and $\frac{\text{in.}}{\text{cm}}$ are reciprocals. Therefore:

$$\frac{1}{2.54\,\frac{\text{cm}}{\text{in.}}} = \frac{1}{2.54}\,\frac{\text{in.}}{\text{cm}} = 0.394\,\frac{\text{in.}}{\text{cm}}.$$

Example 2.9. Convert 2.36 kilograms to milligrams.

Solution. Dimensional analysis shows that the following operations must be carried out in order to obtain the value in milligrams.

$$\cancel{\text{kg}} \times \frac{\cancel{\text{g}}}{\cancel{\text{kg}}} \times \frac{\text{mg}}{\cancel{\text{g}}} = \text{mg}$$

Since there are 1000 $\frac{\text{g}}{\text{kg}}$ and 1000 $\frac{\text{mg}}{\text{g}}$, then:

$$2.36\,\cancel{\text{kg}} \times 1000\,\frac{\cancel{\text{g}}}{\cancel{\text{kg}}} \times 1000\,\frac{\text{mg}}{\cancel{\text{g}}} = 2.36 \times 10^6\ \text{mg}.$$

Example 2.10. Convert $2\frac{7}{16}$ in. to millimeters.

Solution. There are 2.54 $\frac{\text{cm}}{\text{in.}}$ and 10 $\frac{\text{mm}}{\text{cm}}$. Therefore:

$$\frac{39}{16}\,\cancel{\text{in.}} \times 2.54\,\frac{\cancel{\text{cm}}}{\cancel{\text{in.}}} \times 10\,\frac{\text{mm}}{\cancel{\text{cm}}} = 61.9\ \text{mm}.$$

Example 2.11. A car is traveling 60 $\frac{\text{mi}}{\text{hr}}$. What is the speed of the car in $\frac{\text{ft}}{\text{sec}}$?

Solution. To obtain the desired answer it will be necessary to convert mi to ft and hr to sec, and simplify. Since there are 5280 $\frac{\text{ft}}{\text{mi}}$, 60 $\frac{\text{min}}{\text{hr}}$, and 60 $\frac{\text{sec}}{\text{min}}$, then

$$\frac{60\,\cancel{\text{mi}} \times 5280\,\frac{\text{ft}}{\cancel{\text{mi}}}}{1.0\,\cancel{\text{hr}} \times 60\,\frac{\cancel{\text{min}}}{\cancel{\text{hr}}} \times 60\,\frac{\text{sec}}{\cancel{\text{min}}}} = \frac{60 \times 5280}{60 \times 60}\,\frac{\text{ft}}{\text{sec}} = 88\,\frac{\text{ft}}{\text{sec}}.$$

Example 2.12. What is the weight in grams of 2.50 gallons of water, given that there are 231 $\frac{\text{in.}^3}{\text{gal}}$, 2.54 $\frac{\text{cm}}{\text{in.}}$, and that water weighs 1.00 $\frac{\text{g}}{\text{cm}^3}$?

Solution.

$$2.50\ \cancel{\text{gal}} \times 231\ \frac{\cancel{\text{in.}^3}}{\cancel{\text{gal}}} \times \left(2.54\ \frac{\cancel{\text{cm}}}{\cancel{\text{in.}}}\right)^{\cancel{3}} \times 1.00\ \frac{\text{g}}{\cancel{\text{cm}^3}} = 9.46 \times 10^3\ \text{g}.$$

$$\cancel{\text{gal}} \times \frac{\text{in.}^3}{\cancel{\text{gal}}} = \text{in.}^3$$

$$\cancel{\text{in.}^3} \times \frac{\text{cm}^3}{\cancel{\text{in.}^3}} = \text{cm}^3$$

$$\cancel{\text{cm}^3} \times \frac{\text{g}}{\cancel{\text{cm}^3}} = \text{g}.$$

Example 2.13. (a) Derive the dimensional formula for the conversion of $\frac{\text{mi.}}{\text{hr}}$ to $\frac{\text{ft}}{\text{min}}$. (b) Determine the numerical value of the constant relating the two dimensions. (c) Change 25 $\frac{\text{mi}}{\text{hr}}$ to $\frac{\text{ft}}{\text{min}}$.

Solution.

(a) $\frac{\cancel{\text{mi}}}{\cancel{\text{hr}}} \times \frac{\text{ft}}{\cancel{\text{mi}}} \times \frac{\cancel{\text{hr}}}{\text{min}} = \frac{\text{ft}}{\text{min}}$.

(b) Assign unit value to $\frac{\text{mi}}{\text{hr}}$. Then:

$$1.00\ \frac{\text{mi}}{\text{hr}} = 1.00\ \frac{\cancel{\text{mi}}}{\cancel{\text{hr}}} \times 5280\ \frac{\text{ft}}{\cancel{\text{mi}}} \times \frac{1}{60}\ \frac{\cancel{\text{hr}}}{\text{min}} = 88\ \frac{\text{ft}}{\text{min}}.$$

(c) Since $1.00\ \frac{\text{mi}}{\text{hr}} = 88\ \frac{\text{ft}}{\text{min}}$, then:

$$25\ \frac{\text{mi}}{\text{hr}} = (25 \times 88)\ \frac{\text{ft}}{\text{min}} = 2200\ \frac{\text{ft}}{\text{min}}.$$

2.11. Nondimensional Numbers. In the introductory note to this chapter the statement was made that most quantities are dimensional. Nondimensional quantities are called *pure numbers*.

Example 2.14. Two iron rods are 75 ft and 15 ft respectively in length. The longer rod is how many times the length of the shorter rod?

Solution.

$$\frac{75\ \cancel{\text{ft}}}{15\ \cancel{\text{ft}}} = 5.$$

The dimensions cancel, giving the pure number 5. That is, the longer rod is 5 times the length of the shorter rod.

Example 2.15. What is the circumference of a circle having a diameter of 12.0 cm.?

Solution. The circumference, C, of a circle is directly proportional to the diameter, d. That is:

$$C \propto d.$$

The proportionality constant is 3.1416, commonly designated by the Greek letter pi, π. Or:

$$C = 3.1416 \times d.$$

Therefore $C = 3.1416 \times 12.0 \text{ cm.} = 37.7 \text{ cm.}$

Proportionality constants, such as π, are pure numbers.

Problems

Part I

2.50. Evaluate the following numerically and dimensionally:

a. $18 \text{ ft}^2 \div 3 \text{ ft.}$ *Ans.* 6 ft.

b. $(6 \text{ ft} \times 4 \text{ ft}) \div 3 \text{ ft.}$ *Ans.* 8 ft.

c. $350 \text{ cm}^3 \div 7.0 \text{ cm.}$ *Ans.* 50 cm.^{2}

d. $\sqrt[3]{125 \text{ cm}^3}$ *Ans.* 5.0 cm.

e. $12.0 \text{ cm.} \times 15.0 \text{ cm} \times 75.0 \text{ mm.}$ *Ans.* 1.35×10^3 cm.^{3}

2.51. How many milliliters are there in 3.25 *l*? *Ans.* 3.25×10^3 ml.

2.52. How many liters are there in 325 ml? *Ans.* 0.325 *l*.

2.53. How many millimeters are there in 15.0 in.? *Ans.* 381 mm.

2.54. How many feet are there in 1000 mm? *Ans.* 3.28 ft.

2.55. Given that there are 30.5 $\frac{\text{cm}}{\text{ft}}$, determine the number of feet per centimeter $\left(\frac{\text{ft}}{\text{cm}}\right)$. *Ans.* $3.28 \times 10^{-2} \frac{\text{ft}}{\text{cm.}}$

2.56. Convert 1.56 kilograms to milligrams. *Ans.* 1.56×10^6 mg.

2.57. Convert 1.25 meters to angstroms. *Ans.* 1.25×10^{10} A.

2.58. A car is traveling 50.0 $\frac{\text{ft}}{\text{sec}}$. What is the speed of the car in miles per hour $\left(\frac{\text{mi}}{\text{hr}}\right)$? *Ans.* 34.1 $\frac{\text{mi}}{\text{hr}}$.

2.59. Show that $15 \frac{\text{mi}}{\text{hr}} = 22 \frac{\text{ft}}{\text{sec}}$.

2.60. In each of the following derive the dimensional formula for the conversion of the given dimensions to the desired dimensions; then determine the value of the constant relating the two dimensions.

a. mi^3 to ft.^{3} *Ans.* 1.47×10^{11}.

b. mm to km. *Ans.* 10^{-6}.

c. $\frac{\text{ft}^3}{\text{sec}}$ to $\frac{\text{gal}}{\text{min}}$ *Ans.* 449.

d. mm to m *Ans.* 10^{-3}.

e. lb to mg *Ans.* 4.54×10^5.

2.61. What is the weight in grams of 25.0 in.^{3} of water, given that 1.00 cm.^{3} of water weighs 1.00 g? *Ans.* 410 g.

Part II

2.62. Convert 5000 A to in. *Ans.* 1.97×10^{-5} in.

2.63. How many ft^3 are there in 1.00 mi^3? *Ans.* 1.47×10^{11} ft.^{3}

2.64. How many mi^3 are there in 1.00 in.^{3}? *Ans.* 3.93×10^{-15} mi.^{3}

2.65. How many m^3 are there in 1.00 mi^3? *Ans.* 4.17×10^9 m.^{3}

2.66. How many in.^{3} are there in 1.00 mi^3? *Ans.* 2.54×10^{14} in.^{3}

2.67. How many μ are there in 3.17 m? *Ans.* 3.17×10^6 μ.

2.68. How many kilograms are there in 15.0 lb? *Ans.* 6.82 kg.

2.69. How many pounds are there in 15.0 kg? *Ans.* 33.0 lb.

2.70. How many gallons are there in 1.00 yd^3? *Ans.* 202 gal.

2.71. What is the weight in ounces of 25.0 in.^{3} of water, given that 1.00 ft^3 of water weighs 62.4 lb? *Ans.* 14.4 oz.

2.72. What is the numerical constant when converting yards to miles? *Ans.* 5.68×10^{-4}.

2.73. How many A^3 are there in 1.00 cm^3? *Ans.* 10^{24} A.^{3}

2.74. How many pounds are there in 5.0 gallons of water? *Ans.* 41.7 lb.

2.75. How many liters are there in 5.0 gallons of gasoline? *Ans.* 18.9 *l.*

2.76. A car is traveling 60 $\frac{\text{mi}}{\text{hr}}$. What is its speed in $\frac{\text{hr}}{\text{mi}}$?

Ans. 1.67×10^{-2} $\frac{\text{hr}}{\text{mi}}$.

2.77. Classify each of the following as representing length, area, or volume:

a. m.^{2} *Ans.* Area.

b. $\frac{\text{cm}^3}{\text{cm}^2}$ *Ans.* Length.

c. ft^3 *Ans.* Volume.

d. $\frac{\text{mi}}{\text{hr}} \times \frac{\text{min}}{\text{sec}} \times \frac{\text{hr}}{\text{min}} \times \text{sec}$ *Ans.* Length.

e. $\text{gal} \times \frac{\text{in.}^3}{\text{gal}} \times \left(\frac{\text{cm}}{\text{in.}}\right)^3$ *Ans.* Volume.

2.78. In each of the following derive the dimensional formula for the conversion of the given dimensions to the desired dimensions; then determine the value of the constant relating the two dimensions.

a. $\frac{\text{ft}}{\text{sec}}$ to $\frac{\text{mi}}{\text{hr}}$ *Ans.* 0.682.

b. mg to kg *Ans.* 10^{-6}.

c. A to mm *Ans.* 10^{-7}.

d. ft^3 to gal *Ans.* 7.48.

e. $\frac{\text{gal}}{\text{min}}$ to $\frac{\text{yd}^3}{\text{hr.}}$ *Ans.* 0.297.

f. yd to μ. *Ans.* 9.14×10^5.

g. ft^3 to ml *Ans.* 2.83×10^4.

h. in. to A *Ans.* 2.54×10^8.

2.79. The volume, V, of a sphere is given by the formula $V = \frac{4}{3}\pi r^3$, where r is the radius of the sphere.

a. What is the numerical value of the constant relating V and r? *Ans.* 4.1888.

b. What is the numerical value of the constant relating the volume and diameter of a sphere? *Ans.* 0.5236.

c. The diameter of sphere A is twice that of sphere B. How many times greater is the volume of A than that of B? *Ans.* 8.

d. What is the volume of a sphere the radius of which is 5.00 cm ? *Ans.* 524 cm.^3

Density

2.12. Density. *Density* is defined as the mass of a unit volume of a substance. That is:

$$\text{Density} = \frac{\text{mass}}{\text{volume}} \quad \text{or} \quad D = \frac{M}{V}.$$

Any unit of weight and any unit of volume may be chosen. However, in scientific work the densities of solids and liquids are usually given as grams per cubic centimeter $\left(\frac{\text{g}}{\text{cm}^3}\right)$, or as grams per milliliter $\left(\frac{\text{g}}{\text{ml}}\right)$, and the densities of gases are given as grams per liter $\left(\frac{\text{g}}{l}\right)$. English units are commonly used by engineers. In engineering work, therefore, densities are given as pounds per cubic foot $\left(\frac{\text{lb}}{\text{ft.}^3}\right)$. For example, the density of silver is 10.5 $\frac{\text{g}}{\text{cm}^3}$, or 655 $\frac{\text{lb}}{\text{ft.}^3}$. Since the numerical value of density depends upon the units of weight and volume chosen, it is essential that the dimensions be given when expressing density.

Table 2.5 gives the densities of a number of substances.

Example 2.16. Calculate the density of a liquid 17.45 ml of which weighs 16.3 g.

Solution. By definition the density of the liquid is the weight in grams of one milliliter. Therefore:

$$\text{density} = \frac{16.3 \text{ g.}}{17.45 \text{ ml.}} = 0.934 \frac{\text{g.}}{\text{ml.}}.$$

TABLE 2.5

THE DENSITIES OF A NUMBER OF SUBSTANCES

Substance	*Density*		
	$\frac{g}{cm^3}$	$\frac{g}{l}$	$\frac{lb}{ft^3}$
Air	—	1.29	—
Aluminum	2.70	—	168
Bromine	2.93	—	183
Carbon dioxide	—	1.98	—
Carbon tetrachloride	1.60	—	100
Copper	7.92	—	494
Gold	19.30	—	1204
Lead	11.34	—	708
Mercury	13.55	—	846
Oxygen	—	1.43	—
Sulfur	2.06	—	129
Sulfur dioxide	—	2.93	—
Uranium	18.90	—	1189
Water	1.00	—	62.4

Example 2.17. How many milliliters are there in 500 g of mercury, given that the density of mercury is 13.6 $\frac{g}{ml}$?

Solution. Since 1.00 ml of mercury weighs 13.6 g, then the number of milliliters in 500 g. is:

$$\frac{500\ g}{13.6\ \frac{g}{ml}} = \frac{500}{13.6}\cancel{g} \times \frac{ml}{\cancel{g}} = 36.8\ ml.$$

Example 2.18. Given that 140 ml of chlorine gas weighs 0.450 g; find the density of chlorine in grams per liter $\left(\frac{g}{l}\right)$.

Solution. By definition the density of chlorine is the weight in grams of 1.00 liter. Therefore:

$$\frac{140\ ml}{0.450\ g} = \frac{1000\ ml}{x\ g}$$

or $x = 3.21\ g$ = weight of one liter of chlorine.

Example 2.19. A solution of hydrochloric acid has a density of $1.20\ \frac{g}{ml}$ and contains 35.0 per cent HCl by weight. How many grams of HCl are there in 250 ml of the solution?

Solution. Since the per cent by weight of HCl is given, the first step is to find the weight of the 250 ml. of solution.

$$250\ \cancel{ml} \times 1.20\ \frac{g}{\cancel{ml}} = 300\ g\ \text{of solution.}$$

Then $300\ g \times 0.35 = 105\ g$ of HCl.

Example 2.20. What is the weight in grams of a block of silver measuring 2.50 cm by 8.00 cm by 4.00 cm, given that the density of silver is $10.5\ \frac{g}{cm^3}$?

Solution. Since 1.00 cm^3 of silver weighs 10.5 g, the first step is to find the number of cm^3 of silver in the block.

$$2.50\ cm \times 8.00\ cm \times 4.00\ cm = 80.0\ cm.^3$$

Then $$80.0\ \cancel{cm^3} \times 10.5\ \frac{g}{\cancel{cm^3}} = 840\ g = \text{weight of block.}$$

Problems

Part I

2.80. What is the density of ether, given that 300 ml weighs 217.5 g? *Ans.* $0.725\ \frac{g}{ml}$.

2.81. How many milliliters are there per gram of ether $\left(\frac{ml}{g}\right)$? (See problem 2.80) *Ans.* $1.38\ \frac{ml}{g}$.

2.82. How many grams of glycerine, density $1.25\ \frac{g}{cm^3}$, will a 125 ml flask hold? *Ans.* 156 g.

2.83. What is the density of cork if a cube measuring 1.50 cm on a side weighs 1.00 g? *Ans.* $0.296\ \frac{g}{cm^3}$.

2.84. Calculate the density of carbon dioxide gas, given that 450 ml. weighs 0.891 g. *Ans.* $1.98\ \frac{g}{l.}$.

2.85. A carboy contains 41.3 kg of hydrochloric acid, density $1.18\ \frac{g}{ml}$. What is the volume of the carboy in liters? *Ans.* 35.0 *l.*

2.86. A solution of hydrochloric acid has a density of $1.12\ \frac{g}{ml}$. Calculate.

a. The weight of 750 ml of the solution. *Ans.* 840 g.
b. The volume occupied by 750 g. of the solution. *Ans.* 670 ml.

2.87. A solution of sulfuric acid has a density of 1.84 $\frac{\text{g}}{\text{ml}}$ and contains 98.0 per cent acid by weight. What volume of solution would contain 360 grams of acid? *Ans.* 200 ml.

Part II

2.88. What would be the dimensions of a cube of copper weighing 7920 g.? *Ans.* 10 cm.

2.89. What is the weight in grams of 1.00 quart of water? *Ans.* 946 g.

2.90. A cubic foot of iron weighs 428 pounds. What is the density of iron in ounces per cubic inch $\left(\frac{\text{oz.}}{\text{in.}^3}\right)$? *Ans.* 3.96 $\frac{\text{oz}}{\text{in.}^3}$.

2.91. What volume of aluminum would be equal in weight to 100 cm^3 of lead? *Ans.* 420 cm.^3

2.92. Calculate the weight of a block of an alloy measuring 3.62 cm by 1.23 m by 1.00 mm, given that the density of the alloy is 9.63 $\frac{\text{g}}{\text{cm}^3}$. *Ans.* 429 g.

2.93. A block of stone has a density of 428 $\frac{\text{lb}}{\text{ft}^3}$. What is the density in $\frac{\text{g}}{\text{cm}^3}$? *Ans.* 6.86 $\frac{\text{g}}{\text{cm}^3}$.

2.94. A bottle weighs 13.45 g empty and 16.72 g filled with water. The same bottle when filled with a sugar solution weighs 19.01 g. What is the density of the sugar solution? *Ans.* 1.70 $\frac{\text{g}}{\text{ml}}$.

2.95. A graduated cylinder contained 20.0 ml of water. When 100 g. of brass shot were added to the cylinder the water level read 32.6 ml. What is the density of the brass? *Ans.* 7.94 $\frac{\text{g}}{\text{ml}}$.

2.96. What is the weight in grams of 1.00 in.^3 of mercury? *Ans.* 222 g.

2.97. A 100 ml. bulb weighs 67.423 g. when filled with air, and 67.879 g when filled with xenon gas. What is the density of xenon, given that air weighs 1.29 $\frac{\text{g}}{l}$? *Ans.* 5.85 $\frac{\text{g}}{l}$.

2.98. A cube of wood measuring 20.0 cm on a side is placed in water. How much of the wood will extend above the water, given that the density of the wood is 0.80 $\frac{\text{g}}{\text{cm}^3}$? *Ans.* 1600 cm^3

2.99. Given that the density of air is 1.29 $\frac{\text{g}}{l}$, hydrogen 0.09 $\frac{\text{g}}{l}$, and helium 0.18 $\frac{\text{g}}{l}$; what per cent of the lifting power of hydrogen is helium capable of exerting? *Ans.* 93 per cent.

2.100. Five hundred grams of mercury was poured into a 100 ml. graduated cylinder. How much water would have to be added to bring the water up to the 100 ml mark? *Ans.* 63.1 ml.

2.101. Show how the weight of one cubic foot of gold, as given in Table 2.5, p. 43, could be calculated from the density as given in $\frac{g}{cm^3}$.

2.102. A 1.00 in.^{3} of metal weighs 87.7 g when suspended in mercury. What is the metal? *Ans.* Uranium.

2.103. Gold costs approximately $1.12 per gram. What would be the cost of 1.00 in.^{3} of gold? *Ans.* $354.

2.104. What is the constant which, when multiplied by $\frac{lb}{ft^3}$, will give $\frac{g}{cm^3}$? *Ans.* 0.0160.

3

Graphical Treatment of Data

Most experimental data are obtained in an attempt to show a relationship between certain variables. It is often possible to show this relationship by plotting the data on a coordinate system. Quadrant I of the coordinate system is most commonly used since all values in this quadrant are positive. Such a graph of experimental data is helpful in obtaining a mathematical formula expressing the relationship of the variables. When two variables, x and y, are shown to be related then y is said to be a function of x. In mathematical notation, $y = f(x)$.

3.1. Plotting Experimental Data. In Fig. 3.1, values have been plotted for the variables x and y, yielding what appears to be a straight line, ab. The slope, m, of a straight line is defined as the ratio of bc to ac. That is:

$$m = \frac{bc}{ac}.$$

Reference to Sec. 1.13 will show that the slope m is the tangent of the angle Θ. That is: $m = \tan \Theta$.

Example 3.1. Calculate (a) the slope of the line ac, and (b) the value of Θ, in Fig. 3.1.

Solution.

(a) $m = \dfrac{3.25}{9.00} = 0.36 =$ slope of the line ac.

(b) Since $m = \tan \Theta$, then $0.36 = \tan \Theta$, and:

$$\Theta = 19°48' \text{ (From Appendix I by interpolation).}$$

The equation for a straight line is:

$$y = mx + b$$

where m is the slope of the line and b is the value of the intercept of the line with the y-axis.

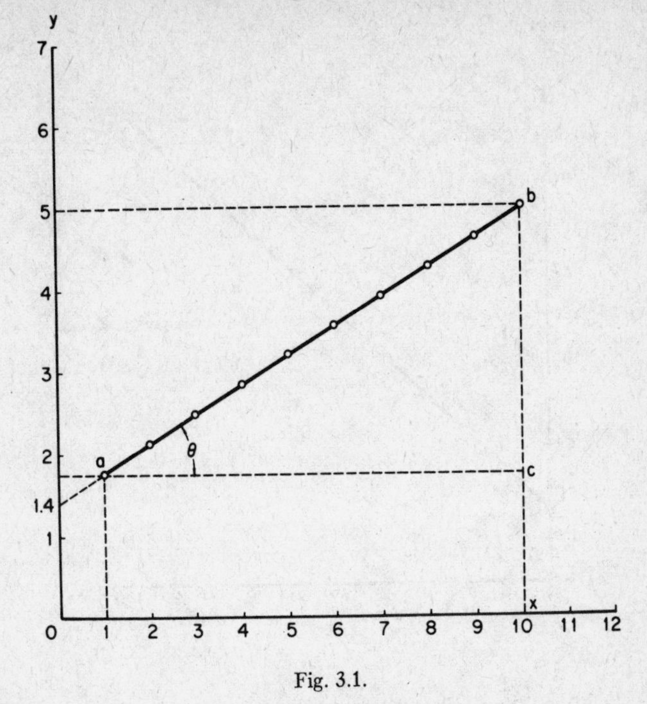

Fig. 3.1.

Example 3.2. (a) What is the equation for the line *ac* in Fig. 3.1? (b) In Fig. 3.1, what is the value of y when $x = 6.4$?

Solution. (a) By inspection, the value of the y-intercept in Fig. 3.1 is equal to 1.4. Therefore, the equation for the line *ac* would be:

$$y = 0.36x + 1.4.$$

(b) $y = (0.36)(6.4) + 1.4 = 3.7.$

Note: The value 3.7 can be checked by inspection of Fig. 3.1.

3.2. Plotting Variables Involving Direct Proportionality. The variables x and y in Fig. 3.1 represent direct proportionality, as shown by the straight-line relationship. That is, $x \propto y$. In Fig. 3.2 corresponding values for Fahrenheit, centigrade, and absolute scales of temperature have been plotted. Again, we have a straight-line (linear) relationship which indicates a direct proportionality involving the three scales of temperature.

Example 3.3. Derive the formula for the straight line in Fig. 3.2 for the variables Fahrenheit and centigrade.

Solution. $y = mx + b$, where $y = {}^\circ\mathrm{F}$, $x = {}^\circ\mathrm{C}$, and $m = \dfrac{180}{100} = 1.8$,

the y-intercept $b = 32$.

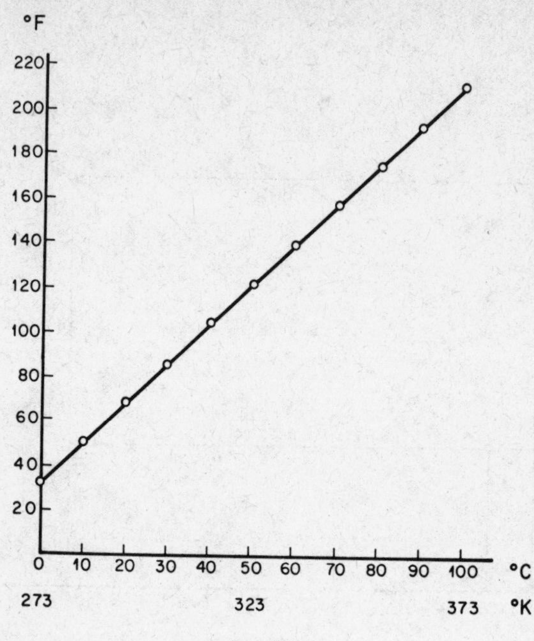

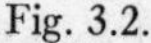

Fig. 3.2.

Substituting the above values in the equation for a straight line we have:

$$°F = (1.8)(°C) + 32.$$

Note: The above formula will be recognized as the one used for the interconversion of Fahrenheit and centigrade temperatures.

3.3. Plotting Variables Involving Inverse Proportionality. In inverse proportionality between two variables x and y, $x \propto \frac{1}{y}$.

Example 3.4. The following data give the time, t (hr), a car must travel at a velocity, v (mi/hr), in order to cover a fixed distance of 600 miles. Also given are the calculated values for $\frac{1}{v}$ (hr/mi).

v (mi/hr)	10	20	40	50	80
t (hr)	60	30	15	12	7.5
$\frac{1}{v}$ (hr/mi)	0.1000	0.0500	0.0250	0.0200	0.0125

a. Plot (1) v and t, and (2) $\frac{1}{v}$ and t.

b. Derive the equation for the straight line obtained in a(2).

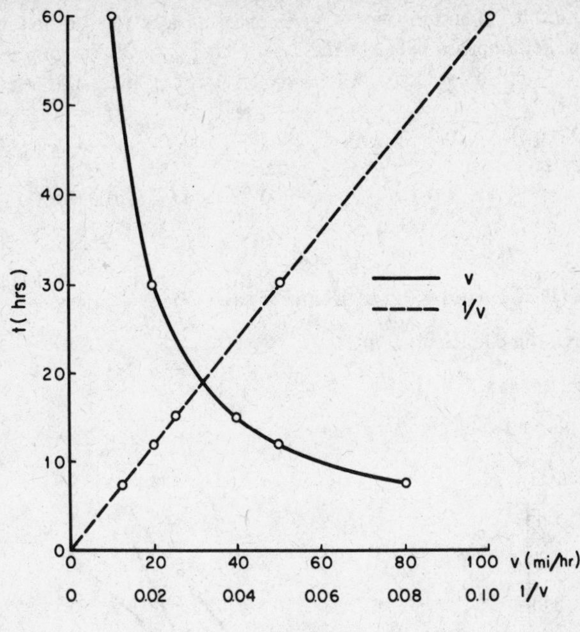

Fig. 3.3.

Solution.

a. See Fig. 3.3. Observe that a straight line results when $\frac{1}{\text{v}}$ and t are plotted. That is, $\text{t} \propto \frac{1}{\text{v}}$.

b. Since $y = mx + b$, where $y = \text{t}$, $x = \frac{1}{\text{v}}$, $m = \frac{60}{0.1} = 600$, and $b = 0$ (since the line passes through the origin):

$$\text{t} = (600)\left(\frac{1}{\text{v}}\right) \quad \text{(the equation for the straight line)}$$

or: $\text{tv} = 600.$

Note: tv = k, the proportionality constant, which in this case is equal to 600 (Sec. 1.11).

3.4. Reducing Data to Linear Plots. In Sec. 3.3, data involving inverse proportionality were reduced to a linear relationship by plotting x and $\frac{1}{y}$. Linear relationships cannot always be obtained for variables in this manner. In order to obtain a linear relationship between some variables it is necessary to plot $x \propto y^n$, $x \propto \sqrt[n]{y}$, $x \propto \log y$, etc. It is mostly a matter of trial and error to determine the relationship between variables.

Example 3.5. The following experimental data were recorded in determining the relationship between the kinetic energy, K.E. (ergs), of a given mass, m (g), and its velocity, v (cm/sec). A 5.00 g mass was used in the experiment.

K.E. (ergs)	10.0	40.0	90.0	160.0	250.0	360.0
v (cm/sec)	2.0	4.0	6.0	8.0	10.0	12.0
v^2	4.0	16.0	36.0	64.0	100.0	144.0
$\frac{\text{K.E.}}{\text{v}}$	5.0	10.0	15.0	20.0	25.0	30.0

a. Plot (1) K.E. and v, (2) K.E. and v^2, and (3) $\frac{\text{K.E.}}{\text{v}}$ and v.

b. Derive the equation for a(2).

Solution.

a. See Fig. 3.4.

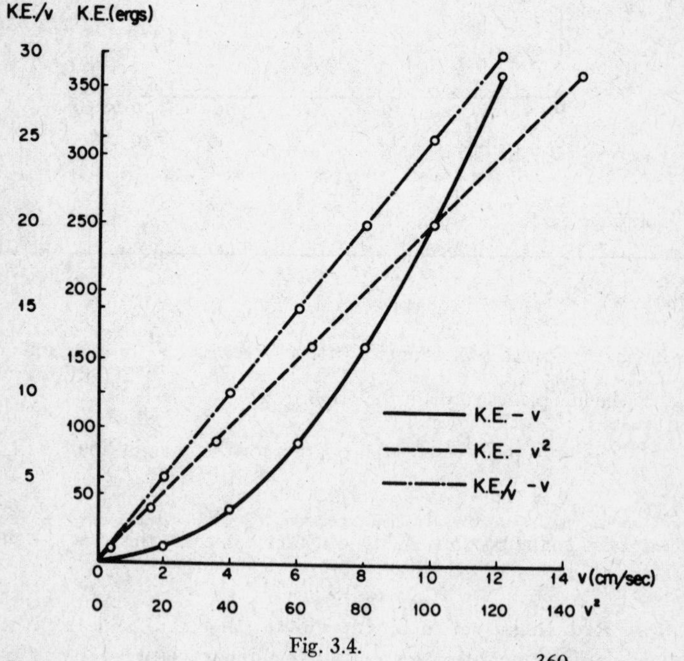

Fig. 3.4.

b. Since $y = mx + b$, where $y = \text{K.E.}$, $x = v^2$, $m = \frac{360}{144} = 2.5$, and $b = 0$, then:

$$\text{K.E.} = 2.5v^2.$$

Since 2.5 is one-half of the 5.0 g mass, we may assume that:

$$\text{K.E.} = \tfrac{1}{2}mv^2.$$

Repeated trials with different masses would show that the constant in the above equation would always be one-half the mass in grams. The above formula will be recognized as the one used to determine the kinetic energy of a moving body.

Example 3.6. The following data were recorded in determining the relationship between the density of a gas, d (g/*l*), and its ability to effuse through a small opening, D (ml/hr). The same sized opening was used for all the gases at 1.00 atm and 0° C.

Gas	*d (g/l)*	$\sqrt{d}$	$1/\sqrt{d}$	*D (ml/hr)*
Hydrogen	0.090	0.300	3.333	450
Neon	0.901	0.950	1.053	142
Oxygen	1.429	1.195	0.837	113
Chlorine	3.214	1.800	0.555	75
Xenon	5.851	2.420	0.413	56

Plot (1) d and D, (2) $\sqrt{d}$ and D, and (3) $1/\sqrt{d}$ and D, derive the equation for (3), and determine how much nitrogen pentoxide gas would escape through the given opening in two hours at 1.00 atm and 0° C. The density of nitrogen pentoxide is 4.821 g/*l*.

Solution. See Fig. 3.5.

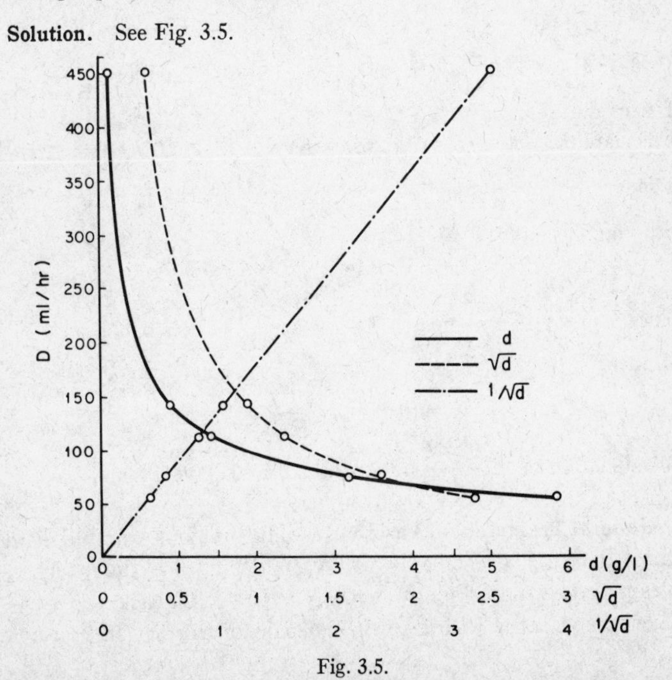

Fig. 3.5.

In the equation for the straight line $y = mx + b$, $y = \mathrm{D}$, $x = 1/\sqrt{\mathrm{d}}$, $b = 0$, and $m = \dfrac{394}{2.92} = 135$. Therefore, the equation for the straight line in Fig. 3.5 would be:

$$\mathrm{D} = (135)\left(\frac{1}{\sqrt{\mathrm{d}}}\right).$$

The amount of nitrogen pentoxide escaping in two hours would be: $\mathrm{D} = (135)\left(\dfrac{1}{\sqrt{4.821}}\right)(2) = 122.8$ ml in two hours. The answer could be checked by reference to Fig. 3.5, where D has a value of approximately 60 ml when $1/\sqrt{\mathrm{d}} = 0.455$.

Example 3.7. When a platinum wire is dipped into a solution containing H^+ ions, and hydrogen gas is bubbled over the immersed portion of wire, an electrical potential, E (volts), is set up in the wire. The following are experimental values for E and $[H^+]$ (moles/l).

E (volts)	0.06	0.12	0.18	0.24	0.30
$[H^+]$ (moles/l)	10^{-1}	10^{-2}	10^{-3}	10^{-4}	10^{-5}
$\dfrac{1}{[H^+]}$	10^{1}	10^{2}	10^{3}	10^{4}	10^{5}
$\log \dfrac{1}{[H^+]}$	1	2	3	4	5

a. Plot (1) E and $[H^+]$, (2) E and $\dfrac{1}{[H^+]}$, and (3) E and $\log \dfrac{1}{[H^+]}$.

b. Derive the equation for a(3).

c. Calculate the potential on the wire when $[H^+] = 5.0 \times 10^{-4}$.

Solution.

a. See Fig. 3.6.

b. Since $y = mx + b$, where $y = \mathrm{E}$, $x = \log \dfrac{1}{[H^+]}$, $b = 0$, and $m = \dfrac{0.30}{5.0} = 0.06$, then:

$$\mathrm{E} = (0.06) \log \frac{1}{[H^+]}.$$

c. $\mathrm{E} = (0.06) \log \dfrac{1}{5.0 \times 10^{-4}} = (0.06)(3.30) = 0.20$ volt.

3.5. General Treatment of Variables. The previous examples have been limited to the use of only two variables at one time. Three variables may be studied simultaneously. In such a case, the x-, y-, and z-axes are used, as with a three-dimensional figure. In Example

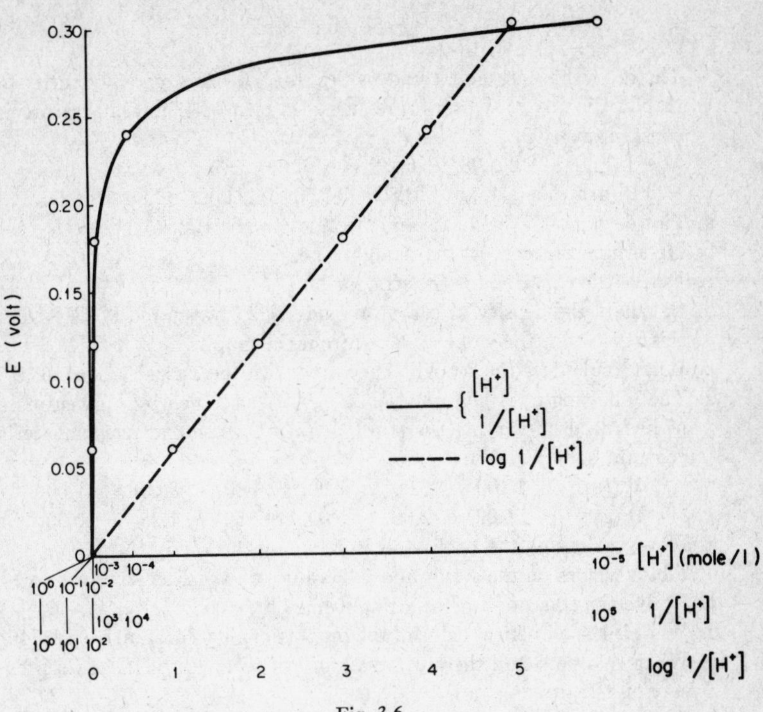

Fig. 3.6.

3.5, kinetic energy and velocity were used as the variables, while the mass was held constant at 5.0 grams. In Example 3.6, density and rate of effusion were used as variables, while the size of the opening, pressure, and temperature were held constant. Any other size opening, temperature, or pressure, when held constant, would merely change the numerical value of the slope of the straight line.

Many techniques, other than those discussed previously, are available for producing linear relationships among variables.

Example 3.8. Derive the equation for the straight line obtained when $\frac{\text{K.E.}}{\text{v}}$ and v are plotted as shown in Fig. 3.4.

Solution. Since $y = mx + b$, where $y = \frac{\text{K.E.}}{\text{v}}$, $x = \text{v}$, $b = 0$, and $m = \frac{25}{10} = 2.5$, the equation for the straight line would be:

$$\frac{\text{K.E.}}{\text{v}} = 2.5\text{v}$$

or, K.E. $= 2.5\text{v}^2$, which is the same equation obtained in Example 3.5.

Problems

3.1. The following data give the densities, d (g/*l*), for a given quantity of carbon dioxide at varying pressures, P (atm). The temperature is constant at 0° C.

d (g/*l*)	1.98	0.99	2.97	1.96	4.95	5.94
P (atm)	1.00	0.50	1.50	2.00	2.50	3.00

a. Plot d values on the y-axis and P values on the x-axis.

b. Determine the slope of the straight line. *Ans.* 1.98.

c. Derive the equation for the straight line.

d. Calculate the density of carbon dioxide at 2.35 atm and 0° C. Does this agree with the value as taken from the graph? *Ans.* 4.65 g/*l*.

e. What angle does the straight line make with the x-axis? *Ans.* 63.2°.

3.2. The following data give the volume, V (*l*), occupied by a given mass of hydrogen at varying pressures, P (atm), when the temperature is constant at 0° C.

V (*l*)	10.00	5.00	2.00	4.00	8.00	1.00
P (atm)	1.00	2.00	5.00	2.50	1.25	10.00

a. Plot V values on the y-axis and P values on the x-axis.

b. Plot V values on the y-axis and 1/P values on the x-axis.

c. Derive the equation for the straight line obtained in *b*.

d. What is the numerical value of the proportionality constant? *Ans.* 10.

e. What volume would the given mass of gas occupy at 0° C and a pressure of 3.75 atm? *Ans.* 2.67 *l*.

3.3. Hydrogen peroxide decomposes according to the equation: $2\ H_2O_2 \rightarrow 2\ H_2O + O_2$ (g). The following data were obtained for the time rate of decomposition, t (sec), of H_2O_2 at 27° C and 1.00 atm.

t (sec)	10.0	15.0	20.0	25.0	30.0	35.0
O_2 (ml)	18.0	21.2	23.4	25.2	26.6	27.8

a. Plot (1) O_2 (ml) and t, and (2) O_2 (ml) and log t.

b. Derive the equation for a(2).

c. How many milliliters of oxygen under the given conditions would be released in 18.0 sec? *Ans.* 22.6 ml.

3.4. The force of attraction, F (dynes), between two oppositely charged electrical particles in a vacuum was determined experimentally as a function of the distance, r (cm), between the particles.

F (dynes)	90.0	22.5	10.0	5.6	3.6
r (cm)	1.0	2.0	3.0	4.0	5.0

a. Plot F as a function of r such that a linear relationship results.

b. Derive the equation involving F and r.

c. What is the value of the proportionality constant relating F and r? *Ans.* 90.

d. What would be the force of attraction in dynes when the two particles are 2.5 cm apart? *Ans.* 14.4 dynes.

3.5. The centripetal force, F, of an object of mass m, revolving in a circular motion of radius r and velocity v, is given by the expression $F = \frac{mv^2}{r}$.

a. Give the proportionality relationship between the variables (1) F and m, (2) F and r, and (3) F and v.
b. What values would have to be plotted on the x- and y-axes in order to show a linear relationship between F and v?
c. What is the expression for the proportionality constant, in terms of F, m, v, and r, when (1) F is a function of v, and (2) F is a function of r?
d. What variables would have to be held constant in order to determine experimentally the relationship between F and v?

3.6. Plot on semilogarithmic paper the values given for E and $\frac{1}{[H^+]}$ in Example 3.7.

3.7. In Fig. 3.6, why are the values for $[H^+]$ lying between 10^0 and 10^{-3} crowded into such a small portion of the x-axis?

3.8. In Problem 3.3, above, plot the values for O_2 (ml) and t on semilogarithmic paper.

4

Statistical Treatment of Data

In carrying out the mathematical operations of addition, subtraction, multiplication, and division, the student is confronted with the problem of the number of digits to retain in the answer. Very definite rules may be followed in such mathematical operations. Fortunately the rules are not difficult to remember and are easy to apply. Much unnecessary work may be avoided in mathematical operations by remembering the rules applying to significant digits.

Statistical techniques are used to study the significance and reliability of experimental data. By means of statistical studies it is often possible to reduce experimental data to a quantitative form. Only the elementary principles of statistics will be discussed.

Significant Digits

4.1. Significant Digits in Experimental Measurements. Experimental data involving numbers are seldom absolutely accurate values. The accuracy of such data is dependent upon the precision of the measuring instrument used. For example, suppose the length of the

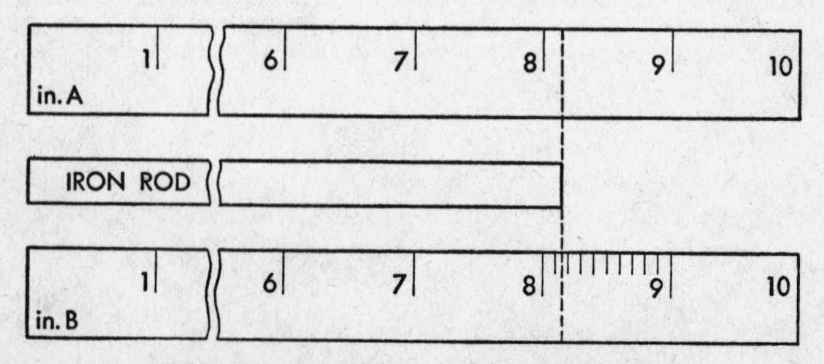

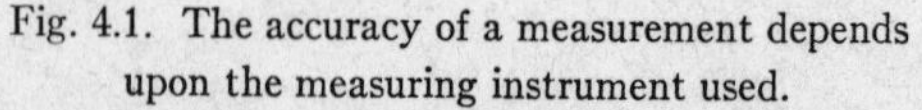
Fig. 4.1. The accuracy of a measurement depends upon the measuring instrument used.

iron rod shown in Fig. 4.1 is determined by the measuring instruments A and B. Using instrument A, in which the centimeter is the smallest scale division, we see that the end of the rod lies between the 8 and 9 centimeter marks. Estimating the length to the nearest tenth of a centimeter gives the value 8.2 cm. Using instrument B, in which the millimeter is the smallest scale division, gives the value 8.16 cm. The underscored figure in each measured value has been estimated, and is therefore not absolutely reliable. Recorded data should represent the greatest accuracy possible with the measuring instrument used.

The discussion which follows deals primarily with the mathematical treatment of data. Such treatment should be consistent with the accuracy of the data used.

Any figure representing a reasonably reliable value should be considered a *significant digit.* The value 8.2 cm, given in Sec. 4.1, represents two significant digits; the value 8.16 cm represents three significant digits. The number of significant digits in a quantity is independent of the location of the decimal point.

From the above it is apparent that the precision of the measuring instrument is a factor in determining the number of significant digits in a number. Another factor is the size of the object measured. For example, two objects were found to weigh 3.1 g and 18.3 g, respectively, on a balance accurate to one-tenth of a gram. The weight of the larger object contains three significant digits, whereas the weight of the smaller object contains only two significant digits.

4.2. Zero as a Significant Digit. Zero occupies a unique position in our number system containing the ten symbols 1, 2, 3, 4, 5, 6, 7, 8, 9, and 0. Depending upon the manner in which it is used, zero may or may not be a significant digit when appearing in a number.

Zero may be used merely to give position value to a combination of one or more of the ten symbols in our number system. When so used, zero is not a significant digit. For example, in the expression 2.54 mm = 0.254 cm = 0.00254 m = 0.00000254 km, none of the zeros are significant digits. In each case the zeros are used to give position value to the significant digits, 254. Zeros may also be used to give position value to integer numbers. For example, the statement that the earth is approximately 8000 miles in diameter represents one significant digit. Actually, the earth may be 7926 miles in diameter on a given axis, which value represents four significant digits. Exponential forms are sometimes used to indicate the number of significant digits in integer values. For example:

$$8000 \text{ miles} = 8 \times 10^3 \text{ miles},$$
$$7900 \text{ miles} = 7.9 \times 10^3 \text{ miles},$$
$$186000 \frac{\text{mi}}{\text{sec}} = 1.86 \times 10^5 \frac{\text{mi}}{\text{sec}}.$$

When used as one of the ten symbols in our number system, zero must be considered as a significant digit, regardless of the position of the decimal point. The zeros in each of the following expressions are significant: 1.06 m, 20.4 cm, 106 cm, 1004 ft, 2.20 lb, 9.600 g, 100.20 cm, and 1.0200 g. Observe that a zero following any of the other nine symbols, when to the right of the decimal point, is significant. One should be extremely careful in the use of zero when recording experimental data.

Some integer values are absolutely accurate values. For example, one egg represents 1.00000—— egg carried to an infinite number of significant digits. It is usually not difficult to recognize such integer values.

Problems

Part I

4.1. Give the number of significant digits in each of the following values.

a. 15 cm. *Ans.* 2.
b. 0.15 cm. *Ans.* 2.
c. 15.0 cm. *Ans.* 3.
d. 0.015 cm. *Ans.* 2.
e. 1.250 g. *Ans.* 4.
f. 13.002 g. *Ans.* 5.
g. 0.0602 g. *Ans.* 3.
h. 6.050×10^2 g. *Ans.* 4.
i. 3.00×10^{-2} m. *Ans.* 3.
j. 8 silver dollars. *Ans.* ∞.

4.2. Express each of the following as a power of ten, the number other than the power of ten containing the significant digits.

a. One mile is equal to approximately 5300 ft. *Ans.* 5.3×10^3.
b. One mile is equal to 5280 ft. *Ans.* 5.280×10^3.
c. Under ordinary room conditions sound travels approximately 1000 ft/sec. *Ans.* 1×10^3 ft/sec.
d. One yard is equal to 0.9144 m. *Ans.* 9.144×10^{-1}.
e. One yard is equal to approximately 900 mm. *Ans.* 9×10^2.

Part II

4.3. Give the number of significant digits in each of the following values.

a. 1.01 *l*. *Ans.* 3.
b. 250.10 m. *Ans.* 5.

c. 0.00101 kg. *Ans.* 3.
d. 9.0900 m. *Ans.* 5.
e. 1.600×10^2 g. *Ans.* 4.
f. 2.0×10^{-3} ml. *Ans.* 2.
g. 80.00×10^2 mm. *Ans.* 4.
h. 3125 nails. *Ans.* ∞.

4.4. Express each of the following as a power of ten, the number other than the power of ten containing the significant digits.

a. One foot is equal to approximately 300 mm. *Ans.* 3×10^2.
b. One foot is equal to 304.8 mm. *Ans.* 3.048×10^2.
c. One cubic foot is equal to approximately 28,000 ml. *Ans.* 2.8×10^4.
d. One cubic foot is equal to 0.028 m^3. *Ans.* 2.8×10^{-2}.
e. At 0° C sound travels with a velocity of 331.36 m/sec. *Ans.* 3.3136×10^2.

4.3. The Use of Significant Digits in Mathematical Operations. Only quantities representing like units and expressed to the same number of decimal places may be added or subtracted, regardless of the number of significant digits in the quantities.

Example 4.1. Add: 12.7 m + 219.31 cm + 332 mm.

Solution. First, change the quantities to like units, in this case the meter. The unit used is a matter of choice. Next, round off the numbers to one decimal place. One decimal place is chosen since this represents the least accurately measured quantity of the three to be added. Then,

12.7 m	=	12.7 m	=	12.7 m
219.31 cm	=	2.1931 m	=	2.2 m
332 mm	=	0.332 m	=	0.3 m
			Add:	15.2 m.

When rounding off numbers, the last figure retained is increased by one if the number immediately following is greater than five. Thus, in Example 4.1, 2.1931 m becomes 2.2 m, and 332 mm becomes 0.3 m.

In multiplication and division, the number of significant digits retained in the product or quotient is the same as the least number of significant digits occurring in either of the quantities involved. Note that the numbers are not rounded off before performing the mathematical operation, and that the units need not be similar.

Example 4.2. A rectangular sheet of platinum measures 5.29 mm by 16.14 mm. What is the area of the sheet?

Solution.

$$\underline{5}.\underline{29} \text{ mm} \times \underline{16}.\underline{14} \text{ mm} = \underline{85}.\underline{3}806 \text{ mm}^2 = \underline{85}.\underline{4} \text{ mm}^2$$

Example 4.3. A car traveled 277.68 mi in 6.21 hr. What was the average speed in mi/hr?

Solution.

$$277.68 \text{ mi} \div 6.21 \text{ hr} = 44.71 \text{ mi/hr} = 44.7 \text{ mi/hr}.$$

In the process of division, it is customary to carry the quotient to one more significant digit than necessary in order to round off the answer to the required number of significant digits.

Example 4.4. Simplify the expression $\frac{5.8 \times 0.0899}{273}$.

Solution.

$$\frac{5.8 \times 0.0899}{273} = 0.00191 = 1.9 \times 10^{-3}.$$

Example 4.5. What is the surface area of a table measuring 2.63 m by 76.5 cm?

Solution. It is customary to use like units when calculating areas and volumes. Such treatment gives areas in terms of unit squares, and volumes in terms of unit cubes. Then:

$$2.63 \text{ m} \times 0.765 \text{ m} = 2.01 \text{ m}^2$$

Only general rules relating to the use of significant digits in mathematical operations have been presented in the foregoing discussions. More comprehensive treatments can be introduced as the need for such arises.

Problems

Part I

4.5. Perform the indicated mathematical operations.

a. 2.1 m + 0.3 m + 2.07 m + 3.224 m. *Ans.* 7.7 m.
b. 21.630 *l* + 844 ml. + 0.036 *l* + 10.196 ml. *Ans.* 22.520 *l*.
c. 3.14 g. + 0.715 g + 10.84 mg. *Ans.* 3.87 g.
d. 13.22 cm − 28.36 mm. *Ans.* 10.38 cm.
e. 0.6260 g − 15 mg. *Ans.* 0.611 g.
f. 32.57 × 7.14 *Ans.* 233.
g. 0.0482 × 0.2134 *Ans.* 1.03×10^{-2}.
h. 7632 ÷ 173 *Ans.* 44.1.
i. 3.438 ÷ 0.988 *Ans.* 3.48.
j. 0.041761 ÷ 32.15 *Ans.* 1.299×10^{-3}.

4.6. What is the area of a rectangle measuring 3.180 cm by 22.4 mm? *Ans.* 7.12 cm.^{2}

4.7. Light travels 1.86×10^5 mi/sec. How far will a beam of light travel in one hour? *Ans.* 6.70×10^8 mi.

4.8. A jar contains 10,000 lead shot averaging 0.2216 g each. What is the weight of the shot? *Ans.* 2216 g.

Part II

4.9. Simplify each of the following expressions.

a. $\frac{2.56 \times 10^3}{454} \times 1.263$. *Ans.* 7.12.

b. $\frac{1566}{1.80} + 32$. *Ans.* 902.

c. $\frac{0.0154 \times 0.1276}{0.00891}$. *Ans.* 2.21×10^{-1}.

4.10. There are 2.54 cm in one inch. How many centimeters are there in 15.0 in.? *Ans.* 38.1 cm.

4.11. What is the cost of 1200 four-cent United States postage stamps? *Ans.* $48.00.

4.12. What is the circumference of a circle of 10.00 in. diameter, given that $\pi = 3.14$? *Ans.* 31.4 in.

4.13. What is the circumference of a circle of 10.00 in. diameter, given that $\pi = 3.1416$? *Ans.* 31.42 in.

4.14. A car travels 1.754×10^2 miles in 191.2 minutes. What is the average speed in miles per hour? *Ans.* 55.04 mi./hr.

4.15. Give the number of feet in a mile to three significant digits. *Ans.* 5.28×10^3 ft.

4.16. How many significant digits are there in the quantity defining the meter in terms of the wave length of the krypton-86 spectral line? *Ans.* 9.

4.17. A circle has a diameter of 10.00 cm. Give the circumference of the circle to four significant digits. *Ans.* 31.42 cm.

4.18. Give the number of miles in one foot to three significant digits. *Ans.* 1.89×10^{-4} mi.

Statistical Methods

4.4 The Arithmetic Mean. The *arithmetic mean* will be designated as $\overline{X}$, and is defined as:

$$\overline{X} = \frac{\Sigma X}{n}$$

where ΣX represents the sum of a series of n measurements.

Example 4.6. Calculate the arithmetic mean for the values 10.42 g, 9.81 g, 9.90 g, 10.35 g, and 9.67 g.

Solution.

$$\overline{X} = \frac{\Sigma X}{n} = \frac{10.42 + 9.81 + 9.90 + 10.35 + 9.67}{5} = 10.03 \text{ g.}$$

The arithmetic mean is commonly referred to as the "mean" or "average."

4.5. The Standard Deviation. The *deviation* of a single value is the difference between that value and the arithmetic mean. It is designated as $(X - \overline{X})$. The variance, S^2, is defined as:

$$S^2 = \frac{(X_1 - \overline{X})^2 + (X_2 - \overline{X})^2 + \ldots\ldots + (X_n - \overline{X})^2}{n - 1}$$

$$= \frac{\Sigma(X - \overline{X})^2}{n - 1}.$$

Expressing the variance as S^2 means that all deviations, whether positive or negative, will have a positive value. The reason for using $n - 1$ rather than n in the expression for the variance involves theory that we need not discuss. It is evident that if n is large there will be little difference between n and $n - 1$.

The *standard deviation*, S, is defined as the square root of the variance. That is:

$$S = \sqrt{\frac{\Sigma(X - \overline{X})^2}{n - 1}}.$$

Example 4.7. Calculate the arithmetic mean, the deviation, the variance, and the standard deviation for the data given in Example 4.6.

Solution.

X (g)	$\overline{X}$ (g)	$(X - \overline{X})$ (g)	S^2 (g^2)	S (g)
10.42		0.39		
9.81		−0.22		
9.90	10.03	−0.13	0.11+	0.33+
10.35		0.32		
9.67		−0.36		

$$S^2 = \frac{(0.39)^2 + (-0.22)^2 + (-0.13)^2 + (0.32)^2 + (-0.36)^2}{4} = 0.11+\ g^2$$

$$S = \sqrt{0.11+} = 0.33+\ g.$$

4.6. Grouped Data. Experimental data sometimes consist of numerous measurements, many of which have the same numerical value. In such cases we may use *grouped data.* The number of times a particular value occurs in grouped data is called the *frequency grouping*, F.

Example 4.8. The following table gives the frequency grouping and ages, X, for 50 people. Calculate $\overline{X}$ and S for the given data.

X (yr)	F	FX	$\overline{X}$ (yr)	S (yr)
11	3	33		
14	6	84		
16	11	176	19.0	4.13
20	15	300		
23	8	184		
25	7	175		

$$\overline{X} = \frac{\Sigma(FX)}{n} = \frac{(33 + 84 + 176 + 300 + 184 + 175)}{50} = 19.0 \text{ yr.}$$

$$S = \sqrt{\frac{\Sigma F(X - \overline{X})^2}{n - 1}}$$

$$= \sqrt{\frac{(3 \times 8^2) + (6 \times 5^2) + (11 \times 3^2) + (15 \times 1^2) + (8 \times 4^2) + (7 \times 6^2)}{49}}$$

$$= 4.13 \text{ yr.}$$

4.7. Probability Distribution. Probability is a measure of the likelihood that an event will occur in a certain way.

Example 4.9. A bag contains 25 red balls, 50 white balls, and 100 blue balls. If one ball is drawn from the bag what is the probability that it will be (a) a red ball, and (b) a red ball or a white ball?

Solution.

(a) Probability $= \dfrac{25}{25 + 50 + 100} = \dfrac{1}{7}$. That is, there is one chance in seven that the ball will be red.

(b) Probability $= \dfrac{25 + 50}{25 + 50 + 100} = \dfrac{3}{7}$. That is, there are three chances in seven that the ball will be either red or white.

4.8. Normal Distributions. The *true value* of a series of measurements cannot be determined. As the number of observations increase, assuming that no gross errors are involved, $\overline{X}$ becomes more and more reliable as the true value. Only when n is infinite, may $\overline{X}$ be assumed to be the true value.

In order to describe distribution curves we will assume the theoretical situation in which n is infinite. Under these conditions the standard deviation, S, will be designated by σ, and the arithmetic mean (average), $\overline{X}$, by μ. If the frequency of occurrence, F, is now plotted on the y-axis, and the standard deviation, σ, on the x-axis, a bell-shaped symmetrical curve results as shown by the solid line in

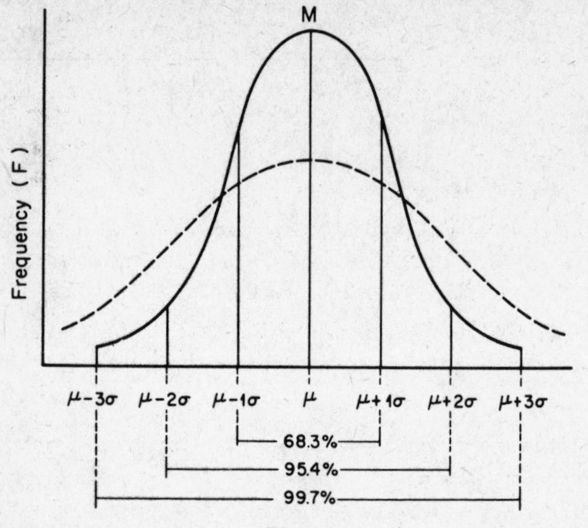

Fig. 4.1.

Fig. 4.1. This particular normal distribution is known as the Gaussian Distribution. The equation for the Gaussian Distribution frequency, F, at X_i is given by the equation:

$$F = \frac{1}{\sigma\sqrt{2\pi}} e^{-\frac{(X_i - \mu)^2}{2\sigma^2}}.$$

In the Gaussian curve (Fig. 4.1) the line $M\mu$ represents the true value of the observations. The Gaussian equation gives a normal distribution curve in which 68.3% of the measurements lie between $\mu - \sigma$ and $\mu + \sigma$, 95.4% between $\mu - 2\sigma$ and $\mu + 2\sigma$, and 99.7% between $\mu - 3\sigma$ and $\mu + 3\sigma$. At the $M\mu$ line, $\mu = 0$.

The total area under the curve, regardless of the value of σ, is always 1.000, a characteristic of all normal distribution curves.

Example 4.10. Out of a total of 5000 measurements which follow a normal distribution curve, how many would lie between $\mu + 2\sigma$ and $\mu + 3\sigma$?

Solution. Between $\mu + 2\sigma$ and $\mu + 3\sigma$ there would lie $\frac{(99.7 - 95.4)}{2}$ $= 2.15\%$ of the measurements. Then:

$$0.0215 \times 5000 = \text{approximately 108 measurements.}$$

It is very unusual to have large numbers of measurements in experimental work. The question now arises as to the variation from the Gaussian distribution for small numbers of measurements (less than 50). The dotted line in Fig. 4.1 shows the nature of the change in the curve for small numbers of measurements. Equations for the frequency of occurrence could be set up for each such curve. As the number of measurements decreases, the distribution curve becomes flatter, thus changing the areas between the different values of σ. For convenience, the areas under the curve have been calculated, and put into the form of a t-table for changing values of n. These values are given in Table 4.1 for various values of n and confidence levels of 50%, 95%, and 99%. For example, by a confidence level of 80% it is meant that 80% of the measurements would lie between certain values of μ representing 80% of the area under the curve. From Fig. 4.1 we see that the 80% confidence level lies between $\mu \pm 1\sigma$ and $\mu \pm 2\sigma$.

4.9. Standard Deviation of the Mean. The equation for the calculation of standard deviation applies to any value of n, large or small. The t-value corrects for sample size. The *standard deviation of the mean* adjusts for n and is defined as $\frac{S}{\sqrt{n}}$. Observe that the standard

TABLE 4.1

t-VALUES FOR VARIOUS VALUES OF n

	Per Cent Confidence Level		
n	50	95	99
2	1.000	12.706	63.657
3	0.816	4.303	9.925
4	0.765	3.182	5.841
5	0.741	2.776	4.604
6	0.727	2.571	4.032
7	0.718	2.447	3.707
8	0.711	2.365	3.499
9	0.706	2.306	3.355
10	0.703	2.262	3.250
11	0.700	2.228	3.169
12	0.697	2.201	3.106
13	0.695	2.179	3.055
14	0.694	2.160	3.012
15	0.692	2.145	2.977
20	0.687	2.086	2.845
25	0.684	2.060	2.787
∞	0.674	1.968	2.576

deviation of the mean decreases as the square root of n. The confidence level gives the probability that any measurement picked at random will fall within the confidence interval $\overline{X} \pm t \frac{S}{\sqrt{n}}$, where the value of t is obtained from a table of t-values (Table 4.1). The true value of $\overline{X}$ is unknown. However, we know that the value of $\overline{X}$ lies within the confidence interval $\overline{X} \pm t \frac{S}{\sqrt{n}}$. In Table 4.1, observe that the confidence interval will get larger as n decreases or as a higher confidence level is desired. Also, as n gets smaller, the value of $\frac{S}{\sqrt{n}}$ gets larger.

Example 4.11. What is the confidence interval for data containing 750 measurements for which there is a 95% probability that any one measurement taken at random will fall within that interval, given that $\overline{X}$ is equal to 340, and the sample standard deviation is 8.5?

Solution. From Table 4.1 we will select the t-value corresponding to $n = \infty$ and the 95% confidence level. The t-value is 1.968. Since the limits are $\overline{X} \pm t \frac{S}{\sqrt{n}}$, then:

$$340 \pm (1.968)(8.5) = 340 \pm 17.$$

That is, there is a 95% probability that any one of the 750 measurements chosen at random will lie within the interval 340 ± 17. Also, approximately 95%, or 713, of the 750 measurements will lie within the interval 340 ± 17.

Example 4.12. Two series of measurements were made in which $\overline{X}$ was the same and equal to 73.46 grams. In one series, n was equal to four with an S value of 4.8; while in the other series, n was equal to 100 with an S value of 1.6. Calculate the 95% confidence interval for each series.

Solution. For $n = 4$, the confidence interval would be:

$$73.46 \pm (3.182)\left(\frac{4.8}{\sqrt{4}}\right) = 73.46 \pm 7.64.$$

For $n = 100$, the confidence interval would be:

$$73.46 \pm (1.968)\left(\frac{1.6}{\sqrt{100}}\right) = 73.46 \pm 0.31.$$

Example 4.13. The following values give the density of aluminum in g/cm^3 as determined by each member of a class of five students: 2.663,

2.714, 2.681, 2.706, and 2.734. Calculate (a) $\overline{X}$, (b) S, and (c) the 95% confidence interval.

Solution.

(a) and (b) We will use a tabular form.

X	$(X - \overline{X})$	$(X - \overline{X})^2$
2.663	−0.037	0.00137
2.714	+0.014	0.00020
2.681	−0.019	0.00036
2.706	+0.006	0.00004
2.734	+0.034	0.00116

$$\overline{X} = \frac{13.498}{5} = 2.700\ \frac{g}{cm^3} \qquad S = \sqrt{\frac{0.00313}{4}} = 0.028\ \frac{g}{cm^3}$$

(c) $$95\%\text{ confidence interval} = 2.700 \pm (2.776)\left(\frac{0.028}{\sqrt{5}}\right) = 2.700 \pm 0.036\ \frac{g}{cm^3}$$

That is, there is a 95% probability that the true value lies within the interval $2.700 \pm 0.036\ \frac{g}{cm^3}$.

4.10. Significance Between Means. The previous discussions were concerned with the statistical treatment of a series of measurements. It is often desirable to determine whether a difference between the means from two series of measurements is significant or merely due to chance. The difference between two means would be $(\overline{X}_1 - \overline{X}_2)$, and the standard deviation would be:

$$\sqrt{\left(\frac{(n_1 - 1)S_1^2 + (n_2 - 1)S_2^2}{(n_1 + n_2 - 2)}\right)\left(\frac{1}{n_1} + \frac{1}{n_2}\right)}$$

The value for the t-ratio:

$$\frac{\overline{X}_1 - \overline{X}_2}{\sqrt{\left(\frac{(n_1 - 1)S_1^2 + (n_2 - 1)S_2^2}{(n_1 + n_2 - 2)}\right)\left(\frac{1}{n_1} + \frac{1}{n_2}\right)}}$$

may then be compared to t-values (Table 4.1) to determine the significance of the difference between the means. The n value in Table 4.1 would be $(n_1 + n_2 - 2)$.

Example 4.14. Two 50 ml burets were compared by weighing the water delivered from each. The following data were obtained.

	n	$\overline{X}$ (g)	S (g)	S^2 (g^2)
Buret 1	40	50.14	0.16	0.0256
Buret 2	40	50.01	0.14	0.0196

Is there a significant difference between the volumes of the two burets?

Solution.

$$\text{t-ratio} = \frac{50.14 - 50.01}{\sqrt{\left(\frac{(39)(0.16^2) + (39)(0.14^2)}{78}\right)\left(\frac{1}{40} + \frac{1}{40}\right)}} = 3.8.$$

$$n = 40 + 40 - 2 = 78.$$

The t-value from Table 4.1 at the 99% confidence level for $n = \infty$ is 2.576. Since the calculated t-ratio, 3.8, is greater than 2.576, the difference in volume between the two burets is significant and not due to chance.

Problems

4.19. Determine (a) the mean and (b) the standard deviation for the numbers 25, 35, 40, 55, and 60. *Ans.* (a) 43, (b) 14.4.

4.20. The following data were recorded for the weight of a test tube as determined by each of five students: 24.34 g, 24.20 g, 24.28 g, 24.41 g, and 24.38 g. Determine the standard deviation. *Ans.* 0.084 g.

4.21. By means of grouped data calculate the mean for the following numbers: 7, 6, 9, 4, 3, 7, 4, 2, 8, 6, 6, 2, 7, 4, 7, 3, 7, 4, 6, 3, 6, 5, 6, 7, 9, 8, 7, 6, 7, 6. *Ans.* 5.7.

4.22. A jar contains 100 pennies, 60 nickels, and 40 dimes. If one coin is taken from the jar what is the probability that it will be (a) a penny, (b) a dime, (c) a nickel or a dime? *Ans.* (a) 1:2, (b) 1:5, (c) 1:2.

4.23. Make a table using the headings X, $\overline{X}$, $(X - \overline{X})$, $(X - \overline{X})^2$, and S for the following percentage grades for a class of 12 students. 98, 97, 86, 84, 81, 79, 76, 74, 73, 68, 64, and 56. *Ans.* S = 12.4%.

4.24. A stock room supply contains 2400 test tubes, of which 200 are defective. Based on the law of probability, how many defective tubes would be found in a desk if each desk had been stocked with 12 test tubes? *Ans.* One defective test tube.

4.25. In a series of measurements, 100 values were recorded. Given that $\overline{X} = 0.214$ g, and that S = 0.046 g, determine (a) the 95% confidence interval, and (b) the standard deviation of the mean.
Ans. (a) 0.214 ± 0.091 g, and (b) 0.0046.

4.26. The following data were obtained by a laboratory class of 10 students in the determination of the per cent of oxygen in potassium chlorate: 38.81, 39.43, 39.04, 38.92, 39.22, 38.76, 39.57, 39.25, 38.98, and 39.54. Determine the 95% confidence interval. *Ans.* 39.13 ± 0.67%.

4.27. A laboratory experiment was designed to determine the effectiveness of two catalysts (A and B) on the decomposition of potassium chlorate. A series of measurements were made to determine the volume of oxygen, in milliliters, released by each catalyst under similar conditions.

Catalyst A 12.6, 12.9, 11.3, 11.7, 12.2, 12.5.
Catalyst B 12.4, 11.5, 12.1, 12.0, 11.6, 12.4.

Is there a significant difference in the effectiveness of the two catalysts? *Ans.* t-ratio = 0.8.

4.28. Two classes (A and B) of 24 students each determined the milliliters of oxygen released by a given quantity of mercuric oxide. Class A carried out the experiment at 300° C, and class B at 350° C. The following are the data:

	n	$\overline{X}$ (ml)	S (ml)
Class A	24	10.16	0.23
Class B	24	10.92	0.21

Is the difference significant? *Ans.* t-ratio = 12.

5

The Crystalline State

Most solids exist in the crystalline state. The molecules of some solids are arranged in a random pattern to form an amorphous solid. In this chapter we are interested in substances existing as crystals. Both covalent and ionic compounds may exist in the crystalline state.

Various experimental techniques have been devised to determine crystalline structure. The most common methods are X-ray diffraction and electron diffraction. To a great extent the gross structure of a crystal is an indication of the arrangement of ions or molecules within the crystal. For this reason molecular geometry is closely associated with crystalline structure.

5.1. Crystal Formation. When ions or molecules change from the gaseous or liquid states to the solid state, the ions or molecules usually orient themselves into a pattern to form a crystalline solid. Ionic solids such as NaCl have a much higher melting point than do molecular solids such as H_2O (Table 5.1).

The higher melting points of ionic solids indicate that the forces holding ions together in an ionic crystal are much greater than are the forces holding molecules together in a molecular crystal. Ionic crystals are held together by electrostatic forces, sometimes called coulombic

TABLE 5.1

THE MELTING POINTS OF SOME IONIC AND MOLECULAR SOLIDS

Ionic		*Molecular*	
Compound	m.p. (°C)	*Compound*	m.p. (°C)
NaCl	801	H_2O	0
CaO	2850	NH_3	−78
NaOH	318	CO_2	−57

forces. Molecular crystals are held together by a number of forces collectively called Van der Waals forces. Evidently electrostatic forces are much stronger than are Van der Waals forces. Therefore, in the process of melting a crystal, more thermal energy is required to overcome the electrostatic forces than the weaker Van der Waals forces. We will now define a crystal as a solid body formed by an element or compound and bounded by plane surfaces that are symmetrically arranged.

5.2. Crystal Geometry. Since they are three-dimensional, crystals may be described in terms of x-, y-, and z-axes, called the *crystallographic axes*, and the angles between the axes. The x-, y-, and z-axes are so chosen as to show a definite relationship to the external features of the crystal. All known crystals may be described in terms of their crystallographic axes and the angles between them, as shown in Fig. 5.1.

There are seven basic crystal systems. Table 5.2 gives the crystal systems, the angles between them, the relative lengths of the x-, y-, and z-axes, the space lattices possible, and examples of each.

The arrangement of ions or molecules in a crystal is called the *space lattice* of the crystal. A crystal, as we see it, consists of a multitude of smaller crystals. The smallest crystal possible that will show the symmetry of the crystal as a whole is called a *unit cell*. The unit cell, of course, is invisible even under a microscope.

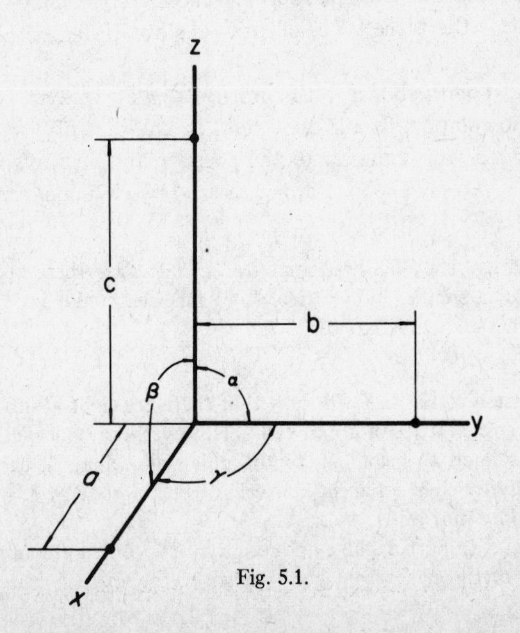

Fig. 5.1.

TABLE 5.2

THE SEVEN BASIC CRYSTAL SYSTEMS

System	*Angles*	*Axes*
1. Cubic	$\alpha = \beta = \gamma = 90°$	a = b = c
2. Tetragonal	$\alpha = \beta = \gamma = 90°$	a = b ≠ c
3. Hexagonal	$\alpha = \beta = 90°; \gamma = 120°$	a = b ≠ c
4. Rhombohedral	$\alpha = \beta = \gamma \neq 90°$	a = b = c
5. Orthorhombic	$\alpha = \beta = \gamma = 90°$	a ≠ b ≠ c
6. Monoclinic	$\alpha = \gamma = 90°; \beta \neq 90°$	a ≠ b ≠ c
7. Triclinic	$\alpha \neq \beta \neq \gamma \neq 90°$	a ≠ b ≠ c

Space Lattices	*Examples*
1. simple, face-centered, body-centered	NaCl, Au
2. simple, body-centered	MgF_2, Sn
3. simple	SiO_2, Zn
4. simple	$CaCO_3$, As
5. simple, face-centered, body-centered	$HgCl_2$, S
6. simple	$KClO_3$, B
7. simple	CuO, $CuSO_4 \cdot 5H_2O$

Of the seven crystal systems, only the cubic system will be discussed. Fig. 5.2 shows the three types of unit cells associated with the cubic system.

The dark spheres shown in the unit cells in Fig. 5.2 are ions in the case of ionic compounds and molecules in the case of covalent compounds. With ionic compounds the positive and negative ions alternate in the crystal in such a manner that no two like ions are adjacent to each other.

Example 5.1. How many atoms of an element are there in a unit cell that forms (a) a simple cubic lattice, (b) a face-centered cubic lattice, and (c) a body-centered cubic lattice?

Solution.

(a) Inspection of Fig. 5.2 will show that there are eight atoms of the element in the space lattice of a unit cell. However, only one-eighth of each atom is considered as belonging to the unit cell. That is, each atom is shared equally by eight adjacent unit cells. Therefore: $8 \times \frac{1}{8} = 1$ atom of the element in a unit cell.

(b) Again, as in part a, the eight corner atoms contribute an equivalent of one atom to the cell. Analogously, each of the six face-centered atoms

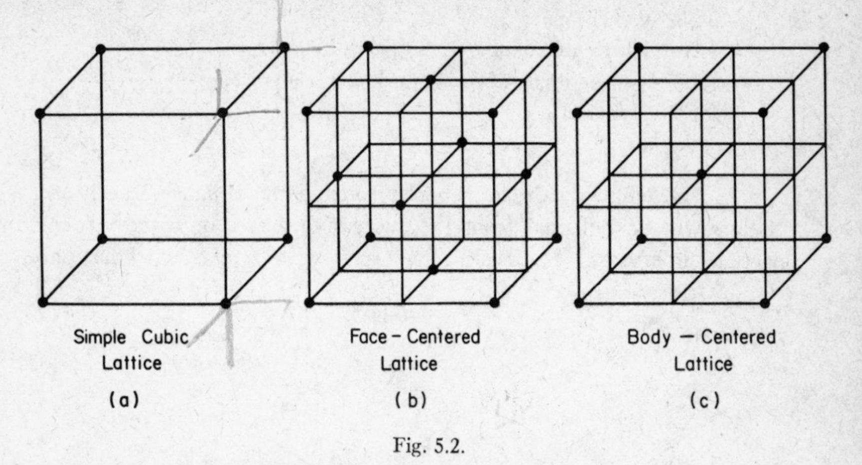

Fig. 5.2.

(shared by two unit cells) contribute one-half atom to the cell, or a total of three atoms. Therefore, each unit cell contains an equivalent of four atoms.

(c) As in part a, the eight corner atoms contribute an equivalent of one atom to the cell. In addition, there is one atom completely within the cell. Therefore, there are two atoms within the unit cell.

When the crystal lattice structure consists of the positive and negative ions of a compound, the composition of a unit cell may be more complex than shown in Fig. 5.2. If the Na^+ and Cl^- ions in a crystal of NaCl are placed alternately in the space lattice, the figure obtained is not symmetrical with respect to Na^+ and Cl^- ions. Therefore, it is not a unit cell. In order to obtain a unit cell that is symmetrical in regard to the positive and negative ions, it is necessary to form a cube consisting of eight of the cubes shown in Fig. 5.2 (a). Each of the six faces of the cube would then have symmetry of ions.

Example 5.2. (a) Draw a diagram of a unit cell of NaCl. (b) How many ions each of Na^+ and Cl^- are there in a unit cell of NaCl?

Solution.

(a) The unit cell of NaCl contains eight Na^+ corner ions, six Na^+ face-centered ions, twelve Cl^- ions on the six faces, and one center Cl^- ion. Draw the structure.

(b) One unit cell of NaCl consists of eight basic cell units as shown in Fig. 5.2 (a). In Example 5.1 we calculated that such a basic cell unit contained an equivalent of one atom or, in this case, one ion. The eight basic cell units contain therefore: $8 \times 1 = 8$ ions (4 Na^+ and 4 Cl^-).

A second solution is to determine the fraction of each ion within a unit cell of NaCl. Then:

$$\left.\begin{array}{lcl} 8\ Na^+ \text{ corner ions} & = & 8 \times \frac{1}{8} = 1 \text{ ion} \\ 6\ Na^+ \text{ face-centered ions} & = & 6 \times \frac{1}{2} = 3 \text{ ions} \end{array}\right\} = 4\ Na^+ \text{ ions.}$$

$$\left.\begin{array}{lcl} 12\ Cl^- \text{ ions on 6 faces} & = & 12 \times \frac{1}{4} = 3 \text{ ions} \\ 1\ Cl^- \text{ center ion} & & = 1 \text{ ion} \end{array}\right\} = 4\ Cl^- \text{ ions.}$$

5.3. Diffraction in Crystals. Electromagnetic radiations such as X-rays will be reflected from the successive layers of space lattice particles in crystals (Fig. 5.3). The wavelength, λ, of such radiation

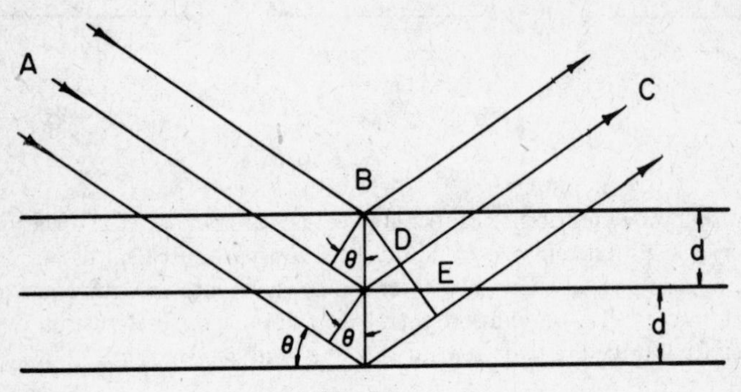

Fig. 5.3.

must be of the same order of magnitude as the distance, d, between successive layers in the crystal. Reference to Fig. 5.3 will show that the entering rays, A, travel increasingly longer distances from successive layers (only three are shown) before forming the emerging wavefront BDE. When the increased distance traveled is not a whole number of wavelengths of the incident radiation, the reflected wave is out of phase and interference results, as a result of which there will be a dark area on the image at C. When the difference in distance is a whole number of wavelengths, the reflected waves will be in phase with little or no diminution of intensity of light. That is, as the angle Θ changes there will be an outward progression of maxima and minima of intensity of light at C similar to water waves. Beginning with the smallest angle Θ they are called first order, second order, etc., reflections. The order of the reflection is designated by n, so that n = 1, 2, 3, etc. By simple geometry it is possible to show that:

$$n\lambda = 2d \sin \Theta$$

where λ is the wavelength of the incident radiation. Since λ and Θ may be determined experimentally, it is possible to use this formula to determine the distance, d, between successive layers in a crystal.

Example 5.3. Diffraction studies of tungsten crystals, using X-rays for which $\lambda = 1.85$ A, showed that the first order reflection (n = 1) occurred at 36° 0′. Calculate the distance between successive layers in a tungsten crystal.

Solution. Since $n\lambda = 2d \sin \Theta$, then:

$$d = \frac{n\lambda}{2 \sin \Theta}$$

and since sin 36° 0′ = 0.588:

$$d = \frac{(1)(1.85 \text{ A})}{(2)(0.588)} = 1.57 \text{ A} = 1.57 \times 10^{-8} \text{ cm.}$$

5.4. The Avogadro Number and Crystal Structure. The gram-atomic weight, abbreviated gram-atom, is the atomic weight of an element expressed in grams. (This will be further discussed in Chapter 6.) The *Avogadro number*, 6.02×10^{23}, represents (1) the number of atoms in one gram-atom of an element, (2) the number of molecules in one mole of a covalent compound, and (3) the number of ions in one mole of an ion. It is so named in honor of the Italian scientist Amedeo Avogadro (1776–1856). If we assume that the atoms, molecules, or ions in a crystal are spheres in direct contact with one another, it should be possible to calculate the Avogadro number from diffraction studies in crystals.

Example 5.4. Silver has a density of 10.5 grams per cubic centimeter, and consists of a face-centered cubic lattice structure. X-ray diffraction studies show the unit cell to measure 4.08 A. Calculate the value of the Avogadro number.

Solution. The atomic weight of silver is 107.9. That is, 107.9 g of Ag would contain 6.02×10^{23} atoms, and:

$$\frac{107.9 \not{g}}{10.5 \frac{\not{g}}{\text{cm}^3}} = 10.3 \text{ cm}^3 = 1.00 \text{ gram-atom of Ag.}$$

One unit cell of Ag contains four Ag atoms (Example 5.1 b). Therefore: $(4.08 \text{ A})^3$ contains four atoms of Ag.

Since $1.00 \text{ A} = 1.0 \times 10^{-8}$ cm, the volume of one atom of Ag is:

$$\frac{[(4.08 \not{\text{A}})(1.0 \times 10^{-8} \text{ cm}/\not{\text{A}})]^3}{4} = 1.70 \times 10^{-23} \text{ cm}^3.$$

Then in 1.00 gram-atom of Ag, 10.3 cm^3, there would be:

$$\frac{10.3\ \cancel{cm^3}}{1.70 \times 10^{-23}\ \cancel{cm^3}/\text{atom}} = 6.06 \times 10^{23} \text{ atoms},$$

which is in close agreement to the accepted value of 6.02×10^{23}.

Example 5.5. Barium crystals exist as a body-centered cubic lattice. Based on X-ray diffraction studies, the unit cell measures 5.02 A. Calculate the density of barium.

Solution. One unit cell of a body-centered cubic lattice contains two atoms (Example 5.1 c). Since 1.0 A = 1.0×10^{-8} cm, then the volume of one Ba atom is:

$$\frac{(5.02 \times 10^{-8}\text{ cm})^3}{2} = 6.30 \times 10^{-23}\text{ cm}^3.$$

Since there are 6.02×10^{23} atoms in one gram-atom of Ba, then:

$$6.30 \times 10^{-23}\,\frac{\text{cm}^3}{\cancel{\text{atom}}} \times 6.02 \times 10^{23}\,\frac{\cancel{\text{atoms}}}{\text{g-atom}} = 38.0\,\frac{\text{cm}^3}{\text{g-atom}}.$$

Since 1.00 gram-atom of Ba = 137.34 g, then:

$$\text{Density of Ba} = \frac{137.34\,\dfrac{\text{g}}{\cancel{\text{g-atom}}}}{38.0\,\dfrac{\text{cm}^3}{\cancel{\text{g-atom}}}} = 3.62\,\frac{\text{g}}{\text{cm}^3}.$$

When crystals are formed, some of the crystal lattice spaces may not be filled. This phenomenon is called a *crystal defect.* Because of such crystal defects the calculated density of a substance (Example 5.5) is sometimes greater than the experimental density. The number of empty spaces are in direct proportion to the density values.

Example 5.6. The experimental density of cesium chloride is 3.907 g/cm^3, and the calculated density 3.920 g/cm^3. What is the per cent of empty lattice spaces in a crystal of cesium chloride?

Solution.

$$\text{Per cent empty lattice spaces} = \frac{3.920 - 3.907}{3.907} \times 100 = 0.33\%.$$

Example 5.7. Gold crystallizes as a face-centered cubic structure. The side of a unit cell is 4.07 A. What is the radius of a gold atom, assuming the atoms to be in contact in the crystal?

Solution. Each face of a unit cell consists of five atoms of gold in contact. The diagonal of the square formed by connecting the centers of the four corner atoms is 4r, where r is the radius of a gold atom. (Draw the diagram.) Also, the diagonal of the square formed would be $(4.07 \times \sqrt{2})$ A. Equating the two values we have:

$$4r = (4.07 \times \sqrt{2}) \text{ A}$$

or

$$r = 1.44 \text{ A} = \text{radius of a gold atom.}$$

Problems

5.1. Lithium fluoride crystallizes as a face-centered cubic structure. How many ions each of lithium and of fluorine are there in a unit cell of lithium fluoride? *Ans.* Four each.

5.2. Aluminum crystallizes as a face-centered cube. How many aluminum atoms are there in a unit cell? *Ans.* Four.

5.3. X-rays of 1.54 A wavelength gave a first order reflection ($n = 1$) of 11.8° for a crystal. (a) What is the distance between layers in the crystal? (b) What would be the angle of the second order reflection? *Ans.* (a) 3.76 A, (b) 24.2°.

5.4. Silver exists as face-centered cubic crystals. A unit cell measures 4.08 A. (a) Calculate the density of silver. (b) Given the experimental density of silver as 10.5 g/cm^3, calculate the per cent of empty lattice spaces in silver crystals. *Ans.* (a) 10.6 g/cm^3, (b) 0.95%.

5.5. Sodium crystallizes as a body-centered cubic lattice structure. The unit cell measures 4.29 A. (a) Calculate the density of sodium. (b) Given the experimental density as 0.97 g/cm^3, calculate the per cent vacant positions in the lattice structure. *Ans.* (a) 1.00 g/cm^3, (b) 3.0%.

5.6. Aluminum forms face-centered cubic crystals. The unit cell measures 4.04 A. (a) What is the weight in grams of a unit cell of aluminum? (b) How many unit cells would there be in a rectangular block of aluminum measuring 6.060×10^{-5} cm by 4.040×10^{-5} cm by 1.616×10^{-4} cm? *Ans.* (a) 17.9×10^{-23} g, (b) 6.00×10^{9} unit cells.

5.7. Barium crystallizes as a body-centered cubic lattice. Its density is 3.05 g/cm^3. What is the length of an edge of a unit cell? *Ans.* 5.07 A.

5.8. What is the diameter of an aluminum atom if aluminum crystals are face-centered cubes? The density of aluminum is 2.70 g/cm^3. *Ans.* 2.90 A.

5.9. From X-ray diffraction studies, the edge of a unit cell of NaCl is 5.63 A. The density is 2.17 g/cm^3. Calculate (a) the density of NaCl, and (b) the per cent of empty lattice spaces. *Ans.* (a) 2.18 g/cm^3, (b) 0.46%.

5.10. The phosphorus molecule, P_4, forms a tetrahedron. How many phosphorus atoms are there in a unit cell consisting of one tetrahedron? *Ans.* One-third atom.

5.11. How many atoms are there in a unit cell of Mg, which forms hexagonal crystals, there being a face-centered atom in each end of the unit cell and three completely enclosed atoms within the unit cell? Suggestion: Draw a diagram of a unit cell. *Ans.* 6 atoms.

5.12. Assume the 12 atoms of magnesium at each corner of the hexagonal unit cell to be in contact, express the volume of a unit cell in terms of r, the radius of a magnesium atom. *Ans.* 33.9 r.

5.13. In hexagonal magnesium crystals the atoms are close-packed (in contact). Calculate the density of magnesium. *Ans.* 1.74 g/cm^3.

5.14. Sodium chloride was purified by recrystallization from a solution containing fluoride ion. Calculate the per cent of fluorine in the NaCl crystals if, on the average, one fluoride ion had replaced a chloride ion in one out of every 1000 unit cells of NaCl. *Ans.* 0.0081%.

5.15. One face of a crystal forms an unsymmetrical triangle ABC with opposite sides abc. Given that angle CAB = 140°, b = 1.25 cm, and c = 2.25 cm, calculate the length of side a. *Ans.* 3.31 cm.

5.16. One face of a crystal is triangular, measuring 1.40 cm by 2.20 cm by 1.90 cm. Calculate the value of each of the three angles on the crystal face. *Ans.* 83°, 59°, and 38°.

6

Chemical Units of Mass

In previous chapters we have discussed some physical units of mass, such as the gram and pound. Units of mass based upon the masses of atoms and molecules have been found more useful by scientists than physical units. Two such chemical units of mass, the *gram-atom* and the *mole*, will be discussed in this chapter.

Relative Units of Mass

6.1. Atomic Weight. A single atom or molecule is too small to weigh, even on the most sensitive balance made. However, scientists are able to determine the weights of single atoms and molecules. This is done by weighing masses containing a known number of atoms or molecules, and dividing the weight obtained by the number of atoms or molecules in the mass. For example, an atom of hydrogen has been found to weigh 1.67×10^{-24} g.

The actual weights of atoms or molecules, in grams, are cumbersome to use because of their extremely small magnitude. The *relative weights* of atoms are less cumbersome to use. By relative weight is meant the number of times heavier an atom is than another atom taken as a standard. The problem is to pick the atom to be used as a standard. This problem is further complicated by the fact that most of the elements exist as *isotopes*. The atom chosen by scientists to be used as the standard of mass was the lighter of the two naturally occurring isotopes of carbon. This atom arbitrarily was assigned a value of exactly 12 *atomic mass units* (a m u). The isotope is designated as C^{12} or carbon-12. When the relative weight of a specific isotope is referred to carbon-12, the value is called the *isotopic weight*. For example, chlorine exists as two naturally occurring isotopes of isotopic weights 34.965 a m u and 36.964 a m u , respectively. Atoms of chlorine therefore are approximately three times as heavy as atoms of carbon-12. The relative weights of the naturally occurring mixtures of the isotopes of the elements, using carbon-12 as the standard, are called the *atomic weights* of the elements (Appendix III). The atomic weight of chlorine is 35.453 amu. In naturally occurring chlorine there are about three times as many atoms of the lighter isotope as there are atoms of the heavier isotope.

The term *isotopic weight* also is called *nuclidic mass*, *atomic mass*, and even *atomic weight*. Each specific kind of atom of an element with a specific mass number is a *nuclide*. That is, collectively, the isotopes of all the elements are called nuclides.

Example 6.1. How many times heavier is an atom of uranium than an atom of carbon-12?

Solution. Uranium occurs as a mixture of isotopes. Therefore, the atoms of the isotopic forms would, on the average, be:

$$\frac{238.03}{12.00} = 19.84 \text{ times as heavy as an atom of carbon-12.}$$

Example 6.2. By mass spectrometric analysis, copper was found to exist as the isotopes Cu^{63} (62.93 amu) and Cu^{65} (64.93 amu). Naturally occurring copper contains 69.4% Cu^{63}. Calculate the atomic weight of naturally occurring copper.

Solution. Naturally occurring copper contains 100.0% − 69.4% = 30.6% Cu^{65}. Let us assume that we have a mass of copper containing 100 atoms. Then 69.4 of the 100 atoms would be Cu^{63} atoms, and 30.6 would be Cu^{65} atoms (the ambiguity of parts of atoms may be assumed for our purpose). The mass of the 100 atoms would be:

$$(69.4 \times 62.93) \text{ amu} + (30.6 \times 64.93) \text{ amu} = 6354 \text{ amu}.$$

The average amu contributed by the 100 atoms would be:

$$\frac{6354 \text{ amu}}{100 \text{ atoms}} = 63.54 \text{ amu} = \text{atomic weight of Cu}.$$

Example 6.3. Naturally occurring boron consists of B^{10} (10.013 amu) and B^{11} (11.009 amu). The atomic weight of boron is 10.811 amu. What is the relative abundance of each of the isotopes?

Solution. Let x = per cent abundance of B^{10}
then: $(100 - x)$ = per cent abundance of B^{11}.

On the basis of 100 atoms of boron:

$$10.811 = \frac{(10.013x) + [11.009\,(100 - x)]}{100}$$

$$x = 19.9\%\ B^{10}$$

and: $(100.0 - 19.9) = 80.1\%\ B^{11}$.

6.2. Formula Weight. A compound contains a definite ratio of the atoms of the constituent elements. In covalent compounds such as water, H_2O, the formula tells us that the ratio of hydrogen atoms to oxygen atoms is 2 : 1. In ionic compounds such as sodium chloride, NaCl, the formula represents the ratio of ions in any given mass of the substance. That is, in any given mass of sodium chloride there are equal numbers of sodium ions, Na^+, and chloride ions, Cl^-.

The sum of the weights of the atoms, as shown in the formula, represents the *formula weight* of a substance. In the case of covalent compounds such as water, the formula weight also may be called the *molecular weight.*

Example 6.4. Calculate the formula weight for (a) H_2O, (b) NaCl, and (c) $Fe_2(SO_4)_3$.

Solution.

(a) $H_2O = (2 \times 1.00797) + 15.9994 = 18.0153$ a.m.u.

(b) $NaCl = 22.9898 + 35.453 = 58.443$ a.m.u.

(c)
$$
\begin{aligned}
Fe_2(SO_4)_3 = \quad 2 \times 55.847 &= 111.694 \\
3 \times 32.064 &= 96.192 \\
12 \times 15.9994 &= \underline{191.993} \\
& 399.879 \text{ a.m.u.}
\end{aligned}
$$

Chemical Units

6.3. The Gram-Atom. The weights in grams of the atoms of the elements must be in the same ratio as their atomic weights. This is really another way of defining atomic weight. On this basis, the weight of a chlorine atom may be calculated, since we know the weight of a hydrogen atom. Then:

$$
\frac{1.008}{35.453} \times (1.67 \times 10^{-24}\text{ g}) = 58.7 \times 10^{-24}\text{ g}
$$
$$
= \text{weight of one atom of chlorine.}
$$

Again, an ambiguity arises in that chlorine consists of two isotopes, having an average atomic weight of 35.453 amu.

From the above we see that 1.008 g of hydrogen will contain the same number of atoms as 35.453 g of chlorine. Continuing this application to all the elements we may conclude that the atomic weights of the elements expressed in grams represent a similar number of atoms for all the elements. This number has been found experimentally to be 6.02×10^{23} atoms, the Avogadro number. (See Sec. 5.4.)

The *gram-atomic weight*, abbreviated *gram-atom*, is the atomic weight of an element expressed in grams. Also, the gram-atom may be defined as the weight in grams of 6.02×10^{23} atoms of an element. The gram-atom concept is valuable to scientists since it enables them to measure out masses of elements containing known numbers of atoms.

Example 6.5. How many gram-atoms of mercury are there in a bottle containing 1000 g. of mercury?

Solution. Since 1.00 gram-atom of mercury is 200.59 g., then 1000 g. of mercury would be:

$$\frac{1000 \cancel{\text{g Hg}}}{200.59 \frac{\cancel{\text{g Hg}}}{\text{g-atom}}} = 4.99 \text{ gram-atoms of Hg.}$$

Example 6.6. A bottle was known to contain 2.50 gram-atoms of sulfur. How many grams of sulfur did the bottle contain?

Solution. Since 1.00 gram-atom of sulfur is 32.064 g., then 2.50 gram-atoms of sulfur would be:

$$32.064 \frac{\text{g S}}{\cancel{\text{g-atom}}} \times 2.50 \cancel{\text{g-atom}} = 80.16 \text{ g of S.}$$

Example 6.7. How many atoms are there in 5.00 g. of oxygen?

Solution. Since 16.00 g. of oxygen (1.00 gram-atom) contains 6.02×10^{23} atoms, then 5.00 g. of oxygen would contain:

$$\frac{5.00 \cancel{\text{g } O_2}}{16.00 \cancel{\text{g } O_2}} \times 6.02 \times 10^{23} \text{ atoms } O_2 = 1.88 \times 10^{23} \text{ atoms of } O_2.$$

Example 6.8. What is the weight in grams of one B^{10} atom? The isotopic weight of B^{10} is 10.013 amu.

Solution. Since 10.013 g of B^{10} contain 6.02×10^{23} atoms, then one atom of B^{10} would weigh:

$$\frac{10.013 \text{ g}}{6.02 \times 10^{23} \text{ atoms}} = 1.66 \times 10^{-23} \frac{\text{g}}{\text{atom}}.$$

6.4. The Mole. The *gram-formula weight* of a substance is the formula weight expressed in grams. For covalent compounds this may also be called the *gram-molecular weight*. For convenience the gram-formula weight, or gram-molecular weight, is abbreviated *mole*.

Just as 1.00 gram-atom of an element contains 6.02×10^{23} atoms, so 1.00 mole of a covalent compound contains 6.02×10^{23} molecules of the compound.

Example 6.9. For convenience, Avogadro's number is often designated by the letter N. Show that one mole of water contains N molecules.

Solution. One mole of water contains:

	2.00 gram-atoms of hydrogen	= 2 N atoms
	1.00 gram-atom of oxygen	= N atoms
Therefore,	1.00 mole of H_2O	= 3 N atoms.

However there are 3 atoms per molecule of water. Therefore,

$$1.00 \text{ mole of } H_2O = \frac{3 \text{ N } \cancel{\text{atoms}}}{3 \frac{\cancel{\text{atoms}}}{\text{molecule}}} = \text{N molecules.}$$

For ionic compounds mole quantities may be interpreted in terms of numbers of ions. For sodium chloride, for example:

$NaCl$	$\rightarrow$ Na^+	+ Cl^-
1.00 mole	1.00 mole	1.00 mole
58.443 g	22.990 g	35.453 g
———	N ions	N ions

Similarly:

$Fe_2(SO_4)_3$	$\rightarrow$ 2 Fe^{+3}	+ 3 SO_4^{-2}
1.00 mole	2.00 mole	3.00 mole
———	2 N ions	3 N ions.

Many of the elements, such as iron and copper, exist as one atom in a molecule. For such elements the gram-atom and mole are identical quantities. That is:

$$1.00 \text{ mole of iron} = 1.00 \text{ gram-atom} = 55.847 \text{ g.}$$

As with the gram-atom, the mole concept enables scientists to measure out masses of substances containing known numbers of ions or molecules.

Example 6.10. How many grams are there in 2.75 moles of $Fe_2(SO_4)_3$?

$$399.9 \frac{\text{g}}{\cancel{\text{mole}}} \times 2.75 \cancel{\text{moles}} = 1100 \text{ g } Fe_2(SO_4)_3.$$

Example 6.11. How many molecules of water are there in 5.00 g of water?

$$\frac{5.00 \cancel{\text{g}}}{18.02 \cancel{\text{g}}} \times 6.02 \times 10^{23} \text{ molecules} = 1.66 \times 10^{23} \text{ molecules } H_2O.$$

Example 6.12. Sulfur has the formula S_8 under average conditions. How many moles of sulfur are there in 500 g of sulfur?

Solution. 1.00 mole of sulfur = $8 \times 32.066 = 256.53$ g. Then:

$$\frac{500 \not{g}\,\not{S}}{256.53 \frac{\not{g}\,\not{S}}{\text{mole}}} = 1.95 \text{ moles of S.}$$

Problems

Part I

6.1. How many atoms are represented by each of the following formulas: (a) Fe_2O_3, (b) $Fe(OH)_3$, (c) $Ca(HCO_3)_2$, (d) $Mg_3(PO_4)_2$, (e) $(NH_4)_2Cr_2O_7$? *Ans.* (a) 5, (b) 7, (c) 11, (d) 13, (e) 19.

6.2. What is the total number of atoms represented in each of the following expressions: (a) 2 Fe_2O_3, (b) 5 $Ca(HCO_3)_2$, (c) 10 $(NH_4)_2SO_4$, (d) 5 $(NH_4)_2Cr_2O_7$, (e) 6 H_2?
Ans. (a) 10, (b) 55, (c) 150, (d) 95, (e) 12.

6.3. Determine the formula weight of each of the following: (a) Fe_2O_3, (b) $Fe(OH)_3$, (c) $Ca(HCO_3)_2$, (d) $Mg_3(PO_4)_2$, (e) $(NH_4)_2Cr_2O_7$.
Ans. (a) 159.70, (b) 106.88, (c) 162.12, (d) 262.91, (e) 252.10.

6.4. How many gram-atoms of sulfur are there in 100 g of sulfur?
Ans. 3.12.

6.5. How many grams are there in 2.75 gram-atoms of iron?
Ans. 154 g.

6.6. How many gram-atoms of iron are there in a rectangular block of iron measuring 4.00 cm by 2.50 cm by 10.00 cm, the density of the iron being 7.90 $\frac{\text{g}}{\text{cm}^3}$? *Ans.* 14.1.

6.7. How many moles are there in each of the following quantities: (a) 250 g of SO_2, (b) 250 g of Fe_2O_3, (c) 250 g of O_2?
Ans. (a) 3.90, (b) 1.57, (c) 7.81.

6.8. How many grams are there in each of the following quantities: (a) 2.65 moles of NaCl, (b) 1.50 moles of sulfur, (c) 5.00 moles of nitrogen? *Ans.* (a) 155 g, (b) 48.10 g, (c) 140 g.

6.9. How many grams are there in 100 millimoles of sugar, $C_{12}H_{22}O_{11}$?
Ans. 34.2 g.

6.10. What is the weight in grams of one atom of lead?
Ans. 344×10^{-24} g.

6.11. What is the weight in grams of 2 N atoms of aluminum?
Ans. 53.96 g.

6.12. How many times heavier is an atom of silver than an atom of iron?
Ans. 1.93 times.

6.13. How many molecules are there in 0.75 mole of CO_2? *Ans.* 0.75 N.

6.14. How many molecules are there in 0.75 mole of Fe? *Ans.* 0.75 N.

6.15. How many atoms are there in 2.50 moles of oxygen? *Ans.* 5 N.

Part II

6.16. Determine the formula weight of (a) oxygen, (b) copper, (c) sugar, $C_{12}H_{22}O_{11}$. *Ans.* (a) 32, (b) 63.54, (c) 342.30.

6.17. How many moles are there in one liter of water? *Ans.* 55.5 moles.

6.18. How many moles are there in a cube of aluminum 10.0 cm on a side? *Ans.* 100.

6.19. How many moles are there in 1.00 kg of iron rust, Fe_2O_3? *Ans.* 6.26.

6.20. How many moles are there in 1.00 kg of iron? *Ans.* 17.9.

6.21. How many moles are there in 1.00 in.^{3} of water? *Ans.* 0.91 mole.

6.22. How many millimoles are there in 1.00 mole of any substance? *Ans.* 1000.

6.23. How many millimoles are there in 1.00 g of water? *Ans.* 55.5.

6.24. How many gram-atoms are there in a cube of platinum 10.0 cm on a side? *Ans.* 110.

6.25. How many gram-atoms are there in 1.00 kg of iron? *Ans.* 17.9.

6.26. How many gram-atoms each of chemically combined hydrogen and oxygen are there in 1.00 mole of water? *Ans.* 2.00 H_2, 1.00 O_2.

6.27. How many moles each of chemically combined hydrogen and oxygen are there in 1.00 mole of water? *Ans.* 1.00 H_2, 0.50 O_2.

6.28. How many grams are there in 1.65 gram-atoms of copper? *Ans.* 105 g.

6.29. What is the weight in grams of 0.35 gram-atom of sodium? *Ans.* 8.0 g.

6.30. A mixture contains 0.25 gram-atom of iron and 1.25 gram-atoms of sulfur. What is the weight of the mixture? *Ans.* 54.1 g.

6.31. An experiment calls for one-half mole of zinc. How many grams would be required? *Ans.* 32.69 g.

6.32. How many moles of H_2SO_4 would be contained in 250 ml of a 95 per cent solution of sulfuric acid, density 1.84 $\frac{g}{ml}$? *Ans.* 4.5 moles.

6.33. One liter of hydrogen chloride gas weighs 1.63 grams. How many liters are there in one mole of hydrogen chloride? *Ans.* 22.4 *l.*

6.34. How many grams are there in one pound-molecular weight of zinc sulfate, $ZnSO_4$? *Ans.* 7.3×10^4 g.

6.35. What is the weight in milligrams of one atom of uranium? *Ans.* 4×10^{-19} mg.

6.36. What is the weight in micrograms of 1,000,000 atoms of gold? *Ans.* 3.3×10^{-10} μg.

6.37. What is the weight in grams of 0.5 N atoms of chlorine? *Ans.* 17.73 g.

6.38. What is the weight in grams of 0.5 N molecules of chlorine? *Ans.* 35.46 g.

6.39. Given that an atom of mercury weighs 333×10^{-24} g ; calculate the atomic weight of mercury by means of N. *Ans.* 200.6.

6.40. How many atoms are there in one cm^3 of lead, density 11.3 g per cm^3? *Ans.* 3.28×10^{22}.

6.41. What is the weight in grams of one molecule of water?
Ans. 3.0×10^{-23} g.

6.42. How many times heavier is a molecule of oxygen than a molecule of hydrogen? *Ans.* 15.9 times.

6.43. How many times heavier is an atom of oxygen than an atom of hydrogen? *Ans.* 15.9 times.

6.44. How many times heavier is an atom of oxygen than a molecule of hydrogen? *Ans.* 7.94 times.

6.45. How many times heavier is a molecule of CO_2 than a molecule of CH_4?
Ans. 2.75 times.

6.46. What is the weight in grams of 10^{30} atoms of oxygen?
Ans. 2.66×10^{7} g.

6.47. How many atoms of sulfur are there in 1.00 g of sulfur?
Ans. 1.9×10^{22}.

6.48. How many atoms of oxygen would be required to weigh 8.00 g ?
Ans. 0.5 N atoms.

6.49. How many moles are there in one pound of sugar, $C_{12}H_{22}O_{11}$?
Ans. 1.33 moles.

6.50. How many significant digits are there in (a) the atomic weight of carbon-12, and (b) the atomic weight of chlorine?
Ans. (a) infinite number, (b) 5.

6.51. Using the Avogadro number, calculate the weight of one calcium atom. *Ans.* 6.64×10^{-23} g.

6.52. Calculate the value in grams of 1.00 a.m u , using the atomic weight of hydrogen and the weight of one atom of hydrogen.
Ans. 1.66×10^{-24} g.

6.53. Using the value 1.00 a m u $= 1.66 \times 10^{-24}$ g , calculate the weight of (a) a mercury atom, and (b) a sulfur atom.
Ans. (a) 3.33×10^{-21} g , (b) 5.32×10^{-22} g.

6.54. How many ions each of Na^+ and Cl^- are there in 12.0 g of NaCl?
Ans. 1.24×10^{23} ions of each.

6.55. How many ions each of Ca^{+2} and Cl^- are there in 25.0 g of $CaCl_2$?
Ans. 1.36×10^{23} Ca^{+2} ions and 2.72×10^{23} Cl^- ions.

6.56. How many grams are there in 1.65 moles of Na^+ ions? *Ans.* 37.9 g.

6.57. How many grams each of Mg^{+2} ions and S^{-2} ions are there in 3.25 moles of MgS? *Ans.* 79.0 g of Mg^{+2}, 104.2 g of S^{-2}.

6.58. Naturally occurring chlorine has an atomic weight of 35.453 amu, and consists of 75.8% Cl^{35} and 24.2% Cl^{37}. The isotopic weight of Cl^{35} is 34.969 amu. What is the isotopic weight of Cl^{37}?
Ans. 36.966 amu.

6.59. How many atoms each of hydrogen and oxygen are there in 25.0 g of water? *Ans.* 16.744×10^{23} hydrogen atoms and 8.372×10^{23} oxygen atoms.

6.60. How many atoms each of iron and oxygen are there in 100 g of Fe_2O_3?
Ans. 1.25 N atoms of iron and 1.88 N atoms of oxygen.

7

The Physical Behavior of Gases

Under suitable conditions most substances may exist either as a gas, liquid, or solid. Many chemical reactions occur involving substances in the gaseous state, both as reactants and as products. It is important, therefore, that the physical laws relating to the gaseous state be understood before discussing problems involving chemical changes of substances existing in the gaseous state. Such physical laws relating to gases may usually be explained in terms of the kinetic energy of molecules.

The Energy of Molecules in the Gaseous State

7.1. The Kinetic-Molecular Theory of Matter. The principal difference between substances in the solid, liquid, and gaseous states lies in the freedom of movement of their molecules. The kinetic-molecular theory of matter deals with the energy associated with molecules due to their motion, the word *kinetic* coming from a Greek word meaning *motion*. Molecules in the liquid and solid states have a somewhat limited freedom of movement, whereas those in the gaseous state are limited in movement only by the walls of the containing vessel.

The kinetic energy of a particle in motion, whether it be a molecule or an aggregate of molecules, is given by the expression:

$$\text{Kinetic energy in ergs} = \tfrac{1}{2}\, mv^2$$

(v: velocity in cm per sec; m: mass in grams)

Since 10^7 ergs = 0.24 cal , the above expression becomes:

$$\text{Kinetic energy in calories} = (1.2 \times 10^{-8})\, mv^2$$

The kinetic energy of all molecules is the same at any given temperature. This means that light molecules, such as hydrogen, must be moving at a greater velocity at any given temperature than are heavier molecules, such as oxygen, in order to compensate for the difference in mass.

7.2. The Pressure Exerted by Gaseous Molecules. Gases are composed of molecules which are vibrating in all directions within the container, as shown in Fig. 7.1.

Gas pressure is the result of the impacts of the molecules on the walls of the containing vessel. The pressure exerted by the gas may be determined by measuring the force necessary to hold the piston, P, in a given position in the cylinder.

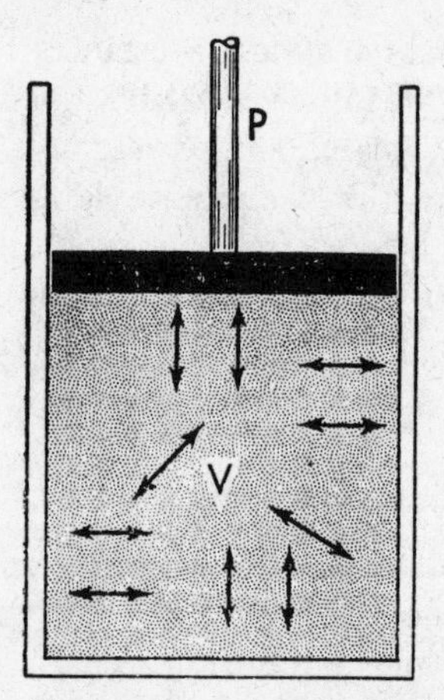

Fig. 7.1. Pressure is produced by the impact of molecules on the walls of the containing vessel.

Pressure is defined as the *force* acting on a *unit surface area*, and may be expressed in any one of several conventional units. Steam pressure is usually measured in pounds per square inch. In scientific work a unit called the *atmosphere* is commonly used. One atmosphere is the pressure exerted by the air on a unit surface area at sea level and is equal to:

1033 grams per square centimeter,
or 14.7 pounds per square inch.

The equivalent of one atmosphere pressure is commonly used in the solution of problems involving gases, and should not be confused with the above absolute values. Considering equal cross-sectional areas, one atmosphere is equivalent to the pressure exerted by:

760 millimeters of mercury,
29.9 inches of mercury,
or 33.9 feet of water.

In the solution of problems involving gas pressure, any of the above absolute or equivalent values may be used, provided the same units are used consistently throughout the problem.

Problems

7.1. Two bullets weighing 4.0 g and 8.0 g, respectively, are traveling at a velocity of 100 m per second. What is the kinetic energy of each (a) in ergs, and (b) in calories?

Ans. (a) 2.0×10^8 ergs, 4.0×10^8 ergs; (b) 4.8 cal, 9.6 cal.

7.2. Two bullets weighing 8.0 g each are traveling at velocities of 100 m per second and 200 m per second, respectively. What is the kinetic energy of each (a) in ergs, and (b) in calories?

Ans. (a) 4.0×10^8 ergs, 16.0×10^8 ergs; (b) 9.6 cal, 38.4 cal.

7.3. The mass of an oxygen molecule is approximately 16 times that of a hydrogen molecule. At room temperature a hydrogen molecule has a velocity of about one mile per second. What is the velocity of an oxygen molecule in feet per second under the same conditions?

Ans. 1320 $\frac{\text{ft}}{\text{sec}}$.

7.4. A pressure of 450 $\frac{\text{lb}}{\text{in.}^2}$ was recorded on a pressure gauge of an oxygen tank. What would this be in terms of atmospheres?

Ans. 30.6 atm.

7.5. A barometer registered 520 mm of Hg on the top of Pikes Peak, Colo. This would be equivalent to how many atmospheres?

Ans. 0.684 atm.

7.6. What is the kinetic energy in (a) ergs, and (b) calories, of a 16.0 lb shot traveling at a velocity of 20.0 $\frac{\text{ft}}{\text{sec}}$?

Ans. (a) 1.35×10^9 ergs; (b) 32.4 cal.

7.7. Under similar conditions, what would be the ratio of velocities of an oxygen molecule and a molecule of sulfur dioxide? A sulfur dioxide molecule has twice the mass of an oxygen molecule. *Ans.* 1.4 : 1.

7.8. What type of variation exists between the kinetic energy of a particle in motion and (a) its mass, and (b) its velocity?

Ans. (a) varies directly; (b) varies as velocity squared.

7.9. How many mm of Hg are equivalent to 0.760 atm?

Ans. 578 mm of Hg.

7.10. A pressure of 2.36 atm would be equivalent to how many pounds per square inch? *Ans.* 34.7 $\frac{\text{lb.}}{\text{in.}^2}$.

Laws Relating to Substances in the Gaseous State

7.3. Introduction. The volume of a given mass of gas is dependent upon the temperature and pressure under which the gas exists. It is, therefore, possible to describe the physical behavior of gases in terms of the three variables: *temperature*, T; *pressure*, P; and *volume*, V. For a given volume of gas under given conditions of temperature and pressure, a change in any one or more of the three variables will result in a change in the remaining variables according to definitely established laws called the *gas laws*. The relationship between any two of the variables may be studied, provided the remaining variable is maintained constant, as shown in Fig. 7.2.

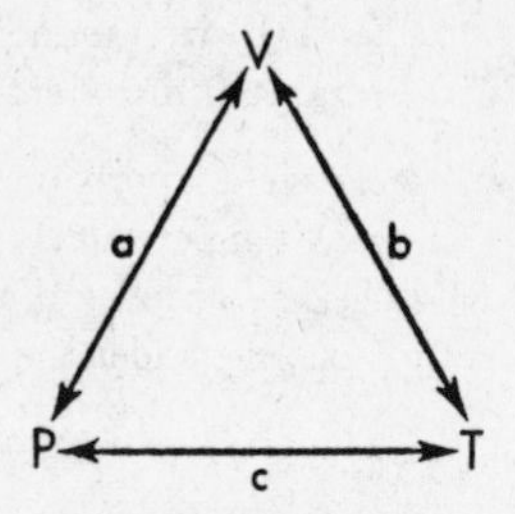

Fig. 7.2. Relationship among the variables pressure, temperature, and volume, for substances in the gaseous state: (a) law of Boyle; (b) law of Charles; (c) law of Gay-Lussac.

7.4. The Law of Charles. The law of Charles expresses the relationship between the two variables, *temperature* and *volume*, for a given mass of gas, the pressure remaining constant. The pressure must remain constant in order that the observed change in volume may be attributed solely to the change in temperature. The relationship is expressed in the law of Charles: pressure constant, the volume of a given mass of gas varies *directly* with the *absolute temperature*. That is:

$$V \propto T \text{ (pressure and mass of gas constant).}$$

The above proportionality indicates that although V and T may not be equal numerically, they vary proportionately with each other. This is shown in Table 7.1.

TABLE 7.1

THE VARIATION IN THE VOLUME OF A GIVEN MASS OF GAS WITH TEMPERATURE, PRESSURE CONSTANT

Temperature (T)	*Volume (V)*
274° K	500 ml
548° K	1000 ml
137° K	250 ml

From Table 7.1 we see that:

$$\frac{V}{T} = \frac{500\text{ ml}}{274°\text{ K}} = \frac{1000\text{ ml}}{548°\text{ K}} = \frac{250\text{ ml}}{137°\text{ K}} = 1.82.$$

That is, for a given mass of gas at constant pressure:

$$\frac{V_1}{T_1} = \frac{V_2}{T_2} = \frac{V_3}{T_3} = K,$$

where K is a constant. The value of the constant for the data given in Table 7.1 is 1.82.

Things equal to the same thing are equal to each other. Therefore:

$$\frac{V_1}{T_1} = \frac{V_2}{T_2}$$

or

$$V_2 = V_1 \times \frac{T_2}{T_1}.$$

$\frac{T_2}{T_1}$: temperature correction fraction
V_1: original volume at T_1
V_2: corrected volume at T_2

Example 7.1. A cylinder contained 600 ml of air at 20° C. What would be the volume of the air at 40° C, pressure constant?

Solution. The four quantities involved are:

Original conditions: $V_1 = 600$ ml.
$T_1 = 20 + 273 = 293°$ K.

Corrected conditions: $V_2 = x$ ml.
$T_2 = 40 + 273 = 313°$ K.

The 600 ml volume will increase proportionately with the temperature. The temperature correction fraction must, therefore, be greater than unity. That is:

$$\frac{T_2}{T_1} = \frac{313° \text{ K}}{293° \text{ K}}.$$

Then:

$$V_2 = 600 \text{ ml} \times \frac{313° \cancel{\text{K}}}{293° \cancel{\text{K}}}$$
$$= 641 \text{ ml.}$$

7.5. The Law of Boyle. The law of Boyle deals with the relationship existing between the two variables, *pressure* and *volume*, for a given mass of gas at constant temperature. The relationship is expressed in the law of Boyle: temperature constant, the volume of a given mass of gas varies *inversely* with the *pressure.* That is:

$$V \propto \frac{1}{P} \text{ (temperature and mass of gas constant).}$$

The variation of volume with pressure, as expressed in the law of Boyle, is shown in Table 7.2.

From Table 7.2 we see that:

$$PV = (760)(500) = (1520)(250) = (380)(1000) = 380{,}000.$$

That is, for a given mass of gas at constant temperature:

$$P_1V_1 = P_2V_2 = P_3V_3 = K,$$

where K is a constant. The value of the constant for the data given in Table 7.2 is 380,000.

Since things equal to the same thing are equal to each other, then:

$$P_1V_1 = P_2V_2$$

or

$$V_2 = V_1 \times \frac{P_1}{P_2}.$$

pressure correction fraction ($\frac{P_1}{P_2}$)
original volume at P_1 (V_1)
corrected volume at P_2 (V_2)

TABLE 7.2

THE VARIATION IN THE VOLUME OF A GIVEN MASS OF GAS WITH PRESSURE, TEMPERATURE CONSTANT

Pressure (P)	*Volume (V)*
760 mm. of Hg	500 ml
1520 mm. of Hg	250 ml
380 mm. of Hg	1000 ml

Example 7.2. A volume of air measuring 380 ml was collected at a pressure of 640 mm of Hg. Calculate the volume the air would occupy at a pressure of 760 mm. of Hg, temperature constant.

Solution. The four quantities involved are:

Original conditions: $V_1 = 380$ ml.
$P_1 = 640$ mm of Hg.

Corrected conditions: $V_2 = x$ ml.
$P_2 = 760$ mm of Hg.

Since the pressure increases, the volume will decrease. This means that the pressure correction fraction is less than unity. That is:

$$\frac{P_1}{P_2} = \frac{640 \text{ mm of Hg}}{760 \text{ mm of Hg}}.$$

Then

$$V_2 = 380 \text{ ml} \times \frac{640 \cancel{\text{mm of Hg}}}{760 \cancel{\text{mm of Hg}}}.$$
$$= 320 \text{ ml}.$$

7.6. The Law of Gay-Lussac. The law of Gay-Lussac deals with the relationship existing between the two variables, *pressure* and *temperature*, for a given mass of gas at constant volume. The relationship is expressed in the law of Gay-Lussac: volume constant, the pressure exerted by a given mass of gas varies *directly* with the *absolute temperature*. That is:

$$P \propto T \text{ (volume and mass of gas constant).}$$

It will be observed that the same type of variation exists between pressure and temperature as exists between volume and temperature (Sec. 7.4). Then:

$$\frac{P_1}{T_1} = \frac{P_2}{T_2}$$

or

$$P_2 = P_1 \times \frac{T_2}{T_1}.$$

- $\frac{T_2}{T_1}$: temperature correction fraction
- P_1: original pressure at T_1
- P_2: corrected pressure at T_2

Example 7.3. The air in a tank was at a pressure of 640 mm of Hg at 23° C. When placed in sunlight the temperature rose to 48° C. What was the pressure in the tank?

Solution. A rise in temperature would result in a rise in pressure. Therefore, the temperature correction fraction must be greater than unity. That is:

$$P_2 = 640 \text{ mm of Hg} \times \frac{(273 + 48)^\circ \text{K}}{(273 + 23)^\circ \text{K}}$$
$$= 694 \text{ mm of Hg.}$$

7.7. A General Gas Law. A general gas law would involve the three variables, temperature, pressure, and volume, simultaneously. Such a general gas law may be obtained by combining any two of the three gas laws discussed previously, since the three variables would be involved. For example, consider the laws of Charles and Boyle combined to give one equation.

$$V_1 \propto T_1, \quad \text{and} \quad V_1 \propto \frac{1}{P_1}.$$

Then $$V_1 \propto \frac{T_1}{P_1} \quad \text{or} \quad V_1 = K\frac{T_1}{P_1}, \quad \text{and} \quad K = \frac{P_1V_1}{T_1}.$$

For a second set of conditions V_2, T_2, and P_2:

$$K = \frac{P_2V_2}{T_2}.$$

Therefore, $$\frac{P_1V_1}{T_1} = \frac{P_2V_2}{T_2} \text{ (for a given mass of gas)}$$

and $$V_2 = V_1 \times \frac{T_2}{T_1} \times \frac{P_1}{P_2}.$$

- $\frac{P_1}{P_2}$: pressure correction fraction
- $\frac{T_2}{T_1}$: temperature correction fraction
- V_1: original volume at T_1 and P_1
- V_2: corrected volume at T_2 and P_2

Example 7.4. A volume of 250 ml of oxygen was collected at 20° C and 785 mm of Hg. The next day the temperature was 37° C and the pressure 770 mm of Hg. Calculate the resultant volume of the oxygen.

Solution. The quantities involved are:

Original conditions: $P_1 = 785$ mm of Hg.
$V_1 = 250$ ml.
$T_1 = 293^\circ$ K.

Corrected conditions: $P_2 = 770$ mm of Hg.
$V_2 = x$ ml.
$T_2 = 310^\circ$ K.

Both the increase in temperature and the decrease in pressure will result in an increase in volume of the oxygen. Therefore:

$$V_2 = 250 \text{ ml} \times \frac{310^\circ \cancel{\text{K}}}{293^\circ \cancel{\text{K}}} \times \frac{785 \cancel{\text{mm of Hg}}}{770 \cancel{\text{mm of Hg}}}$$

$$= 270 \text{ ml.}$$

7.8. A General Equation of State. Equations that express the relationships among the temperature, pressure, volume, and mass of a substance are called *equations of state.* The mathematical expressions for the laws of Charles and Boyle are equations of state. In each of the previous equations of state given, one or two of the four variables involved — pressure, temperature, volume, and mass of gas — were held constant. We will now derive a formula involving all four variables.

For one mole of any gas at standard conditions:

$$\frac{PV}{T} = \frac{(22.4\ l)(1.00 \text{ atm})}{(273^\circ \text{ K})}$$

$$= 0.0821\ l\text{-atm per degree per mole.}$$

Then, for any number of moles of gas, n:

$$\frac{PV}{T} = 0.0821 \text{ n.}$$

Let the constant 0.0821 = R, then:

$$\frac{PV}{T} = nR \quad \text{or} \quad PV = nRT,$$

where R is called the *molar gas constant.*

The choice of liter, atmosphere, °K , and mole as units in which to represent the variables is an arbitrary one. In solving problems, the information given in the problem determines the choice of the particular equation of state to be used.

Example 7.5. What would be the volume occupied by 7.31 g of carbon dioxide, CO_2, at 720 mm of Hg and 27° C ?

Solution. A study of the problem shows that we are given the mass, pressure, and temperature of the gas. The volume is the unknown variable. We can therefore use the general equation of state, PV = nRT, to solve the problem. Solving the equation for the unknown variable, V, we have:

$$V = \frac{nRT}{P}.$$

In order to use the value 0.0821 for R, we must measure n in moles, T in °K., and P in atmospheres. The value of V will then represent liters. Then:

$$n = \frac{7.31\,\not{g}}{44.01\,\not{g}\ \text{mole}^{-1}} = 0.166 \text{ mole}$$

$$T = 273 + 27 = 300° \text{ K}$$

$$P = \frac{720\ \not{\text{mm of Hg}}}{760\ \not{\text{mm of Hg}}\ \text{atm}^{-1}} = 0.947 \text{ atm.}$$

We are now ready to solve for the volume, V, in liters.

$$V = \frac{0.166\ \not{\text{mole}} \times 0.0821\ \frac{l\text{-}\not{\text{atm}}}{\not{\text{mole}}\ \not{°\text{K}}} \times 300°\ \not{\text{K}}}{0.947\ \not{\text{atm}}} = 4.32\ l \text{ of } CO_2$$

at 27° C and 720 mm of Hg.

Since $n = \frac{g}{M}$, where g is the number of grams of a compound of molecular weight M, we may write:

$$PV = \frac{g}{M}RT \quad \text{or} \quad M = \frac{gRT}{PV}$$

Example 7.6. What is the molecular weight of a gas, 350 ml of which weighs 1.069 g at 40° C and 785 mm of Hg?

Solution.

$$M = \frac{gRT}{PV} = \frac{1.069 \text{ g} \times 0.0821\ \frac{\not{l}\text{-}\not{\text{atm}}}{\text{mole}\ \not{°\text{K}}} \times 313°\ \not{\text{K}}}{\frac{785}{760}\ \not{\text{atm}} \times 0.350\ \not{l}} = 76.0\ \frac{\text{g}}{\text{mole}}$$

7.9. The Change in Density of a Gas with Changes in Temperature and Pressure. Since the volume of a given mass of gas is dependent upon the temperature and pressure under which it exists, it is apparent that the density of a gas will be dependent upon these same variables.

From Fig. 7.3 we see that the density of a gas varies directly with the pressure. That is, $d \propto P$.

From Fig. 7.4 we see that the density of a gas varies inversely with the absolute temperature. That is, $d \propto \frac{1}{T}$. From the relationship that $d \propto P$ and that $d \propto \frac{1}{T}$, for a given mass of gas, it can be shown that (Secs. 7.4 and 7.5):

$$\frac{d_1T_1}{P_1} = \frac{d_2T_2}{P_2}$$

or
$$d_2 = d_1 \times \frac{T_1}{T_2} \times \frac{P_2}{P_1}.$$

Example 7.7. A liter of oxygen weighs 1.43 grams at 0° C and 760 mm of Hg. Calculate the weight of one liter of oxygen at 23° C and 720 mm of Hg.

Solution. The quantities involved are:

Original conditions:
$$d_1 = 1.43\,\frac{g}{l}$$
$$T_1 = 273°\text{ K}$$
$$P_1 = 760\text{ mm of Hg.}$$

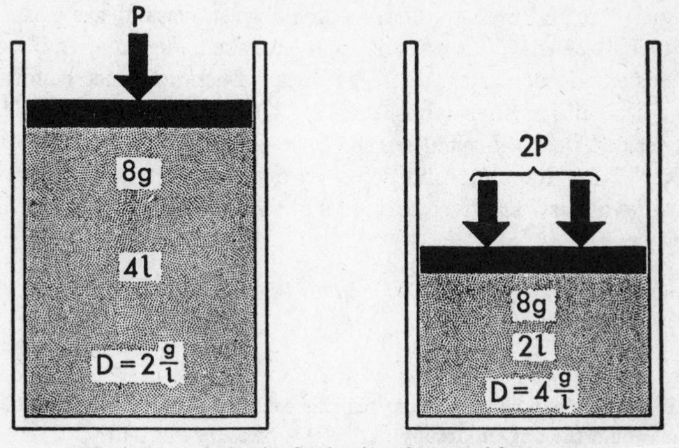

Fig. 7.3. Variation in the density of a gas with pressure.

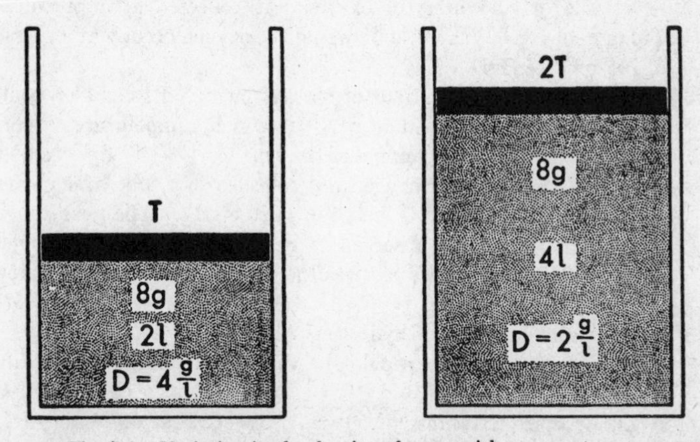

Fig. 7.4. Variation in the density of a gas with temperature.

Corrected conditions: $d_2 = x \frac{g}{l}$

$T_2 = 296° K$

$P_2 = 720$ mm of Hg.

The increase in temperature and decrease in pressure both tend to increase the volume of the gas and therefore decrease its density. Therefore:

$$d_2 = 1.43 \frac{g}{l} \times \frac{273° \cancel{K}}{296° \cancel{K}} \times \frac{720 \cancel{\text{mm of Hg}}}{760 \cancel{\text{mm of Hg}}}$$

$$= 1.25 \frac{g}{l}.$$

7.10. Standard Conditions of Temperature and Pressure. Since the temperature and pressure under which a given mass of gas exists determine its density, it is essential that gases be under comparable conditions when comparing their densities. Such reference conditions have been arbitrarily established as 0° C and 760 mm of Hg. These are termed *standard conditions*, S.C., or *normal temperature and pressure*, N.T.P. Standard conditions may be assumed to be implied when densities of gases are given without stating the conditions under which they exist.

Problems

Part I

7.11. A volume of 473 ml of oxygen was collected at 27° C. What volume would the oxygen occupy at 173° C, pressure constant?
Ans. 703 ml.

7.12. A volume of 2.45 liters of oxygen was collected at a pressure of 740 mm of Hg. What volume would the oxygen occupy at a pressure of 765 mm of Hg? *Ans.* 2.37 *l.*

7.13. The pressure on a cubic foot of air was increased from 14.7 pounds per square inch to 231 pounds per square inch, temperature constant. Calculate the resultant volume of the air. *Ans.* 0.064 ft.^{3}

7.14. A volume of 21.5 ml of oxygen was collected in a tube over mercury at a temperature of 17° C and 740 mm of Hg. The next day the volume of oxygen was observed to be 22.1 ml with the barometer still at 740 mm of Hg. What was the temperature of the laboratory?
Ans. 25° C.

7.15. A volume of 84.0 ml of hydrogen was collected at standard conditions. At what pressure would the volume be 100 ml, temperature constant? *Ans.* 638 mm of Hg.

7.16. A gas occupies a volume of 50 ml at 30° C and 680 mm of Hg. Calculate the volume the gas would occupy at standard conditions.
Ans. 40 ml.

7.17. Air weighs 1.29 grams per liter at S.C. Calculate the density of the air on Pikes Peak when the pressure is 450 mm of Hg and the temperature 17° C. *Ans.* 0.719 $\frac{g}{l}$.

7.18. One liter of nitrogen weighs 1.25 grams at S.C. At what temperature would the density be one-half this value, pressure constant? *Ans.* 273° C.

7.19. A closed metal cylinder contains air at a pressure of 930 mm of Hg and a temperature of 27° C. To what temperature would the air in the cylinder have to be raised in order to exert a pressure of 1500 mm. of Hg? *Ans.* 211° C.

7.20. At standard conditions one liter of ammonia, NH_3, weighs 0.771 g. What is the density of ammonia at 640 mm of Hg and 27° C ? *Ans.* 0.591 $\frac{g}{l}$.

7.21. What volume would one mole of a gas occupy at 100° C and 640 mm of Hg? *Ans.* 36.3 *l.*

7.22. What volume would 1.31 g of hydrogen occupy at 2.00 atm and 400° K ? *Ans.* 10.76 *l.*

7.23. What is the formula weight of a gas, 1.15 g of which occupies a volume of 2800 ml at standard conditions? *Ans.* 92.

Part II

7.24. A room is 16 ft by 12 ft by 12 ft. Would air enter or leave the room and how much if the temperature changed from 27° C to −3° C, pressure remaining constant? *Ans.* 230 ft^3 entering.

7.25. A bottle of nitrogen was collected at 0° C. Assuming the pressure to remain constant, at what temperature would the volume be doubled? *Ans.* 273° C.

7.26. An automobile tire contains air at 38 lb per $in.^2$ How many times the original volume would the air in the tire occupy if released at 15 lb per $in.^2$, temperature constant? Note: a tire gauge registers excess over atmospheric pressure. *Ans.* 3.5 times.

7.27. The temperature of a tire rose from 50° F to 120° F as the result of traveling on a hot pavement. Assuming the volume of the tire to be constant, what was the resulting pressure in the tire if the initial pressure was 40 lb per $in.^2$? *Ans.* 46 $\frac{lb}{in.^2}$.

7.28. A volume of 385 ml of air at 760 mm of Hg and 27° C was carried to a mountaintop where the temperature was −23° C and the pressure 470 mm of Hg. Calculate the resultant volume of the air. *Ans.* 519 ml.

7.29. At the place Piccard started his ascent in the stratosphere balloon, the temperature was 17° C and the pressure 640 mm of Hg. At the highest altitude reached, the temperature was $-48°$ C and the pressure 310 mm of Hg. To what fractional part of its total capacity was the balloon filled before ascending in order that it would be fully expanded at the highest altitude reached? *Ans.* 0.62.

7.30. A given mass of chlorine occupies a volume of 130 ft^3 at 726 mm of Hg. Calculate the volume the chlorine would occupy at 2.00 atmospheres, temperature constant. *Ans.* 62.1 $ft.^3$

7.31. What pressure would be required to compress 250 l of carbon dioxide at 1.00 atmosphere into a 15.0 l cylinder, temperature constant? *Ans.* 16.7 atm.

7.32. A given mass of nitrogen occupies a volume of 25.0 ft^3 at 70° F At constant pressure, what will be the volume of the nitrogen at 212° F? *Ans.* 31.7 $ft.^3$

7.33. In a laboratory experiment a student collected 186 ml of carbon dioxide over mercury. The barometer registered 74.5 cm and the thermometer 68° F. What would be the volume of the carbon dioxide collected when reduced to standard conditions? *Ans.* 170 ml.

7.34. A sealed glass bulb contained helium at a pressure of 750 mm of Hg and 27° C. The bulb was packed in dry ice at $-73°$ C. What was the resultant pressure of the helium? *Ans.* 500 mm of Hg.

7.35. A 2.00 $ft.^3$ cylinder of nitrogen is under a pressure of 2000 $\frac{lb}{in.^2}$. What volume would the nitrogen occupy if released in a room in which the pressure was 745 mm of Hg, there being no change in temperature? *Ans.* 278 $ft.^3$

7.36. By means of a vacuum pump it is possible to obtain a pressure of 10^{-6} mm of Hg. What would this be equivalent to in pounds per square inch? *Ans.* $2 \times 10^{-8} \frac{lb}{in.^2}$.

7.37. At standard conditions nitrogen weighs 1.25 $\frac{g}{l}$. What is the weight of 650 ml of nitrogen at 725 mm of Hg and 23° C.? *Ans.* 0.72 $\frac{g}{l}$.

7.38. What weight of oxygen would be contained in a 2.00 ft^3 cylinder at a pressure of 2000 $\frac{lb}{in.^2}$ and 68° F? At S.C. one liter of oxygen weighs 1.43 g. *Ans.* 10.3 kg.

7.39. The temperature of a given mass of gas is changed from 23° C to 319° C, and the pressure from 420 mm of Hg to 840 mm of Hg. What is the resultant volume? *Ans.* No change.

7.40. At standard conditions 1.00 l of carbon dioxide weighs 1.98 g. What is the weight per liter at 15° C and 675 mm of Hg? *Ans.* 1.67 $\frac{g}{l}$.

7.41. A given mass of neon occupies a volume of 125 ml at 75.0 cm of Hg and 68° F. What volume would the neon occupy at 3.75 atmospheres and 300° K? *Ans.* 33.7 ml.

7.42. A given quantity of gas occupies a volume of 875 ml at 68° F and 73 cm of Hg. Calculate the volume the gas would occupy at 350° K. and 12 atmospheres pressure. *Ans.* 84 ml.

7.43. A tank containing 4 ft^3 of butane gas at 15 atm pressure is connected to a tank containing 6 ft^3 of the gas at 5 atm pressure. Calculate the resultant pressure in the connected tanks, assuming no temperature change. *Ans.* 9 atm.

7.44. One liter of a gas weighs 1.33 grams at 750 mm of Hg and 17° C. Calculate the weight of 500 ml of the gas at 640 mm. of Hg and 37° C. *Ans.* 0.53 g.

7.45. An automobile tire gauge registered 40 lb with the barometer reading 760 mm of Hg and the temperature 23° C. After driving on a hot pavement the gauge registered 43 lb. What was the temperature of the tire, assuming its volume to remain constant? *Ans.* 39° C.

7.46. It was desired to obtain a volume of 1000 ml of oxygen at 100° C and 640 mm of Hg. How many moles of oxygen would be required? *Ans.* 0.0275 mole.

7.47. What volume would 2.00 g of methane, CH_4, occupy at 27° C and 710 mm of Hg? *Ans.* 3.30 *l.*

Laws Relating to Mixtures of Gases

7.11. Dalton's Law of Partial Pressures. Dalton's law states that the total pressure exerted by a mixture of gases is equal to the sum of the partial pressures of each of the gases constituting the mixture. By *partial pressure* is meant the pressure each gas would exert if it alone occupied the volume occupied by the mixture of gases. For example, if one liter of hydrogen and one liter of oxygen, each at 0° C and 380 mm of Hg, are forced into a third container of one liter capacity as shown in Fig. 7.5, (page 98), then the resultant pressure of the mixture would be:

$$\underbrace{380 \text{ mm of Hg}}_{\text{partial pressure of hydrogen}} + \underbrace{380 \text{ mm of Hg}}_{\text{partial pressure of oxygen}} = \underbrace{760 \text{ mm of Hg.}}_{\text{total pressure}}$$

Dalton's law finds its most useful application in general chemistry to calculations involving the collection of gases over water, where

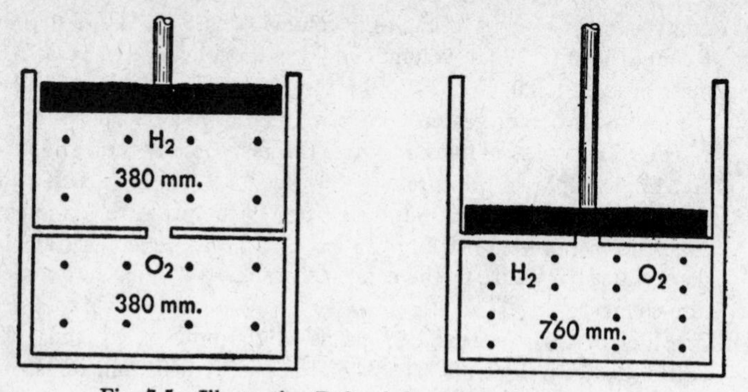

Fig. 7.5. Illustrating Dalton's law of partial pressures.

water vapor is always present with the collected gas. In such calculations a correction must be made for the water vapor present. The higher the temperature of the water over which the gas is collected, the greater the amount of water vapor present. The actual amount of gas collected is therefore less than the measured volume. In Fig. 7.6 the gas, G, has been collected over the water, L. The space G therefore contains both water vapor and gas. If the level of the liquid is the same inside and outside the tube, T, then the pressure of the gases in the tube is equal to the barometric pressure of the room in which the gas is collected. Then, from Dalton's law:

$$P_g + P_w = P_a.$$

- P_a: atmospheric pressure
- P_w: partial pressure of water vapor
- P_g: partial pressure of the collected gas

Solving the above for P_g gives:

$$P_g = P_a - P_w.$$

Therefore, to find the pressure, P_g, under which the collected gas would exist if it alone occupied the total volume in the tube, subtract the vapor pressure of the water at the given temperature from the atmospheric pressure. A table giving the vapor pressure of water for temperatures ranging from 0° C to 100° C will be found in the Appendix.

Example 7.8. Hydrogen was collected over water at 27° C and 807 mm of Hg. The volume of gas above the water was 124 ml. Calculate the volume the hydrogen would occupy dry at S.C.

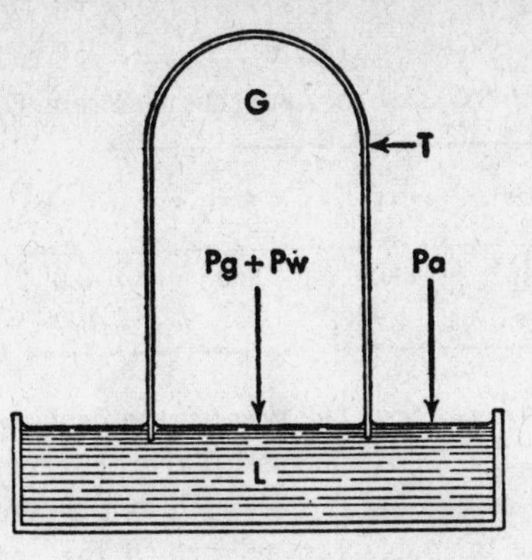

Fig. 7.6. Dalton's law of partial pressures applied to gases collected over water.

Solution. First calculate the partial pressure of the hydrogen. At 27° C the partial pressure of the water vapor (vapor pressure) would be 27 mm of Hg. Therefore:

$$P_{H_2} = 807 \text{ mm of Hg} - 27 \text{ mm of Hg} = 780 \text{ mm of Hg.}$$

This means that, if the hydrogen alone occupied the 124 ml, it would exert a pressure of 780 mm of Hg. We may now apply the general gas law (Sec. 7.7) to change the hydrogen to standard conditions (Sec. 7.10). Then:

$$V_{H_2} = 124 \text{ ml} \times \frac{273 \cancel{° K}}{300 \cancel{° K}} \times \frac{780 \cancel{\text{mm of Hg}}}{760 \cancel{\text{mm of Hg}}}$$
$$= 116 \text{ ml at S.C.}$$

7.12. Graham's Law of Gaseous Diffusion. The law of Graham deals with the relationship existing between the *rates of diffusion*, D, of gases, and their *densities*, d. By diffusion is meant the ability of gaseous molecules to pass through small openings, such as in the fabric of balloons or unglazed porcelain; or the intermingling of molecules in a mixture. On the basis of the kinetic-molecular theory it is evident that increasing the temperature of a gas will increase its rate of diffusion, since increasing the temperature increases the velocity of the molecules, thus producing a greater number of impacts on the walls of the containing vessel in any given unit of time. It is also apparent that increasing the pressure of a gas will produce more impacts on the walls of the vessel, thus increasing the rate of diffusion.

TABLE 7.3

DIFFUSION OF GASES AS RELATED TO THEIR DENSITIES

Gas	$d\left(\frac{g}{l}\right)$	$\sqrt{d}$	$D\left(\frac{ml}{hr}\right)$
H_2	0.09	0.3	400
O_2	1.43	1.2	100

From Table 7.3 we see that the greater the density of the gas, the slower the rate of diffusion of the gas. This is in accord with the law of Graham: under similar conditions such as temperature, pressure, and the size of the openings through which the gases diffuse, the rates of diffusion of gases vary *inversely* as the *square roots* of their *densities*. That is:

$$D \propto \frac{1}{\sqrt{d}}.$$

Treating these variables in the same manner as the variables pressure and volume in Sec. 7.5, we have:

$$D_1\sqrt{d_1} = D_2\sqrt{d_2}$$

or

$$D_2 = D_1 \times \frac{\sqrt{d_1}}{\sqrt{d_2}} = D_1 \times \sqrt{\frac{d_1}{d_2}}.$$

When one considers that light molecules are moving with a greater velocity than are heavier molecules (Sec. 7.1), it becomes evident why the lighter molecules will diffuse the faster.

Example 7.9. Two balloons of the same size and like material are filled respectively with hydrogen and oxygen at the same temperature and pressure. If the oxygen escapes at the rate of 65 ml per hour, calculate the rate of escape of the hydrogen.

Solution. Using the values of the densities given in Table 7.3, we have:

$$D_{H_2} = 65\,\frac{ml}{hr} \times \sqrt{\frac{1.43}{0.09}}$$

$$= 260\,\frac{ml}{hr}.$$

Since the hydrogen is the lighter of the two gases it will diffuse the faster. Also, the quantity $\sqrt{\frac{1.43}{0.09}} = 4$ tells us that hydrogen will diffuse four times faster than oxygen.

The densities of gases are proportional to their molecular weights. For example, in the case of oxygen and hydrogen, we have:

$$\frac{1.43\,\frac{g}{l}}{0.09\,\frac{g}{l}} = \frac{32.00}{2.016} = 15.9.$$

It is therefore possible to substitute the molecular weights of the gases, M, for their densities in the above formula. That is:

$$D_2 = D_1 \times \sqrt{\frac{M_1}{M_2}}.$$

Example 7.10. Compare the rates of diffusion of methane, CH_4, and sulfur dioxide, SO_2.

Solution. The molecular weight of $CH_4 = 16$ and of $SO_2 = 64$. Therefore the CH_4 will diffuse the faster. Let the rate of diffusion of SO_2 be unity. Then:

$$\text{Rate of diffusion of } CH_4 = 1 \times \sqrt{\tfrac{64}{16}} = 2.$$

That is, CH_4 diffuses twice as fast as SO_2.

Problems

Part I

7.48. Hydrogen was collected over water, the volume of hydrogen and water vapor being one liter at 25° C and 640 mm of Hg.

a. What is the partial pressure of the hydrogen? *Ans.* 616 mm of Hg.

b. Calculate the volume the hydrogen would occupy dry under the given conditions. *Ans.* 0.963 *l*.

c. Calculate the volume the hydrogen would occupy dry at standard conditions. *Ans.* 0.742 *l*.

d. What fractional part of the original liter was water vapor? *Ans.* 0.037.

7.49. A volume of 1.43 liters of dry hydrogen at 27° C and 760 mm of Hg was bubbled through water. What was the volume of the mixture of hydrogen and water vapor, temperature and pressure constant? *Ans.* 1.48 *l*.

7.50. Nitrogen weighs 1.25 g per *l* and chlorine 3.21 g per *l* at standard conditions. Which will escape the faster and by how much if enclosed in an unglazed porcelain container in equal amounts? *Ans.* N_2, 1.6 times faster.

7.51. Chlorine will escape through a small opening one-sixth as fast as hydrogen under similar conditions. Given that one liter of hydrogen weighs 0.0899 g, calculate the density of chlorine. *Ans.* 3.24 g per *l*.

7.52. Two porous containers were filled respectively with hydrogen and oxygen at S.C. At the end of one hour 880 ml of hydrogen had escaped. How much oxygen had escaped during this same period of time? *Ans.* 220 ml.

7.53. Calculate the partial pressures of oxygen and nitrogen in the atmosphere, given that air contains 21 per cent oxygen and 78 per cent nitrogen by volume. Barometric pressure, 746 mm of Hg.
Ans. O_2 = 157 mm ; N_2 = 582 mm.

7.54. Arrange the following gases in order of their increasing rates of diffusion: N_2, He, H_2, CH_4, CO_2, O_2. *Ans.* CO_2, O_2, N_2, CH_4, He, H_2.

7.55. It required 16 sec for 250 ml of CH_4 to diffuse through a small opening. Under the same conditions of temperature and pressure, how long would be required for 1500 ml of SO_2 to diffuse through the same opening? *Ans.* 192 sec.

Part II

7.56. In a laboratory experiment 763 ml of gas was collected over water at 35° C and 748 mm of Hg. Calculate the volume the gas would occupy dry at standard conditions. *Ans.* 628 ml.

7.57. In an experiment hydrogen was collected over water at 27° C and 725 mm of Hg. The volume of hydrogen and water vapor measured 350 ml Upon standing the temperature changed to 17° C and the pressure to 750 mm of Hg. What was the resultant volume?
Ans. 332 ml.

7.58. What volume will 2.500 g of oxygen occupy at 27° C and 760 mm of Hg, when collected over (a) mercury, and (b) water?
Ans. (a) 1.92 *l* (b) 1.99 *l*.

7.59. A mixture of gases consists of 20% N_2, 30% O_2, and 50% He at 760 mm of Hg. What is the partial pressure of each gas?
Ans. N_2 = 152 mm ; O_2 = 228 mm ; He = 380 mm.

7.60. Two liters of oxygen and eight liters of nitrogen, at S.C., are mixed in a 25.0 *l*. tank. At 0° C (a) what is the pressure in the tank, and (b) what is the partial pressure of each gas?
Ans. (a) 304 mm ; (b) O_2 = 61 mm., N_2 = 243 mm.

7.61. Three similar balloons were filled respectively with O_2, CO_2, and Cl_2 under similar conditions. In ten hours one-half the CO_2 had escaped. How much of each of the other gases escaped during the same period of time? *Ans.* O_2 = 0.6; Cl_2 = 0.4.

7.62. A porous container was filled with equal amounts of oxygen and a gas of unknown molecular weight. The oxygen escaped 1.77 times faster than the unknown gas. Calculate the molecular weight of the unknown gas. *Ans.* 99.

7.63. What are the relative rates of diffusion of the gases H_2 and $COCl_2$ under similar conditions? *Ans.* 7 : 1.

7.64. Methane, CH_4, diffuses through an opening at the rate of 135 ml per sec. At what rate will argon diffuse through the same opening under similar conditions? *Ans.* 85.4 ml per sec.

7.65. Arrange the following in order of increasing time required to diffuse through a given opening under similar conditions: 150 ml $COCl_2$, 500 ml. H_2, 375 ml. CO_2. *Ans.* H_2, $COCl_2$, CO_2.

8

Chemical Equations

The chemist expresses the behavior of atoms, ions, and molecules in the form of chemical equations. A chemical equation must satisfy three conditions. (1) *It must represent the experimental facts.* That is, the equation must represent the reaction that actually occurs under the specified conditions. This means that, in order to write an equation, the reactants and products must be known. If you do not know the products you must rely on your knowledge of chemistry to predict them. (2) *The same number and kind of atoms must appear on the reactant and product sides of the equation.* By "kind of atoms" is meant the atoms of any given element regardless of oxidation state. (3) *The net electrical charge must be the same on both sides of the equation.*

Many of the equations that follow will show only the atoms, ions, or molecules that are actually involved in the chemical reaction. For example, when water solutions containing silver nitrate, $AgNO_3$, and sodium chloride, NaCl, are mixed, insoluble silver chloride, AgCl (s), precipitates from the solution. That is:

$$AgNO_3 + NaCl \rightarrow AgCl\downarrow + NaNO_3.$$

In the ionic form this equation becomes:

$$Ag^+ + \cancel{NO_3^-} + \cancel{Na^+} + Cl^- \rightarrow AgCl\,(s) + \cancel{Na^+} + \cancel{NO_3^-}.$$

In the above reaction observe that the Na^+ and NO_3^- ions remain unchanged during the course of the reaction. They may therefore be canceled as shown, reducing the equation to:

$$\begin{array}{ccccc} Ag^+ & + & Cl^- & \rightarrow & AgCl\,(s). \\ | & & | & & |\quad| \\ \text{1 atom} & & \text{1 atom} & & \text{1 atom}\ \ \text{1 atom} \end{array}$$

The third condition is satisfied since the net electrical charge is zero for both sides of the equation.

General Types of Reactions

8.1. Combination Reactions Involving Elements. When a metal reacts with a nonmetal the product usually is an ionic compound. For example:

$$2\,\overset{e^-}{\overbrace{Na + Cl_2}} \rightarrow 2\,NaCl.$$

In the above reaction electrons have been transferred from the sodium atoms to the chlorine atoms to form Na^+ and Cl^- ions. The above therefore is an *oxidation-reduction reaction* in which the sodium atoms have been *oxidized* (lose electrons), and the chlorine atoms have been *reduced* (gain electrons).

When a nonmetal reacts with a nonmetal the product usually is a *covalent* compound, that is, a compound in which electrons are *shared* between atoms. Such compounds exist as molecules rather than ions. For example, phosphorus and oxygen react to form $\mathbf{P_2O_5}$ molecules.

$$P + O_2 \rightarrow P_2O_5. \quad \text{(skeleton equation)}$$

A *skeleton equation*, such as shown above, usually is balanced by trial and error methods. The equation would be

$$4\,P + 5\,O_2 \rightarrow 2\,P_2O_5.$$

8.2. Decomposition Reactions. High temperatures will bring about the decomposition of many compounds. In such reactions a single reactant is decomposed into two or more products. For example, mercuric oxide will decompose into mercury and oxygen when heated.

$$2\,\overset{e^-}{\overbrace{HgO}} \rightarrow 2\,Hg + O_2\uparrow$$

Mercuric oxide is an *ionic* compound. In the above reaction mercury has been *reduced* (gains electrons) and oxygen has been *oxidized* (loses electrons). Most of the oxides of the precious metals such as gold, silver, and platinum decompose when heated.

Many salts of oxyacids have a tendency to produce oxygen as a decomposition product.

$$2\,KNO_3 \xrightarrow{\Delta} 2\,KNO_2 + O_2\uparrow.$$

$$2\,Pb(NO_3)_2 \xrightarrow{\Delta} 2\,PbO + 4\,NO_2\uparrow + O_2\uparrow.$$

$$2\,KClO_3 \xrightarrow{\Delta} 2\,KCl + 3\,O_2\uparrow.$$

Many carbonates decompose when heated to yield carbon dioxide.

$$CaCO_3 \xrightarrow{\Delta} CaO + CO_2\uparrow .$$

$$2\ NaHCO_3 \xrightarrow{\Delta} Na_2CO_3 + H_2O + CO_2\uparrow .$$

Decomposition reactions may be quite complex. When heated, ammonium dichromate undergoes the following reaction:

$$2\ (NH_4)_2Cr_2O_7 \xrightarrow{\Delta} 4\ NH_3\uparrow\ + 2\ H_2O + 2\ Cr_2O_3 + 3\ O_2\uparrow .$$

8.3. Combustion Reactions. Reactions which take place with the emission of *heat* and *light* are called *combustion reactions.* Most combustion reactions involve burning with atmospheric oxygen, such as the burning of coal, wood, natural gas, and oils. In reactions of this type the carbon of the fuel is burned to carbon dioxide and the hydrogen to water.

$$C + O_2 \rightarrow CO_2.$$

$$2\ H_2 + O_2 \rightarrow 2\ H_2O.$$

Natural gas is principally methane, CH_4.

$$CH_4 + 2\ O_2 \rightarrow CO_2 + 2\ H_2O.$$

One of the principal constituents of gasoline is heptane, C_7H_{16}.

$$C_7H_{16} + 11\ O_2 \rightarrow 7\ CO_2 + 8\ H_2O.$$

Alcohol is C_2H_5OH.

$$2\ C_2H_5OH + 7\ O_2 \rightarrow 4\ CO_2 + 6\ H_2O.$$

8.4. Reactions Involving Water as a Reactant. Three types of reactions will be given in which water functions as a reactant.

(1) *The reaction of oxides with water.* Water-soluble nonmetallic oxides react with water to form *acids.*

$$P_2O_5 + 3\ H_2O \rightarrow 2\ H_3PO_4 \text{ (phosphoric acid)}.$$

Acids are covalent compounds.

Water-soluble metallic oxides react with water to form *bases.*

$$CaO + H_2O \rightarrow Ca(OH)_2 \text{ (calcium hydroxide)}.$$

Most bases are ionic compounds.

(2) *The reaction of metals with water.* Some of the lighter metals in Groups IA and IIA of the periodic table react with cold water to produce hydrogen.

$$Ca + 2\ H_2O \rightarrow Ca(OH)_2 + H_2 \uparrow .$$

Some of the heavier metals such as aluminum, zinc, and iron require superheated steam in order to react.

$$3\ Fe + 4\ H_2O \xrightarrow{\Delta} Fe_3O_4 + 4\ H_2 \uparrow .$$

(3) *Hydrate formation.* Most ions combine with water to form *hydrates.*

$$H^+ + H_2O \rightarrow \underset{\text{hydronium ion}}{H_3O^+} \quad (\text{or } H^+ \cdot H_2O).$$

$$Cu^{+2} + 4\ H_2O \rightarrow Cu(H_2O)_4^{+2} \quad (\text{or } Cu^{+2} \cdot 4\ H_2O).$$

Sometimes the water of hydration of the ions is indicated in the formula, such as $CuSO_4 \cdot 5\ H_2O$. This could be written $Cu(H_2O)_4^{+2}SO_4(H_2O)^{-2}$.

8.5. Reactions in Which Water Functions as a Solvent. Four types of reactions will be given in which water functions as a solvent.

(1) *Ionic dissociation in water solution.* Ionic compounds that are soluble in water, and some covalent compounds such as acids and weak bases, undergo ionic dissociation when dissolved in water. Salts and bases ionize according to the equations:

$$CaCl_2 \rightarrow Ca^{+2} + 2\ Cl^-$$

and

$$KOH \rightarrow K^+ + OH^-.$$

Molecules of acids are in equilibrium with their ions in water solution. Acids such as acetic acid, $HC_2H_3O_2$, that are only slightly ionized are called *weak acids.*

$$HCl \rightleftharpoons H^+ + Cl^-.$$
$$HC_2H_3O_2 \rightleftharpoons H^+ + C_2H_3O_2^-.$$

Polyprotic acids, such as H_2SO_4, undergo step-by-step ionization.

$$H_2SO_4 \rightleftharpoons H^+ + HSO_4^-. \quad \text{Step 1.}$$
$$HSO_4^- \rightleftharpoons H^+ + SO_4^{-2}. \quad \text{Step 2.}$$

The primary ionization (Step 1) occurs to a much greater extent than do subsequent steps.

(2) *Neutralization.* In the process of *neutralization* an acid and a base react to produce a salt and water.

$$NaOH + HCl \rightarrow NaCl + H_2O.$$

Writing the equation in the ionic form, and canceling nonfunctional ions we have:

$$\cancel{Na^+} + OH^- + H^+ + \cancel{Cl^-} \rightarrow \cancel{Na^+} + \cancel{Cl^-} + H_2O$$

or

$$OH^- + H^+ \rightarrow H_2O.$$

The last equation is the basic equation for all neutralization reactions. In the above reaction the NaCl could be recovered in crystalline form by evaporating the water.

In the neutralization of a polyprotic acid the reaction occurs in steps, as with NaOH and H_2SO_4.

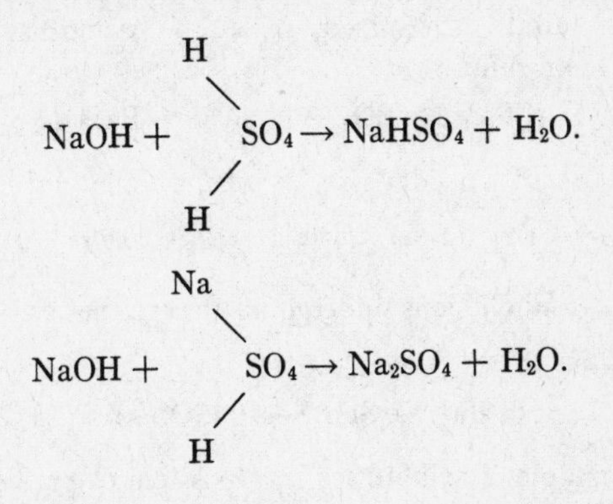

Conditions such as temperature and concentration of the reactants determine which neutralization step will predominate.

(3) *Displacement reactions.* These are reactions in which an atom and an ion exchange electrons. If a strip of zinc is placed in a solution of copper sulfate the following reaction occurs:

$$\overset{e^-}{\widehat{Fe + Cu}}{}^{+2} \rightarrow Fe^{+2} + Cu.$$

In the above reaction iron atoms lose electrons (are oxidized) and the copper ions gain electrons (are reduced). With nonmetals the electron exchange is reversed.

$$\overset{e^-}{\widehat{Cl_2 + 2\,I}}{}^- \rightarrow 2\,Cl^- + I_2.$$

The preparation of hydrogen by the action of certain metals on acid represents displacement.

$$Zn + H_2SO_4 \rightarrow ZnSO_4 + H_2 \uparrow .$$

or

$$\overset{e^-}{\frown} \quad Zn + 2\,H^+ \rightarrow Zn^{+2} + H_2 \uparrow .$$

The *electromotive series* of metals and nonmetals may be used to predict whether or not a replacement reaction will take place between an atom and an ion.

(4) *Precipitation from solution.* A substance precipitates because it is insoluble in the solvent. When solutions of two ions that form an insoluble compound are mixed, the compound will precipitate from the solution. For example, when solutions of barium chloride and sodium sulfate are mixed, insoluble barium sulfate, $BaSO_4$, precipitates from solution.

$$BaCl_2 + Na_2SO_4 \rightarrow 2\,NaCl + BaSO_4 \downarrow .$$

Since in water solution salts exist as ions, we may write:

$$Ba^{+2} + \cancel{2\,Cl^-} + \cancel{2\,Na^+} + SO_4^{-2} \rightarrow \cancel{2\,Na^+} + \cancel{2\,Cl^-} + BaSO_4\,(s).$$

Since the Na^+ and Cl^- ions undergo no change they may be canceled, and the equation becomes:

$$Ba^{+2} + SO_4^{-2} \rightarrow BaSO_4\,(s).$$

By using a table of solubilities it is possible to predict whether or not precipitation will occur when two or more solutions are mixed.

Balancing Chemical Reactions

8.6. Writing Chemical Equations.

Example 8.1. Write the equation for the burning of propane, C_3H_8, in air.

Solution. When an organic compound such as propane burns, the carbon forms either free carbon, CO, or CO_2, depending on the relative amount of oxygen present and the temperature, while the hydrogen forms water. Then:

$$a\,C_3H_8 + b\,O_2 \rightarrow c\,CO_2 + d\,H_2O.$$

where a, b, c, and d are numerical constants to be determined. The three carbon atoms will form 3 CO_2, and the eight hydrogen atoms will form 4 H_2O. We must now evaluate b. We now know that:

$$C_3H_8 + b\,O_2 \rightarrow 3\,CO_2 + 4\,H_2O.$$

Since there are 10 atoms of oxygen in the products, then $b = 5$. Or:

$$C_3H_8 + 5\ O_2 \rightarrow 3\ CO_2 + 4\ H_2O \text{ (balanced).}$$

Example 8.2. The compound $KMnO_4$ is prepared commercially by two consecutive reactions.

1. $MnO_2 + KOH + O_2 \rightarrow K_2MnO_4 + H_2O.$
2. $K_2MnO_4 + CO_2 + H_2O \rightarrow KMnO_4 + MnO_2 + KHCO_3.$

a. Write the equation for each step.
b. What are the by-products of the process?

Solution. a. 1. The elements that appear in only one compound as a reactant or as a product should first be balanced — in this case, Mn, K, and H. We see that Mn is balanced in the given reaction. We can balance K and H simultaneously by taking 2 KOH on the reactant side. Oxygen remains to be balanced. As the reaction now stands we have:

$$MnO_2 + 2\ KOH + a\ O_2 \rightarrow K_2MnO_4 + H_2O.$$

We now have 6 atoms of oxygen as reactants and 5 atoms of oxygen as products. If $a = \frac{1}{2}$ then the oxygen atoms will balance. Then:

$$MnO_2 + 2\ KOH + \tfrac{1}{2}\ O_2 \rightarrow K_2MnO_4 + H_2O \text{ (balanced).}$$

The fractional coefficient $\frac{1}{2}$ could be eliminated by multiplying the equation by two.

a. 2. The key elements to be balanced in this reaction are carbon and hydrogen. The reactant H_2O would yield 2 $KHCO_3$. This in turn would require 2 CO_2 in order to get two atoms of carbon. If we now balance the manganese atoms, then the oxygen atoms automatically will balance. Since there are two Mn atoms in the products, we must have 2 K_2MnO_4 on the reactant side. This in turn would require 4 $KHCO_3$ and 2 H_2O on the product side. Then:

$$2\ K_2MnO_4 + 4\ CO_2 + 2\ H_2O \rightarrow KMnO_4 + MnO_2 + 4\ KHCO_3 \text{ (balanced).}$$

b. From the given reactions we see that MnO_2 is recycled back through reaction one. Also, the intermediate K_2MnO_4 is used in reaction two. It is usually possible to determine the by-products of a series of reactions by adding the equations algebraically. Before adding we will multiply equation one by 2 in order to be able to cancel out K_2MnO_4. Then:

$$\cancel{2}\,MnO_2 + 4\ KOH + O_2 \rightarrow \cancel{2\ K_2MnO_4} + \cancel{2\ H_2O}$$

$$\cancel{2\ K_2MnO_4} + 4\ CO_2 + \cancel{2\ H_2O} \rightarrow KMnO_4 + \cancel{MnO_2} + 4\ KHCO_3.$$

Add: $$MnO_2 + 4\ KOH + O_2 + 4\ CO_2 \rightarrow KMnO_4 + 4\ KHCO_3.$$

After adding the equations we see that $KHCO_3$ appears to be the only by-product of the process.

Problems

Part I

8.1. Write equations for the following reactions:

a. $Ca + O_2 \rightarrow CaO$.
b. $Al + N_2 \rightarrow AlN$.
c. $P + Cl_2 \rightarrow PCl_3$.
d. $P + Cl_2 \rightarrow PCl_5$.
e. $KClO_4 \xrightarrow{\Delta} KCl + O_2$.
f. $NaNO_3 \xrightarrow{\Delta} NaNO_2 + O_2$.
g. $PbCO_3 \xrightarrow{\Delta} PbO + CO_2$.
h. $Fe_2(C_2O_4)_3 \xrightarrow{\Delta} FeC_2O_4 + CO_2$.
i. $Fe(OH)_3 \xrightarrow{\Delta} Fe_2O_3 + H_2O$.
j. $HNO_2 \xrightarrow{\Delta} HNO_3 + NO + H_2O$.

8.2. Write equations for the following reactions in which water acts as a reactant:

a. $SO_2 + H_2O \rightarrow H_2SO_3$.
b. $N_2O_5 + H_2O \rightarrow HNO_3$.
c. $ClO_2 + H_2O \rightarrow HClO_2 + HClO_3$.
d. $Na_2O + H_2O \rightarrow NaOH$.
e. $MgO + H_2O \rightarrow Mg(OH)_2$.
f. $Na + H_2O \rightarrow NaOH + H_2$.
g. $Mg + H_2O \xrightarrow{\Delta} Mg(OH)_2 + H_2$.
h. $Al^{+3} + H_2O \rightarrow Al(H_2O)_6^{+3}$.
i. $Na_2CO_3 + H_2O \rightarrow Na_2CO_3 \cdot 10\ H_2O$.
j. $Cl_2 + H_2O \rightarrow Cl_2 \cdot 8\ H_2O$.

8.3. Write equations for the following ionic dissociation reactions:

a. $MgCl_2 \rightarrow Mg^{+2} + Cl^-$.
b. $Ca(NO_3)_2 \rightarrow Ca^{+2} + NO_3^-$.
c. $Al_2(SO_4)_3 \rightarrow Al^{+3} + SO_4^{-2}$.
d. $Mg(OH)_2 \rightarrow Mg^{+2} + OH^-$.
e. $Na_2SO_4 \rightarrow Na^+ + SO_4^{-2}$.

8.4. Write the equations for the three successive ionization steps that occur when phosphoric acid, H_3PO_4, is dissolved in water.

8.5. Write the equations of neutralization for the following reactions:

a. $Al(OH)_3 + HCl \rightarrow AlCl_3 + H_2O$.
b. $Ca(OH)_2 + H_3PO_4 \rightarrow Ca(H_2PO_4)_2 + H_2O$.
c. $Ca(OH)_2 + H_3PO_4 \rightarrow CaHPO_4 + H_2O$.
d. $Ca(OH)_2 + H_3PO_4 \rightarrow Ca_3(PO_4)_2 + H_2O$.
e. $NaOH + HNO_3 \rightarrow NaNO_3 + H_2O$.

8.6. Write equations for the following displacement reactions in solution:

a. $Al + CuSO_4 \rightarrow Al_2(SO_4)_3 + Cu$.

b. $Mg + HCl \rightarrow MgCl_2 + H_2$.

c. $Zn + AgNO_3 \rightarrow Zn(NO_3)_2 + Ag$.

d. $Cl_2 + NaBr \rightarrow NaCl + Br_2$.

e. $Br_2 + KI \rightarrow KBr + I_2$.

8.7. Write equations for the combustion of:

a. Benzene, C_6H_6.

b. Pentane, C_5H_{12}.

c. Methyl alcohol, CH_3OH.

d. Carbon monoxide, CO.

e. Sugar, $C_{12}H_{22}O_{11}$.

8.8. The following compounds are insoluble in water: Ag_2CO_3, $PbCO_3$, $BaCrO_4$, HgI_2, Ag_3PO_4, Bi_2S_3, FeS, Ag_2S, PbS, $BaSO_4$. Write the ionic equations for the reactions that take place to produce precipitates when water solutions of the following compounds are mixed:

a. $Pb(NO_3)_2$ and Na_2CO_3.

b. $BaCl_2$ and K_2CrO_4.

c. $AgNO_3$ and Na_3PO_4.

d. $BiCl_3$ and H_2S.

e. $Hg(NO_3)_2$ and KI.

Part II

8.9. Balance the following skeleton equations:

a. $MnO_2 + HCl \rightarrow MnCl_2 + H_2O + Cl_2$.

b. $Cu + HNO_3 \rightarrow Cu(NO_3)_2 + NO + H_2O$.

c. $AgCl + NH_4OH \rightarrow Ag(NH_3)_2Cl + H_2O$.

d. $C + HNO_3 \rightarrow NO_2 + CO_2 + H_2O$.

e. $CO + Fe_2O_3 \rightarrow Fe + CO_2$.

f. $HCl + HClO_3 \rightarrow H_2O + Cl_2$.

g. $HCl + Fe_2O_3 \rightarrow FeCl_3 + H_2O$.

h. $KI + H_2O \rightarrow KOH + I_2$.

i. $SO_3 + HNO_3 \rightarrow H_2SO_4 + N_2O_5$.

j. $FeCl_3 + H_2S \rightarrow FeCl_2 + HCl + S$.

8.10. Show that the ionic equations written in Problem 8.8 satisfy the condition of net electrical charge for an equation.

8.11. Predict the products, and then write the equation for each of the following:

a. $Al + Br_2$.

b. $AgNO_3$ + a water solution of Na_2CO_3.

c. BaO + a water solution of SO_3.

d. CaO + a water solution of HNO_3.

e. $C_{10}H_{16} + O_2$.
f. $Al + H^+$.
g. $Fe^{+2} + S^{-2}$.
h. $Li_2O + H_2O$.
i. $CO_2 + H_2O$.
j. $Zn + Ag^+$.

8.12. Write the equation for each of the following reactions, then reduce each to the simplest form of equation involving atoms, ions, and molecules:
a. Aluminum chloride + ammonium hydroxide → aluminum hydroxide ↓ + ammonium chloride.
b. Silver nitrate + sodium phosphate → silver phosphate ↓ + sodium nitrate.
c. Chlorine + potassium iodide → potassium chloride + iodine ↓.
d. Ferric sulfate + calcium hydroxide → ferric hydroxide ↓ + calcium sulfate ↓.
e. Barium nitrate + ammonium carbonate → barium carbonate ↓ + ammonium nitrate.
f. Potassium hydroxide + sulfuric acid → potassium sulfate + water.
g. Ferrous hydroxide + hydrochloric acid → ferrous chloride + water.
h. Cupric sulfate + hydrogen sulfide → cupric sulfide ↓ + sulfuric acid.
i. Silver nitrate + potassium chromate → silver chromate ↓ + potassium nitrate.

8.13. Write equations for the following reactions:
a. Heat decomposes sodium bicarbonate to form sodium carbonate, carbon dioxide, and X.
b. Ammonia reacts with oxygen to form nitric acid and X.
c. When $NaClO$ is heated the products are $NaClO_3$ and X.
d. Copper reacts with concentrated nitric acid to yield copper (II) nitrate, NO_2, and X.
e. Phosphorus trichloride reacts with water to yield phosphorous acid and X.

8.14. Write equations for the following reactions:
a. $Ca_3(PO_4)_2 + SiO_2 + C \rightarrow CaSiO_3 + CO + P_4$
b. $H_3PO_4 + (NH_4)_2MoO_4 + HNO_3 \rightarrow (NH_4)_3PO_4 \cdot 12MoO_3 + NH_4NO_3 + H_2O$
c. $HClO_4 + P_4O_{10} \rightarrow H_3PO_4 + Cl_2O_7$
d. $Ca_3P_2 + H_2O \rightarrow Ca(OH)_2 + PH_3$

8.15. Sulfuric acid is prepared commercially by the following series of reactions:
1. Iron pyrites, FeS_2, is roasted to form iron (III) oxide and sulfur dioxide.
2. The sulfur dioxide is oxidized to sulfur trioxide.
3. The sulfur trioxide is dissolved in H_2SO_4 to form $H_2S_2O_7$.
4. The $H_2S_2O_7$ is dissolved in water to form H_2SO_4.

a. Write the equation for each step.
b. Write the overall equation for the process.
c. What are the by-products of the process?

8.16. Sodium bicarbonate, $NaHCO_3$, is prepared commercially by the following series of reactions:

1. Calcium carbonate is heated to produce calcium oxide and carbon dioxide.
2. Ammonium chloride is heated with calcium oxide to produce ammonia, sodium chloride, and water.
3. Carbon dioxide, ammonia, and water react to yield ammonium hydrogen carbonate.
4. Ammonium hydrogen carbonate and sodium chloride react to produce sodium hydrogen carbonate and ammonium chloride.

a. Write the equation for each step.
b. Write the overall equation for the process.
c. What are the by-products of the process?

9

Equivalent Weights of the Elements

The previous chapters have dealt with fundamentals which the student must understand in order to speak the language of the chemist. We are now ready to use some of these fundamental ideas in studying the mass relationships involved when the various elements combine to form compounds. In order for the chemist to explain quantitatively the laws controlling the formation of compounds from elements, a unit of mass called the *gram-equivalent weight* has been devised. This represents the third unit of mass discussed, based upon the actual masses of atoms and molecules, the gram-atom and the mole having been presented in Chapter 6.

9.1. Combining Proportions among the Elements. Under similar conditions, such as temperature and pressure, the atoms of two or more elements which react to form a compound will always combine in the same ratio. For example, when hydrogen burns in oxygen the product is H_2O, indicating that two atoms of hydrogen combine with one of oxygen. Also, excluding variations due to isotopes, the atoms of a given element all possess the same mass. The above facts account for the *law of constant composition:* the proportions by weight of the elements constituting a compound are always the same.

Table 9.1 gives the percentage composition by weight of several compounds.

TABLE 9.1

PERCENTAGE COMPOSITION BY WEIGHT OF SEVERAL COMPOUNDS

Compound	*Percentage Composition*
H_2O	hydrogen 11.19%, oxygen 88.81%
NaCl	sodium 39.34%, chlorine 60.66%
CaO	calcium 71.47%, oxygen 28.53%
H_2S	hydrogen 5.92%, sulfur 94.08%
HCl	hydrogen 2.76%, chlorine 97.24%

By means of the experimental values given in Table 9.1 we can show that elements combine by weight in the ratios of their atomic weights or some multiple thereof.

Example 9.1. Using the data given in Table 9.1, calculate (a) the number of grams of hydrogen, and (b) the number of grams of calcium that will combine with 16.00 g of oxygen (1.00 gram-atom).

Solution. (a) From Table 9.1 we see that 11.19 g of hydrogen will combine with 88.81 g of oxygen to form 100 g of H_2O.

Since $\frac{11.19 \text{ g } H_2}{88.81 \text{ g } O_2}$ is the number of grams of H_2 combined with 1.00 g of O_2,

then $16.00 \text{ g } O_2 \times \frac{11.19 \text{ g } H_2}{88.81 \text{ g } O_2} = 2.02 \text{ g } H_2$ combined with 16.00 g O_2.

(b) From Table 9.1 we see that 71.47 g of Ca will combine with 28.53 g of O_2 to form 100 g of CaO.

Since $\frac{71.47 \text{ g Ca}}{28.53 \text{ g } O_2} =$ g of Ca combined with 1.00 g of O_2, then $16.00 \text{ g } O_2 \times$ $\frac{71.47 \text{ g Ca}}{28.53 \text{ g } O_2} = 40.08$ g Ca will combine with 16.00 g O_2.

By means of calculations such as the above it is possible to determine the combining powers of all the elements in terms of gram-atoms. Fig. 9.1 shows the results of such calculations for the six elements listed in Table 9.1.

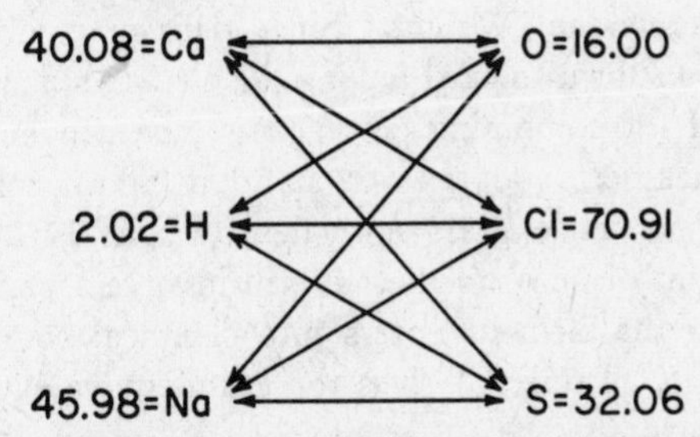

Fig. 9.1. Combining proportions among several of the elements.

Observe in Fig. 9.1 that the combining masses given for calcium, oxygen, and sulfur are equal numerically to their atomic weights; for sodium, hydrogen, and chlorine, the combining masses are twice their atomic weights.

Example 9.2. A piece of calcium weighing 2.16 grams was exposed to the air until oxidation was complete. (a) How much oxygen combined with the calcium? (b) What was the weight of the calcium oxide formed?

Solution. (a) From Fig. 9.1 we see that 40.1 g of calcium combines with 16.0 g of oxygen.

Since $\frac{16.00 \text{ g } O_2}{40.08 \text{ g Ca}}$ = g of O_2 combined with 1.00 g of Ca, then 2.16 g Ca $\times$

$\frac{16.00 \text{ g } O_2}{40.08 \text{ g Ca}}$ = 0.862 g O_2 combined with 2.16 g Ca.

(b) The weight of the calcium oxide is the sum of the weights of the calcium and oxygen reacting. That is:

Weight of calcium oxide formed = **2.16 g + 0.862 g = 3.02 g.**

9.2. The Avogadro Number and Combining Power. The combining proportions of the elements, in terms of gram-atoms, are related directly to the Avogadro number, N. For example, equal numbers of sodium atoms and chlorine atoms combine to form any given mass of sodium chloride, NaCl. When sodium atoms and chlorine atoms combine, each sodium atom loses one electron and each chlorine atom gains one electron. This is supported in Fig. 9.1 where we see that 2.00 gram-atoms of sodium (45.98 g or 2 N atoms) combine with 2.00 gram-atoms of chlorine (70.91 g or 2 N atoms). To form calcium chloride, $CaCl_2$, the ratio of atoms would be one calcium atom to two chlorine atoms. This must be true since a calcium atom loses two electrons when it reacts to form an ionic compound. Therefore, 1.00 gram-atom of calcium (40.08 g or N atoms) combines with 2.00 gram-atoms of chlorine (70.91 g or 2 N atoms).

9.3. Gram-equivalent Weight. Since definite combining proportions have been shown to exist among the elements, it is evident that some standard for combining power would be convenient. Such a unit of mass has been adopted as a standard and is called the *gram-equivalent weight.* The gram-equivalent weight is defined as the number of grams of an element that will involve a gain or loss of N electrons when the element enters into chemical combination with another element. Table 9.2 gives the gram-equivalent weights of a number of elements.

Table 9.2 could be extended to include all the elements. The usefulness of gram-equivalent weight values is that they enable us to determine the amount of one element that will combine with a given weight of another element. That is, the combining ratios of elements by weight will be in the same ratio as their gram-equivalent weights.

Example 9.3. How many grams of aluminum will combine with 7.63 g of chlorine?

TABLE 9.2

THE GRAM-EQUIVALENT WEIGHTS OF SOME ELEMENTS

Element	*Atomic Weight*	*Oxidation Number*	1.00 *Gram-equivalent Weight*
Oxygen	15.9994	−2	8.000 g
Hydrogen	1.00797	+1	1.008 g
Chlorine	35.453	−1	35.453 g
Sodium	22.9898	+1	22.990 g
Calcium	40.08	+2	20.04 g
Sulfur	32.064	−2	16.032 g
Aluminum	26.9815	+3	8.994 g
Silver	107.870	+1	107.870 g

Solution. The two elements will combine in amounts proportional to their gram-equivalent weights. From Table 9.2 we see that 8.994 g of Al will combine with 35.453 g of Cl_2.

Since $\frac{8.994 \text{ g Al}}{35.453 \text{ g } Cl_2}$ = g Al combined with 1.00 g Cl_2, then 7.63 g Cl_2 × $\frac{8.994 \text{ g Al}}{35.453 \text{ g } Cl_2}$ = 1.93 g Al combined with 7.63 g Cl_2.

Example 9.4. An analysis of magnesium oxide showed that 2.099 g of the oxide contained 0.833 g of oxygen and 1.266 g of magnesium. From the above data calculate the gram-equivalent weight of magnesium.

Solution. The gram-equivalent weight of magnesium is the amount of the element that will combine with 8.0000 g of oxygen.

Since $\frac{1.266 \text{ g Mg}}{0.833 \text{ g } O_2}$ = g Mg combined with 1.00 g O_2, then 8.000 g O_2 × $\frac{1.266 \text{ g Mg}}{0.833 \text{ g } O_2}$ = 12.16 g Mg combined with 8.000 g O_2. That is, the gram-equivalent weight of Mg is 12.16 g.

9.4. Gram-equivalent Weight and Oxidation Number. A study of Table 9.2 will show that the gram-equivalent weight of an element may be obtained by dividing the gram-atomic weight of an element by its oxidation number. That is:

$$\text{Gram-equivalent weight} = \frac{\text{gram-atomic weight}}{\text{oxidation number}}.$$

The relationship expressed in the above formula exists because each unit value of the oxidation number represents a gain or loss of one electron per atom. Therefore, one gram-equivalent weight would involve a gain or loss of N electrons.

Example 9.5. Determine the gram-equivalent weight of the metal as given in each of the following formulas: (a) NaCl, (b) $CaCl_2$, (c) Fe_2O_3.

Solution. First determine the oxidation number of the metal; then use the formula given above.

(a) NaCl. Oxidation number of Na = +1. Therefore:

$$\text{Gram-equivalent weight of Na} = \frac{22.990 \text{ g}}{1} = 22.990 \text{ g}.$$

(b) $CaCl_2$. Oxidation number of Ca = +2. Therefore:

$$\text{Gram-equivalent weight of Ca} = \frac{40.08 \text{ g}}{2} = 20.04 \text{ g}.$$

(c) Fe_2O_3. Oxidation number of Fe = +3. Therefore:

$$\text{Gram-equivalent weight of Fe} = \frac{55.85 \text{ g}}{3} = 18.62 \text{ g}.$$

It so happens that the atoms of two or more elements sometimes combine in more than one ratio, each combination occurring only under certain specified conditions. That is, some elements exhibit more than one oxidation state. For example, iron may exist in either the ferrous state, Fe^{+2}, or the ferric state, Fe^{+3}. It follows, therefore, that an element will have a gram equivalent weight for each oxidation state it assumes.

Example 9.6. Determine the equivalent weight of iron in (a) $FeCl_2$, and in (b) $FeCl_3$.

Solution. Following the same procedure as in Example 9.6, we have:

(a) $\text{Gram-equivalent weight of Fe in FeCl}_2 = \frac{55.85 \text{ g}}{2} = 27.93 \text{ g}.$

(b) $\text{Gram-equivalent weight of Fe in FeCl}_3 = \frac{55.85 \text{ g}}{3} = 18.62 \text{ g}.$

Observe that the gram-equivalent weights for an element in its various oxidation states are in a ratio of small integer values. For example, for iron, as calculated in Example 9.6:

$$\frac{18.62}{27.93} = \frac{2}{3}.$$

The above small-integer ratio relationship is sometimes referred to as the *law of multiple proportions:* when two elements combine to form more than one compound, the weights of one combined with a fixed weight of the other element are in a ratio of small whole numbers. In Example 9.6 a fixed weight of chlorine, 35.453 g , is combined with 18.62 g of Fe in $FeCl_3$, and with 27.93 g of Fe in $FeCl_2$.

Example 9.7. Tin and oxygen combine to form two different oxides. One contains 78.77 per cent tin, and the other 88.12 per cent tin. Determine the equivalent weight of tin in each of the two oxides and show that the values are in accord with the law of multiple proportions.

Solution. The gram-equivalent weight of tin is the number of grams of tin which will combine with 8.0000 g of oxygen.

Oxide Number 1. From the data given in the problem, $\frac{78.77 \text{ g Sn}}{21.23 \text{ g } O_2} =$ g of Sn combined with 1.00 g of O_2. Therefore, $8.00 \text{ g } O_2 \times \frac{78.77 \text{ g Sn}}{21.23 \text{ g } O_2}$ $= 29.68$ g Sn combined with 8.00 g O_2.

That is, the gram-equivalent weight of Sn is 29.68 g.

Oxide Number 2. From the data given in the problem, $\frac{88.12 \text{ g Sn}}{11.88 \text{ g } O_2} =$ g of Sn combined with 1.00 g of O_2. Therefore, $8.00 \text{ g } O_2 \times \frac{88.12 \text{ g Sn}}{11.88 \text{ g } O_2}$ $= 59.34$ g Sn combined with 8.00 g O_2.

That is, the gram-equivalent weight of Sn is 59.34 g.

Problems

Part I

9.1. How much oxygen will combine with 1.00 g of calcium? *Ans.* 0.400 g.

9.2. How much calcium will combine with 1.00 g of oxygen? *Ans.* 2.50 g.

9.3. How much sulfur will combine with 15.0 g of sodium? *Ans.* 10.5 g.

9.4. In a laboratory experiment it was found that 0.562 g of aluminum combined with 0.500 g of oxygen. Calculate the gram-equivalent weight of aluminum. *Ans.* 8.99 g.

9.5. Zinc oxide contains 80.3 per cent zinc. Calculate the gram-equivalent weight of zinc. *Ans.* 32.6 g.

9.6. Calculate the gram-equivalent weight of bismuth if 3.96 g of the element forms 4.71 g of oxide. *Ans.* 42.2 g.

9.7. What is the (a) oxidation number, and (b) gram-equivalent weight of tin in $SnCl_4$? *Ans.* (a) 4; (b) 29.68 g.

9.8. The oxidation number of an element is +3 and the gram-equivalent weight 69.67 g. (a) What is the atomic weight of the element, and (b) what is the symbol of the element? *Ans.* (a) 209.0; (b) Bi.

9.9. How many electrons are lost by 1.00 g of Mg when the element combines with chlorine? *Ans.* 4.95×10^{22}.

9.10. Two oxides of copper contain respectively 20.1 per cent and 11.2 per cent oxygen. (a) What is the gram-equivalent weight of copper in each of the two oxides? (b) Show that the data are in accord with the law of multiple proportions. *Ans.* (a) 31.8 g , 63.4 g ; (b) 1 : 2.

9.11. Nitrogen forms five oxides: N_2O, NO, N_2O_3, NO_2, and N_2O_5. (a) What is the gram-equivalent weight of nitrogen in each of the oxides? (b) Which oxide contains 53.3 per cent oxygen?

Ans. (a) 14.0 g , 7.00 g , 4.67 g , 3.50 g , 2.80 g ; (b) NO.

9.12. On the basis of the percentage composition of H_2S, as given in Table 9.1, calculate the gram-equivalent weight of sulfur. *Ans.* 16.02 g.

9.13. Given that the oxidation state of sulfur is −2, calculate the gram-equivalent weight of sulfur. *Ans.* 16.032 g.

9.14. The following questions refer to Example 19.3. (a) How many atoms are there in (1) 1.93 grams of aluminum, and (2) 7.63 grams of chlorine? (b) How many electrons will be (1) lost by 1.93 grams of aluminum, and (2) gained by 7.63 grams of chlorine when the two elements combine? *Ans.* (a–1) 4.30×10^{22} atoms, (a–2) 1.30×10^{23} atoms, (b–1 and b–2) 1.30×10^{23} electrons.

Part II

9.15. When 5.81 g of silver oxide was heated, 0.401 g of oxygen was liberated. Calculate the gram-equivalent weight of silver.

Ans. 108 g.

9.16. It was found that 4.90 g of a monovalent element combined with 1.00 g of oxygen. What was the element? *Ans.* K.

9.17. A metallic oxide contains 52.9 per cent metal. What is the gram-equivalent weight of the metal? *Ans.* 9.0.

9.18. A mass of metallic oxide weighing 2.59 g contained 0.401 g of oxygen. Calculate the gram-equivalent weight of the metal. *Ans.* 43.8.

9.19. The gram-equivalent weight of a trivalent metal is 17.34 g. What is the metal? *Ans.* Cr.

9.20. The mole weight of R_2O_3 is 326. What is the gram-equivalent weight of R? *Ans.* 46.3.

9.21. A 4.00 g sample of cupric oxide, CuO, was reduced to free copper by passing hydrogen over the hot oxide. The reduced copper weighed 3.20 g. Calculate the gram-equivalent weight of copper.

Ans. 32.0 g.

9.22. An iron nail weighs 6.34 grams. What weight of rust, Fe_2O_3, would be formed by the nail? *Ans.* 9.06 g.

9.23. How much sodium and chlorine could be obtained by the decomposition of 12.0 grams of salt, NaCl? *Ans.* 7.28 g Cl_2; 4.72 g Na.

9.24. Calculate the gram-equivalent weight of copper in $Cu_3(PO_4)_2$.
Ans. 31.77 g.

9.25. How many electrons would be liberated by 1.00 g of the element in the process of formation of the following ions: (a) Cs^+, (b) Ca^{+2}, (c) Al^{+3}? *Ans.* (a) 4.5×10^{21}; (b) 3.0×10^{22}; (c) 6.7×10^{22}.

9.26. How many grams of magnesium would have to react in order to liberate 4 N electrons? *Ans.* 48.62 g.

9.27. Mercury and chlorine form two compounds. In one compound 0.669 g of mercury combines with 0.118 g of chlorine; in the other compound 1.00 g of mercury combines with 0.355 g of chlorine. What is the gram-equivalent weight of mercury in each compound?
Ans. 200; 100.

9.28. Iron exists in two possible oxidation states, +2 and +3. How many grams of iron will combine with 25.0 g of iodine?
Ans. 5.50 g ; 3.67 g.

9.29. On the basis of electrons gained and lost, show why calcium and oxygen combine in the ratio of 40.08 g of calcium to 16.00 g of oxygen.

9.30. On the basis of electrons gained and lost, show why one gram-atom of iron in the +3 oxidation state will be combined with 1.5 gram-atoms of oxygen.

10

The Quantitative Significance of Chemical Formulas

Formulas are the expressions used by chemists to designate the exact chemical composition of an element or compound. The quantitative interpretation of formulas is basic in the study of chemistry, and gives one a better appreciation of the science of chemistry.

For example, the formula H_2O may be interpreted in terms of atoms and molecules by stating that H_2O represents one molecule of water. The one molecule of water contains two atoms of hydrogen and one atom of oxygen. Again, the formula H_2O may be interpreted as representing one mole of water, or 18.02 g. One mole of water contains 6.02×10^{23} molecules. Or, we may say that one mole of water consists of two gram-atoms of hydrogen (2.02 g or 12.04×10^{23} atoms) and one gram-atom of oxygen (16.00 g or 6.02×10^{23} atoms).

Some Quantitative Relationships Involving Molecules in the Gaseous State

10.1. The Mole Volume of Substances in the Gaseous State. An interesting relationship is shown when we compare the volumes occupied by one mole of gaseous substances under the same conditions of temperature and pressure. This relationship is shown for five gases in Table 10.1.

TABLE 10.1

THE MOLE VOLUME OF A NUMBER OF GASEOUS SUBSTANCES AT STANDARD CONDITIONS

Gas	*One Mole*	*Weight of One Liter*	*Mole Volume at S.C.*
O_2	32.00 g	1.43 g	32.00 ÷ 1.43 = 22.4 *l*
N_2	28.02 g	1.25 g	28.02 ÷ 1.25 = 22.4 *l*
H_2	2.016 g	0.0899 g	2.016 ÷ 0.0899 = 22.4 *l*
HCl	36.46 g	1.63 g	36.46 ÷ 1.63 = 22.4 *l*
CO_2	44.01 g	1.965 g	44.01 ÷ 1.965 = 22.4 *l*

From the table we see that the volume occupied by one mole of each of the five gases at standard conditions is 22.4 liters, or 22,400 ml. The quantity 22.4 liters is called the *gram-molecular volume* (G.M.V.), and represents the volume occupied by one mole of any gas at standard conditions. Since one mole represents the weight in grams of 6.02×10^{23} molecules (Sec. 6.4), it follows that use may be made of the gram-molecular volume principle to determine the molecular weight of substances when in the gaseous state.

Example 10.1. It was found that 326 ml. of a gas weighed 0.492 g at standard conditions. Calculate the molecular weight of the gas.

Solution. The molecular weight of the gas is the weight in grams of 22,400 ml at standard conditions. Then:

$$\frac{326 \text{ ml.}}{0.492 \text{ g.}} = \frac{22{,}400 \text{ ml.}}{x \text{ g.}}$$

or

$$x = 33.8 = \text{molecular weight of the gas.}$$

$$\frac{0.492 \text{ g}}{326 \text{ ml}} = \text{weight in grams of 1.00 ml of the gas at S.C.}$$

And:

$$22{,}400 \frac{\cancel{\text{ml}}}{\text{mole}} \times \frac{0.492 \text{ g}}{326 \cancel{\text{ml}}} = 33.8 \frac{\text{g}}{\text{mole}}.$$

That is, the molecular weight of the gas is 33.8.

Example 10.2. It was found that 426 ml of a gas weighed 0.492 g at 27° C and 640 mm. of Hg. Calculate the molecular weight of the gas.

Solution. Again, the molecular weight of the gas is the weight of 22,400 ml at standard conditions. However, the given volume of gas is not at standard conditions. We must, therefore, reduce the given volume to standard conditions. Then:

$$V_{\text{S.C.}} = 426 \text{ ml} \times \frac{273 \cancel{°\text{K}}}{300 \cancel{°\text{K}}} \times \frac{640 \cancel{\text{mm of Hg}}}{760 \cancel{\text{mm of Hg}}}$$

$$= 326 \text{ ml.}$$

The problem is now identical to Example 10.1. Compressing the 426 ml of gas to 326 ml volume would not change the weight of the given mass of gas. The molecular weight of the gas is, therefore, 33.8 as in Example 10.1.

The gram-molecular volume principle may also be used to determine the weight of any given volume of a gaseous substance, providing its molecular weight is known.

Example 10.3. Determine the weight of one liter of hydrogen sulfide, H_2S, at standard conditions.

Solution. We know that one mole of H_2S, or

$$(2 \times 1.008) + 32.064 = 34.080 \text{ g},$$

occupies a volume of 22.4 liters at standard conditions. Therefore:

$$H_2S = \frac{34.080 \text{ g}}{22.4\ l} = 1.52 \frac{\text{g}}{l}.$$

Example 10.4. Determine the weight of 375 ml of oxygen at 23° C and 740 mm of Hg.

Solution. Since the gram-molecular volume principle applies only to standard conditions, it will be necessary to reduce the given volume of oxygen to the volume it would occupy at standard conditions. Then:

$$V_{\text{S.C.}} = 375 \text{ ml} \times \frac{273 \cancel{°\text{K}}}{296 \cancel{°\text{K}}} \times \frac{740 \cancel{\text{mm of Hg}}}{760 \cancel{\text{mm of Hg}}}$$
$$= 337 \text{ ml}.$$

We know that 22,400 ml of oxygen weighs 32.00 g at S.C. Therefore:

$$\frac{32.00 \text{ g } O_2}{22{,}400 \text{ ml } O_2} = \text{weight in grams of 1.00 ml } O_2 \text{ at S.C.}$$

And: $337 \cancel{\text{ml } O_2} \times \frac{32.00 \text{ g } O_2}{22{,}400 \cancel{\text{ml } O_2}} = 0.481 \text{ g } O_2.$ Weight of 337 ml of O_2.

10.2. The Gram-equivalent Volume of Substances in the Gaseous State. From Sec. 10.1 we see that one mole of hydrogen, or 2.02 g, occupies a volume of 22.4 l at S.C. Therefore, one gram-equivalent weight of hydrogen, or 1.01 g, would occupy a volume of 11.2 l at S.C. Similarly, one mole of oxygen, or 32.00 g, occupies a volume of 22.4 l at S.C., and one gram-equivalent weight of oxygen, or 8.00 g, will occupy a volume of 5.6 l at S.C.

We can now extend the concept of gram-equivalent weight to replacement reactions involving elements. In a replacement reaction equal gram-equivalent weights (g-eq. wt.) of elements are involved, as shown in the following reactions.

$$\begin{array}{lcr} 2\,Na & + 2\,HCl \rightarrow 2\,NaCl + & H_2 \\ \text{2 moles} & & \text{1 mole} \\ \text{2 g-eq. wt.} & & \text{2 g-eq. wt.} \\ 2\,Al & + 6\,HCl \rightarrow 2\,AlCl_3 + & 3\,H_2 \\ \text{2 moles} & & \text{3 moles} \\ \text{6 g-eq. wt.} & & \text{6 g-eq. wt.} \end{array}$$

From the above equations we see that one gram-equivalent weight of a metal will displace one gram-equivalent weight of hydrogen (1.008 g or 11,200 ml at S.C.) from an acid solution.

Example 10.5. When 0.684 g of zinc reacted with sulfuric acid there was liberated 234 ml of hydrogen, measured at standard conditions. Determine the gram-equivalent weight of zinc.

Solution. By definition, the gram-equivalent weight of zinc is the number of grams required to liberate 11,200 ml of H_2 at S.C. Therefore:

$$\frac{0.684 \text{ g Zn}}{234 \text{ ml } H_2} = \text{g of Zn required to release 1.00 ml } H_2 \text{ at S.C.},$$

and the gram-equivalent weight of Zn is:

$$11{,}200 \frac{\cancel{\text{ml } H_2}}{\text{g-eq.}} \times \frac{0.684 \text{ g Zn}}{234 \cancel{\text{ml } H_2}} = 32.7 \frac{\text{g Zn}}{\text{g-eq.}}.$$

Example 10.6. It was found that 0.450 g of Al liberated 760 ml of hydrogen from a solution of sulfuric acid, the gas being collected over water at 27° C and 640 mm of Hg. Calculate the gram-equivalent weight of aluminum from the data given in the problem.

Solution. One gram-equivalent weight of aluminum will displace 11,200 ml of hydrogen at standard conditions. Evidently the first step in the solution will be to convert the collected gas to the volume it would occupy at standard conditions. Then:

$$V_{\text{S.C.}} = 760 \text{ ml} \times \frac{273 \cancel{°\text{K}}}{300 \cancel{°\text{K}}} \times \frac{(640 - 27) \cancel{\text{mm of Hg}}}{760 \cancel{\text{mm of Hg}}}$$

$$= 558 \text{ ml of } H_2\text{, dry, at standard conditions.}$$

The amount of aluminum required to liberate 11,200 ml of hydrogen can now be calculated.

Since: $\frac{0.450 \text{ g Al}}{558 \text{ ml } H_2}$ = g of Al required to release 1.00 ml H_2 at S.C., then

the gram-equivalent weight of Al is:

$$11{,}200 \frac{\cancel{\text{ml } H_2}}{\text{g-eq.}} \times \frac{0.450 \text{ g Al}}{558 \cancel{\text{ml } H_2}} = 9.00 \frac{\text{g Al}}{\text{g-eq.}}.$$

Example 10.7. In a laboratory experiment it was found that 1.324 g of zinc reacted with 226.8 ml of oxygen, measured at S.C., to form zinc oxide, ZnO. Calculate the gram-equivalent weight of zinc.

Solution. The number of grams of zinc combined with 5600 ml of oxygen measured at S.C. would be the gram-equivalent weight of zinc.

Since: $\frac{1.324 \text{ g Zn}}{226.8 \text{ ml } O_2}$ = g of Zn reacting with 1.00 ml O_2 at S.C., then the gram-equivalent weight of Zn is:

$$5600\,\frac{\cancel{\text{ml } O_2}}{\text{g-eq.}} \times \frac{1.324\text{ g Zn}}{226.8\,\cancel{\text{ml } O_2}} = 32.69\,\frac{\text{g Zn}}{\text{g-eq.}}.$$

Problems

Part I

10.1. One liter of chlorine gas at standard conditions weighs 3.214 g. Calculate the molecular weight of chlorine. *Ans.* 72.0.

10.2. Calculate the molecular weight of a gas, 225 ml of which weighs 0.281 g at standard conditions. *Ans.* 28.0.

10.3. Calculate the molecular weight of a gas, 642 ml of which weighs 1.61 g at 100° C and 740 mm of Hg. *Ans.* 78.8.

10.4. Calculate the weight of one liter of (a) ammonia, and (b) helium at standard conditions. *Ans.* (a) 0.76 g; (b) 0.18 g.

10.5. Calculate the weight of one liter of carbon dioxide at (a) standard conditions, and (b) 27° C and 730 mm of Hg. *Ans.* (a) 1.96 g; (b) 1.71 g.

10.6. A compound has the formula $COCl_2$. Calculate the weight of one liter of the gas at standard conditions. *Ans.* 4.42 g.

10.7. The molecular weight of a gaseous substance is 80. Calculate the volume occupied by one gram of the substance at standard conditions. *Ans.* 280 ml.

10.8. Calculate the weight of 350 ml of CO_2 at S.C. *Ans.* 0.687 g.

10.9. How many moles of water will contain 1.25 moles of (a) hydrogen, and (b) oxygen? *Ans.* (a) 1.25 moles, (b) 2.50 moles.

10.10. In a laboratory experiment the decomposition of 2.364 g of mercuric oxide produced 2.189 g of mercury and 122.3 ml of oxygen at S.C. Calculate the gram-equivalent weight of mercury. *Ans.* 100.2 g.

10.11. When 1.391 g. of mercuric oxide was heated, 71.8 ml. of oxygen was liberated at S.C. Calculate the gram-equivalent weight of mercury. *Ans.* 100 g.

10.12. How many liters of hydrogen at standard conditions would be obtained by the action of 0.100 gram-equivalent weight of a metal reacting with an acid? *Ans.* 1.12 *l.*

10.13. When 0.723 g of iron reacted with a solution of H_2SO_4, and the released hydrogen was collected by displacement of water, there was obtained 340 ml of hydrogen and water vapor at 27° C and 740 mm of Hg. Calculate the gram-equivalent weight of iron. *Ans.* 27.9 g.

Part II

10.14. When 0.700 g of a substance was heated to 300° C, it formed 350 ml of gaseous vapor. The barometric pressure was 72.4 cm. Calculate the molecular weight of the substance. *Ans.* 98.7.

10.15. Calculate the molecular weight of a substance, 2.810 g of which formed 850 ml of vapor at 200° C and a reduced pressure of 60 mm of Hg. *Ans.* 163.

10.16. What is the weight of 1500 ml of argon at S.C.? *Ans.* 2.67 g.

10.17. What volume would 12.0 g of neon occupy at standard conditions? *Ans.* 13.3 *l*.

10.18. What volume would 12.0 g of neon occupy when collected over water at 22° C and 720 mm of Hg? *Ans.* 15.6 *l*.

10.19. A mass of gas weighing 22.50 g occupies a volume of 17.5 *l* at 23° C and 740 mm of Hg. What would be the weight of 1.00 liter of the gas at 100° C and 780 mm of Hg? *Ans.* 1.08 $\frac{g}{l}$.

10.20. The substance in problem 10.14 is an element. What is the element? *Ans.* Oxygen.

10.21. A mass 0.475 g of a gaseous substance occupied a volume of 131.5 ml at 40° C and 740 mm of Hg. Calculate the molecular weight of the substance. *Ans.* 95.

10.22. How many molecules are there in one liter of oxygen at S.C.? *Ans.* 2.69×10^{22}.

10.23. How many molecules are there in 1.00 ml of hydrogen at 27° C and 640 mm of Hg? *Ans.* 2.06×10^{19}.

10.24. Which of the three gases — CO_2, NH_3, N_2 — has the smallest weight per liter under similar conditions? *Ans.* NH_3.

10.25. A tank has a volume of 100 *l*. How many grams of oxygen will the tank hold at S.C.? *Ans.* 143 g.

10.26. How many grams of oxygen will the tank given in problem 10.20 hold at 27° C and 720 mm of Hg? *Ans.* 123 g.

10.27. It was found that 0.336 *l* of a gas weighed 0.240 g at S.C. Was the gas CH_4, O_2, NH_3, or F_2? *Ans.* CH_4.

10.28. What is the difference in weight between 50.0 *l* each of the gases PH_3 and CH_4 at S.C.? *Ans.* 40.2 g.

10.29. What is the density of oxygen at 86° F and 64 cm of Hg? *Ans.* 1.09 $\frac{g}{l}$.

10.30. One of the principal constituents of gasoline is heptane, C_7H_{16}, the density of which is 0.68 g per $cm.^3$ What volume would one liter of heptane occupy if vaporized at 200° C and 740 mm of Hg? *Ans.* 271 *l*.

10.31. How many moles of aluminum are associated with 2.00 moles of sulfur in the form of ammonium sulfide? *Ans.* 1.33 moles.

10.32. How many gram-atoms of oxygen are there in 2.50 moles of Fe_2O_3? *Ans.* 7.5 gram-atoms.

10.33. How many moles of oxygen are there in 2.50 moles of Fe_2O_3? *Ans.* 3.25 moles.

10.34. How many moles of iron are there in 2.50 moles of Fe_2O_3? *Ans.* 5.00 moles.

10.35. When 0.590 g of sodium reacted with water, there was obtained 314 ml of hydrogen and water vapor, collected by displacement of water at 17° C and 755 mm of Hg. Calculate the gram-equivalent weight of sodium. *Ans.* 23 g.

10.36. How many liters of hydrogen, measured over water at 24° C and 720 mm of Hg, would be obtained by the action of 2.50 g of calcium on water? *Ans.* 1.65 *l.*

Some Quantitative Relationships Involving Formulas

10.3. Determination of the Formula of a Compound. In order to determine the molecular formula of a compound, the following information must be available: (1) the elements constituting the compound; (2) the atomic weights of the constituent elements; (3) the percentage composition of the compound; and (4) the molecular weight of the compound. The above information must be obtained experimentally. However, details of the experimental procedure need not be known in order to understand the calculations involved.

Example 10.8. A compound upon analysis was found to have the following percentage composition: carbon, 81.82 per cent; and hydrogen, 18.18 per cent. The molecular weight was found to be 44. Determine the formula of the compound.

Solution. The solution lies in finding the values of A and B in the expression C_AH_B. Since 81.82 per cent of the molecular weight is contributed by carbon and 18.18 per cent by hydrogen, then:

$$0.8182 \times 44 = 36 \text{ units contributed by carbon}$$

and

$$0.1818 \times 44 = 8 \text{ units contributed by hydrogen.}$$

Next we must find how many carbon atoms are required to contribute 36 units to the molecular weight. This will be the value of A. Since one carbon atom contributes 12 units, its atomic weight, then:

$$36 \div 12 = 3 \text{ atoms of carbon} = \text{A.}$$

Each atom of hydrogen in the molecule contributes 1 unit to the molecular weight. Therefore:

$$8 \div 1 = 8 \text{ atoms of hydrogen} = \text{B.}$$

The formula for the compound is, therefore, C_3H_8.

Example 10.9. A compound was found to contain 40.01 per cent carbon, 6.67 per cent hydrogen, and 53.32 per cent oxygen. The molecular weight was found to be 178. What is the formula of the compound?

Solution. Following the same procedure as in Example 10.8, we will first determine the contribution of each element to the molecular weight, 178. Then:

Carbon $= 0.4001 \times 178 = 71.2$ units of molecular weight;
Hydrogen $= 0.0667 \times 178 = 11.9$ units of molecular weight;
Oxygen $= 0.5332 \times 178 = 94.9$ units of molecular weight.

The number of atoms of each in a molecule would be:

$71.2 \div 12 = 6$ carbon atoms;
$11.9 \div 1 = 12$ hydrogen atoms;
$94.9 \div 16 = 6$ oxygen atoms.

The formula of the compound is, therefore, $C_6H_{12}O_6$.

As will be observed later, most methods for the determination of the molecular weight of a compound give only approximate values. The correct molecular weight of $C_6H_{12}O_6$ would be 180.16. This means that the contribution of each atom as calculated above may be somewhat in error and may, therefore, not represent an exact multiple of the atomic weight of the element. However, since a molecule must contain an integer number of atoms of each of the elements constituting the molecule, the inherent error in the experimental value must be taken into consideration. Observe that the nearest integer value was used in the second step above. On the other hand, methods of analysis have enabled chemists to determine percentage composition and atomic weights with a high degree of accuracy.

The term molecular weight has no significance when referred to ionic compounds (Sec. 6.2). The formulas for such compounds must, therefore, be determined without this information.

Example 10.10. A compound was found to contain 88.80 per cent copper and 11.20 per cent oxygen. What is the formula of the compound?

Solution. Since the molecular weight has no meaning, the contribution of each atom to the molecular weight cannot be determined. It is possible, however, to determine the simplest ratio of atoms in the substance. First, divide the percentages given by the respective atomic weights of the elements. Then:

$$\left.\begin{aligned} \text{Copper} &= \frac{88.80}{63.54} = 1.40 \\ \text{Oxygen} &= \frac{11.20}{16.00} = 0.70 \end{aligned}\right\} \text{Ratio of copper to oxygen atoms.}$$

Next, reduce the above ratio to the simplest integer ratio of atoms. These values represent the number of atoms of each element as expressed in the formula. Then:

$$1.40 : 0.70 = \underset{\text{copper}}{2} : \underset{\text{oxygen}}{1}.$$

The formula is, therefore, Cu_2O.

Example 10.11. Red lead is composed of 90.65 per cent lead and 9.35 per cent oxygen. What is the formula for red lead?

Solution. The ratio of atoms would be:

$$\left.\begin{aligned} \text{Lead} &= \frac{90.65}{207.19} = 0.438 \\ \text{Oxygen} &= \frac{9.35}{16.00} = 0.584 \end{aligned}\right\} \text{Ratio of lead to oxygen atoms.}$$

Reducing the ratio 0.438 : 0.584 to small integer values will evidently require more careful consideration than in Example 10.10. In this case trial and error is the best method. Evidently, one of the following integer ratios is the correct one:

$$\frac{0.438}{0.584} = \frac{1}{2} = \frac{1}{3} = \frac{2}{3} = \frac{1}{4} = \frac{3}{4} = \frac{4}{5} = \frac{5}{6}.$$

A few moments' inspection will show that:

$$\left.\begin{aligned} & \frac{0.438}{0.584} = \frac{3}{4} \\ \text{or} \quad & 0.438 \times 4 = 0.584 \times 3 \\ \text{and} \quad & 1752 = 1752. \end{aligned}\right\} \text{Proof for the validity of a proportion.}$$

The formula for red lead is, therefore, Pb_3O_4.

Example 10.12. A 1.224 g sample of an organic compound containing carbon, hydrogen, and oxygen, was burned in an atmosphere of oxygen to CO_2 and H_2O. There was formed 1.794 g of CO_2 and 0.729 g of H_2O. What was the empirical formula for the organic compound?

Solution. All of the carbon in the CO_2 and all of the hydrogen in the H_2O came from the compound being analyzed. Some of the oxygen originated from the atmosphere of oxygen. Therefore, we will determine the amounts of carbon and hydrogen and obtain the oxygen by difference.

$$CO_2 \text{ contains } \tfrac{12}{44} \times 100 = 27.3\% \text{ C}.$$

Then: $1.794 \cancel{\text{g } CO_2} \times 0.273 \dfrac{\text{g C}}{\cancel{\text{g } CO_2}} = 0.490$ g C in 1.224 g of the compound.

Again:

$$H_2O \text{ contains } \frac{2.02}{18.02} \times 100 = 11.2\% \ H_2.$$

And:

$$0.729 \ \cancel{g\ H_2O} \times 0.112 \ \frac{g\ H_2}{\cancel{g\ H_2O}} = 0.082 \ g\ H_2 \text{ in } 1.224 \text{ g of the compound.}$$

Then by difference:

$$1.224 \text{ g} - (0.490 \text{ g} + 0.082 \text{ g}) = 0.652 \text{ g } O_2 \text{ in } 1.224 \text{ g of the compound.}$$

To determine the empirical formula:

$$C = \frac{0.490}{1.224} \div 12.0 = 0.00333;$$

$$H = \frac{0.082}{1.224} \div 1.01 = 0.00666;$$

$$O = \frac{0.652}{1.224} \div 16.0 = 0.00333.$$

The simplest empirical formula for the compound is therefore:

$$C_{\frac{0.00333}{0.00333}} H_{\frac{0.00666}{0.00333}} O_{\frac{0.00333}{0.00333}} = CH_2O.$$

10.4. Determination of the Percentage Composition of a Compound from the Formula. The percentage composition of a compound for which the formula is not known must be determined experimentally by standard methods of chemical analysis. If the formula is known, then the percentage composition may be calculated.

Example 10.13. Calculate the percentage composition of water.

Solution.

$$\text{One mole of water} = (2 \times 1.008) + 15.999 = 18.015 \text{ g.}$$

That is, in 18.015 grams of water there are 2.016 grams of hydrogen and 15.999 grams of oxygen. Therefore:

$$\frac{2.016 \text{ g}}{18.015 \text{ g}} \times 100 = 11.19 \text{ per cent hydrogen}$$

and

$$\frac{15.999 \text{ g}}{18.015 \text{ g}} \times 100 = 88.81 \text{ per cent oxygen.}$$

Example 10.14. What is the per cent of water in washing soda, $Na_2CO_3 \cdot 10\ H_2O$?

Solution. First calculate the weight of one mole. Then:

$$\begin{array}{ccccccc} Na_2 & & C & & O_3 & & 10 \quad H_2O \\ (2 \times 22.990) & + & 12.011 & + & (3 \times 15.999) & + & (10 \times 18.015) = 286.14 \text{ g.} \end{array}$$

The weight of one mole is 286.14 grams, of which 180.15 grams is water. Therefore:

$$\frac{180.15 \text{ g}}{286.14 \text{ g}} \times 100 = 62.96 \text{ per cent water.}$$

Problems

Part I

10.37. A compound was found to contain 32.00 per cent carbon, 42.66 per cent oxygen, 18.67 per cent nitrogen, and 6.67 per cent hydrogen. Calculate the formula if the molecular weight is known to be about 75. *Ans.* $C_2O_2NH_5$.

10.38. A compound was found to contain 42.11 per cent carbon, 51.46 per cent oxygen, and 6.43 per cent hydrogen. The molecular weight was found to be approximately 340. What is the formula of the compound? *Ans.* $C_{12}H_{22}O_{11}$.

10.39. An oxide of iron contained 30.0 per cent oxygen. What was the formula of the oxide? *Ans.* Fe_2O_3.

10.40. Calculate the formula of a compound which contains 31.80 per cent potassium, 29.00 per cent chlorine, and 39.20 per cent oxygen. *Ans.* $KClO_3$.

10.41. A gaseous compound was found to contain 75 per cent carbon and 25 per cent hydrogen. It was found that 22.4 liters of the gas at standard conditions weighed 16 grams. What is the formula of the compound? *Ans.* CH_4.

10.42. A compound was found to have the formula CH_2O, as calculated without knowledge of the molecular weight. Later the molecular weight was found to be approximately 177. What was the formula of the compound? *Ans.* $C_6H_{12}O_6$.

10.43. A compound was found to contain 20.00 per cent hydrogen and 80.00 per cent carbon. It was found that 250 ml. of the gas weighed 0.256 g at 27° C and 640 mm of Hg. What was the formula of the compound? *Ans.* C_2H_6.

10.44. What is the per cent of aluminum in Al_2O_3? *Ans.* 52.9%.

10.45. Determine the percentage composition of formaldehyde, CH_2O, and glucose, $C_6H_{12}O_6$. *Ans.* C = 40%; H = 6.7%; O = 53.3%.

Part II

10.46. One liter of a gas was found to weigh 1.25 grams at standard conditions. Analysis showed it to contain 42.85 per cent carbon and 57.15 per cent oxygen. What is the formula of the compound? *Ans.* CO.

10.47. A sample of iron weighing 0.763 g was burned in oxygen. The product weighed 0.982 g. What was the formula of the oxide formed? *Ans.* FeO.

10.48. A compound A upon analysis showed the following composition: potassium 38.67 per cent; nitrogen, 13.85 per cent; and oxygen, 47.48 per cent. When heated, a compound B was formed having the composition: potassium 45.85 per cent; nitrogen 16.47 per cent; and oxygen 37.66 per cent. Write the equation for the reaction.

Ans. $2\ KNO_3 \rightarrow 2\ KNO_2 + O_2$.

10.49. Calculate the formulas of the following inorganic compounds:

(a) Na = 39.3%; Cl = 60.7%.

(b) Al = 15.8%; S = 28.1%; O_2 = 56.1%.

Ans. (a) NaCl; (b) $Al_2(SO_4)_3$.

10.50. A gas is known to be one of the five oxides of nitrogen. Upon analysis it was found to contain 36.8 per cent nitrogen. What is the formula for the oxide? *Ans.* N_2O_3.

10.51. A gas had the composition 71.72 per cent chlorine, 16.16 per cent oxygen, and 12.12 per cent carbon. Oxygen diffused through a small opening 1.76 times faster than the gas analyzed. What is the formula of the gas? *Ans.* $COCl_2$.

10.52. The molecular weight of chlorine, as determined experimentally, is 72. What is the formula for chlorine? *Ans.* Cl_2.

10.53. The molecular weight of argon has been found to be 40. What is the formula for argon? *Ans.* A.

10.54. What is the per cent of CaO in $CaCO_3$? *Ans.* 56%.

10.55. The bones of an adult person weigh about 24 pounds and they are 50 per cent calcium phosphate, $Ca_3(PO_4)_2$. How many pounds of phosphorus are there in the bones of the average adult?

Ans. 2.4 lb.

10.56. Washing soda is sold in two forms, as the anhydrous salt Na_2CO_3, and as the hydrated salt $Na_2CO_3 \cdot 10\ H_2O$. Considering that the active constituent in washing soda is Na_2CO_3, which would be the cheaper to the consumer, the anhydrous salt at ten cents per pound or the hydrated salt at five cents per pound? *Ans.* Anhydrous salt.

10.57. How much iron is there in one ton of iron ore containing 80 per cent hematite, Fe_2O_3? *Ans.* 1100 lb.

10.58. What is the per cent of copper in an ore containing 5.0 per cent of malachite, $CuCO_3 \cdot Cu(OH)_2$? *Ans.* 2.88%.

10.59. Calculate the per cent of available chlorine in bleaching powder, $CaOCl_2$, assuming that only one of the two chlorine atoms in the formula is liberated as free chlorine. *Ans.* 28%.

10.60. A compound was known to be either $CuCl_2$ or $CuBr_2$. A 5.00 g sample yielded 2.36 g of copper upon reduction. What was the compound? *Ans.* $CuCl_2$.

10.61. Lead dioxide, PbO_2, liberates one-half the oxygen atoms as free oxygen when heated. How much PbO_2 would be required to produce 10.0 g of oxygen? *Ans.* 150 g.

10.62. How much oxygen could be obtained from one pound of mercuric oxide, HgO, assuming all the oxygen to be liberated upon heating? *Ans.* 0.074 lb.

10.63. Calculate the formula of a compound, given that 55.85 g of iron combines with 32.06 g of sulfur. *Ans.* FeS.

10.64. A sample of copper weighing 3.18 g formed 3.98 g of oxide when made to react with oxygen. What is the formula of the oxide? *Ans.* CuO.

10.65. How many gram-atoms of oxygen are there in 500 g of Fe_2O_3? *Ans.* 9.40.

10.66. Blood hemoglobin contains 0.33 per cent iron. Assuming that there are two atoms of iron per molecule of hemoglobin, calculate the approximate molecular weight of hemoglobin. *Ans.* 34,000.

10.67. One liter of a gas weighs 1.34 g at S.C. The simplest formula for the substance is known to be CH_3. What is the correct molecular formula of the substance? *Ans.* C_2H_6.

10.68. Upon analysis an organic compound was found to consist of carbon and hydrogen. When 0.781 g of the compound was burned there was formed 2.64 g of CO_2 and 0.540 g of H_2O. The molecular weight was found to be approximately 78. Determine the formula of the substance. *Ans.* C_6H_6.

10.69. A compound upon analysis was found to contain potassium, chromium, and oxygen. There was present in the compound 26.57 per cent potassium and 35.36 per cent chromium. Determine the formula of the compound. *Ans.* $K_2Cr_2O_7$.

10.70. How much iron could be obtained from one ton of iron ore containing 45.0 per cent Fe_2O_3? *Ans.* 630 lb.

11

The Quantitative Significance of Chemical Equations

Chemical equations show how atoms are rearranged during the course of a chemical reaction. Since an equation consists of formulas, the quantitative interpretation of an equation is a summation of the interpretation of the formulas in the equation.

A chemical equation expresses a definite weight relationship among the reactants and products in a reaction. This statement is supported by three fundamental facts: (1) atoms are the smallest parts of an element that are involved in a chemical change, (2) excluding variations due to isotopes, the atoms of each of the elements have fixed weights, and (3) under similar conditions atoms will always combine in the same ratio to form a compound.

General Interpretation of Equations

11.1. Molecular Interpretation. In an equation the reactants and products contain the same number and kind of atoms. However, since an equation shows the rearrangement of atoms as the result of a chemical change, the reactants and products need not contain the same number of molecules. The preceding statements are summarized in the following equations.

	Zn	+	$2\,HCl$	→	$ZnCl_2$	+	H_2 (g)
Molecules	1		2		1		1
Atoms	1		4		3		2

or	Zn	+	$2\,H^+$	→	Zn^{+2}	+	H_2 (g)
Molecules	1		—		—		1
Atoms	1		2		1		2
Ions	—		2		1		—

Observe that, in the above reactions, there are the same number and kind of atoms on each side of the equations. By "kind of atoms" we mean atoms or ions that represent the same element, such as H_2 and 2 H^+.

11.2. Mole Interpretation. Since the symbol of an element represents a definite weight of that element, an equation can be interpreted in terms of weights of reactants and products. Reactants or products existing as gases may be interpreted in an equation in terms of volume relationships. It is usually more practical to express an equation in terms of weight and volume than in terms of atoms, ions, and molecules. The first equation given in Sec. 11.1 will now be interpreted in terms of mole quantities.

	Zn	+	2 HCl	→	$ZnCl_2$	+	H_2 (g)
Moles	1		2		1		1
Grams	65.37		72.922		136.28		2.016
Liters (S.C.)	—		—		—		22.4
Pounds	65.37		72.922		136.28		2.016

The molecular interpretation and the mole interpretation are interrelated in that one mole of a substance contains 6.023×10^{23} molecules. That is, one mole of zinc (65.37 g) and one mole of hydrogen (2.016 g) each contain 6.023×10^{23} molecules.

Observe that, in the above equation:

Weight of reactants = weight of products.

Since chemical equations indicate weight and volume relationships, three types of problems will be discussed:

1. Problems that involve weight only.
2. Problems that involve volume only.
3. Problems that involve weight and volume simultaneously.

Mole quantities, as expressed in equations, involve relatively small integer values (1, 2, 3, etc.). Therefore, calculations involving mole quantities require less arithmetic than do problems involving weight units such as grams and pounds. For this reason the general plan of solution for the problems that follow will be through the use of mole quantities.

Mass and Volume Relationships in Reactions

11.3. Problems Involving Weight Relationships among the Reactants and Products.

Example 11.1. How many grams of zinc sulfate, $ZnSO_4$, would be formed by the action of 4.31 grams of zinc on sulfuric acid?

Solution. Observe the following procedure in solving problems based upon chemical equations. (1) Write the equation involved. (2) Underscore the substances about which the problem is concerned. (3) Designate the quantities of these substances as expressed in the equation. That is:

$$\begin{array}{lcccccc} \underline{Zn} & + & H_2SO_4 & \rightarrow & \underline{ZnSO_4} & + & H_2. \\ 1.00 \text{ mole} & & & & 1.00 \text{ mole} & & \\ 65.37 \text{ g} & & & & 161.43 \text{ g} & & \end{array}$$

1. Change 4.31 g of Zn to moles of Zn.

Then:

$$\frac{4.31 \text{ g } \cancel{Zn}}{65.37 \dfrac{\cancel{\text{g } Zn}}{\text{mole Zn}}} = \frac{4.31}{65.37} \text{ mole Zn.}$$

2. Since 1.00 mole of Zn yields 1.00 mole of $ZnSO_4$, then:

$$\frac{4.31}{65.37} \cancel{\text{mole Zn}} \times 1.00 \frac{\text{mole } ZnSO_4}{\cancel{\text{mole Zn}}} = \frac{4.31}{65.37} \text{ mole } ZnSO_4.$$

3. Since there are 161.43 g of $ZnSO_4$ in 1.00 mole, then:

$$\frac{4.31}{65.37} \cancel{\text{mole } ZnSO_4} \times 161.43 \frac{\text{g } ZnSO_4}{\cancel{\text{mole } ZnSO_4}} = 10.6 \text{ g } ZnSO_4.$$

Combining the above steps gives:

$$\frac{4.31 \cancel{\text{ g Zn}}}{65.37 \dfrac{\cancel{\text{g Zn}}}{\cancel{\text{mole Zn}}}} \times 1.00 \frac{\cancel{\text{mole } ZnSO_4}}{\cancel{\text{mole Zn}}} \times 161.43 \frac{\text{g } ZnSO_4}{\cancel{\text{mole } ZnSO_4}} = 10.6 \text{ g } ZnSO_4.$$

That is, 4.31 g of Zn will yield 10.6 g of $ZnSO_4$.

Note: Having solved the above problem let us look back over the steps involved and the logic of the process. We are going to work with mole quantities. Therefore, we changed 4.31 g of Zn to moles of Zn (step 1). The equation tells us that 1.00 mole of Zn yields 1.00 mole of $ZnSO_4$. We now can calculate the yield in moles of $ZnSO_4$ (step 2). Since the answer must be in grams of $ZnSO_4$, we must change moles of $ZnSO_4$ to grams of $ZnSO_4$ (step 3).

The first step in working a problem is to outline the method of solution.

Example 11.2. How many moles of Fe_2O_3 would be formed by the action of oxygen on one kilogram of iron?

Solution. The solution to the problem is based upon the equation:

$$\begin{array}{lcccc} \underline{4\ Fe} & + & 3\ O_2 & \rightarrow & \underline{2\ Fe_2O_3.} \\ 4.00 \text{ moles} & & & & 2.00 \text{ moles} \\ 223.4 \text{ g} & & & & 319.4 \text{ g} \end{array}$$

1. Convert 1000 g of Fe into moles of Fe.

Then:
$$\frac{1000\ \cancel{\text{g Fe}}}{55.8\ \dfrac{\cancel{\text{g Fe}}}{\text{mole Fe}}} = \frac{1000}{55.8}\ \text{moles Fe.}$$

2. Since 4.00 moles of Fe yield 2.00 moles of Fe_2O_3, then:

$$\frac{1000}{55.8}\ \cancel{\text{mole Fe}} \times \frac{2.00\ \text{moles } Fe_2O_3}{4.00\ \cancel{\text{moles Fe}}} = \frac{2000}{223.2}\ \text{moles } Fe_2O_3$$
$$= 8.95\ \text{moles } Fe_2O_3.$$

Combining the above steps gives:

$$\frac{1000\ \cancel{\text{g Fe}}}{55.8\ \dfrac{\cancel{\text{g Fe}}}{\cancel{\text{mole Fe}}}} \times \frac{2.00\ \text{moles } Fe_2O_3}{4.00\ \cancel{\text{moles Fe}}} = 8.95\ \text{moles } Fe_2O_3.$$

Example 11.3. The decomposition by heat of 2000 pounds of limestone, $CaCO_3$, will produce how much quicklime, CaO?

Solution. The unit of weight involved is the pound. Therefore:

$CaCO_3$	→	CaO	+	CO_2
1.00 lb-mole		1.00 lb-mole		
100.1 lb		56.1 lb		

1. Change 2000 lb of $CaCO_3$ to pound-moles of $CaCO_3$.

Then:
$$\frac{2000\ \cancel{\text{lb } CaCO_3}}{100.1\ \dfrac{\cancel{\text{lb } CaCO_3}}{\text{lb-mole } CaCO_3}} = \frac{2000}{100.1}\ \text{lb-mole } CaCO_3.$$

2. Since 1.00 lb-mole of $CaCO_3$ yields 1.00 lb-mole of CaO, then:

$$\frac{2000}{100.1}\ \cancel{\text{lb-mole } CaCO_3} \times 1.00\ \frac{\text{lb-mole CaO}}{\cancel{\text{lb-mole } CaCO_3}} = \frac{2000}{100.1}\ \text{lb-mole CaO.}$$

3. Since there are 56.1 lb of CaO in 1.00 lb-mole:

$$\frac{2000}{100.1}\ \cancel{\text{lb-mole CaO}} \times 56.1\ \frac{\text{lb CaO}}{\cancel{\text{lb-mole CaO}}} = 1121\ \text{lb CaO.}$$

Combining the above steps gives:

$$\frac{2000\ \cancel{\text{lb } CaCO_3}}{100.1\ \dfrac{\cancel{\text{lb } CaCO_3}}{\cancel{\text{lb-mole } CaCO_3}}} \times 1.00\ \frac{\cancel{\text{lb-mole CaO}}}{\cancel{\text{lb-mole } CaCO_3}} \times 56.1\ \frac{\text{lb CaO}}{\cancel{\text{lb-mole CaO}}}$$
$$= 1121\ \text{lb CaO.}$$

It is not always necessary to write the complete equation to solve a problem. For example, when silver reacts with nitric acid, the silver is converted to silver nitrate, $AgNO_3$, with a number of other products, the reaction being somewhat complex. The products other than silver nitrate need not be indicated, provided they are not involved in the problem.

Example 11.4. Wire silver weighing 3.48 g was dissolved in nitric acid. What weight of silver nitrate was formed?

Solution. Since the silver was converted to silver nitrate, we are interested only in the fact that one gram-atom of silver, 107.87 g, will form one mole of $AgNO_3$, 169.88 g. That is:

Ag	→	$AgNO_3$.
1.00 mole		1.00 mole
107.87 g		169.88 g

Note that 1.00 gram-atom of Ag and 1.00 mole of Ag are the same quantity, 107.88 g.

1. Change 3.48 g of Ag to moles of Ag.

$$\frac{3.48\ \cancel{\text{g Ag}}}{107.88\ \dfrac{\cancel{\text{g Ag}}}{\text{mole Ag}}} = \frac{3.48}{107.88}\ \text{mole Ag.}$$

2. Since 1.00 mole Ag yields 1.00 mole $AgNO_3$, then:

$$\frac{3.48}{107.88}\ \cancel{\text{mole Ag}} \times 1.00\ \frac{\text{mole AgNO}_3}{\cancel{\text{mole Ag}}} = \frac{3.48}{107.88}\ \text{mole AgNO}_3.$$

3. Since there are 169.88 g $AgNO_3$ in 1.00 mole, then:

$$\frac{3.48}{107.88}\ \cancel{\text{mole AgNO}_3} \times 169.88\ \frac{\text{g AgNO}_3}{\cancel{\text{mole AgNO}_3}} = 5.48\ \text{g AgNO}_3.$$

Combining the above steps gives:

$$\frac{3.48\ \cancel{\text{g Ag}}}{107.88\ \dfrac{\cancel{\text{g Ag}}}{\cancel{\text{mole Ag}}}} \times 1.00\ \frac{\cancel{\text{mole AgNO}_3}}{\cancel{\text{mole Ag}}} \times 169.88\ \frac{\text{g AgNO}_3}{\cancel{\text{mole AgNO}_3}}$$
$$= 5.48\ \text{g AgNO}_3.$$

Example 11.5. How many pounds of water of hydration are there in 100 pounds of washing soda, $Na_2CO_3 \cdot 10\ H_2O$?

Solution. One pound-mole of washing soda contains 10 pound-moles of water. That is:

$$\begin{array}{lcl} \underline{Na_2CO_3 \cdot 10H_2O} & \rightarrow & \underline{10\ H_2O.} \\ 1.00 \text{ lb-mole} & & 10.00 \text{ lb-moles} \\ 286 \text{ lb} & & 180 \text{ lb} \end{array}$$

1. Change 100 lb of washing soda to pound-moles of washing soda.

$$\frac{100\ \cancel{\text{lb washing soda}}}{286\ \dfrac{\cancel{\text{lb washing soda}}}{\text{lb-mole washing soda}}} = \frac{100}{286} \text{ lb-mole washing soda.}$$

2. Since 1.00 lb-mole of washing soda yields 10.00 lb-moles of water, then:

$$\frac{100}{286}\ \cancel{\text{lb-mole washing soda}} \times 10.00\ \frac{\text{lb-mole } H_2O}{\cancel{\text{lb-mole washing soda}}} = \frac{1000}{286} \text{ lb-mole } H_2O.$$

3. Since there are 18.0 lb of H_2O in 1.00 lb-mole, then:

$$\frac{1000}{286}\ \cancel{\text{lb-mole } H_2O} \times 18.0\ \frac{\text{lb } H_2O}{\cancel{\text{lb-mole } H_2O}} = 63.0 \text{ lb } H_2O.$$

Combining the above steps gives:

$$\frac{100\ \cancel{\text{lb soda}}}{286\ \dfrac{\cancel{\text{lb soda}}}{\cancel{\text{lb-mole soda}}}} \times 10.00\ \frac{\cancel{\text{lb-mole } H_2O}}{\cancel{\text{lb-mole soda}}} \times 18.0\ \frac{\text{lb } H_2O}{\cancel{\text{lb-mole } H_2O}} = 63.0 \text{ lb } H_2O.$$

Example 11.6. Iodic acid, HIO_3, is converted to iodine pentoxide, I_2O_5, by heat. Assuming that I_2O_5 is the only product of the reaction containing iodine, how much I_2O_5 could be obtained from 250 g of HIO_3?

Solution. Only iodine need be balanced. Therefore:

$$\begin{array}{lcl} \underline{2\ HIO_3} & \rightarrow & \underline{I_2O_5.} \\ 2.00 \text{ moles} & & 1.00 \text{ mole} \\ 351.8 \text{ g} & & 333.8 \text{ g} \end{array}$$

1. Change 250 g of HIO_3 to moles of HIO_3.

$$\frac{250\ \cancel{\text{g } HIO_3}}{175.9\ \dfrac{\cancel{\text{g } HIO_3}}{\text{mole } HIO_3}} = \frac{250}{175.9} \text{ moles } HIO_3.$$

2. Since 2.00 moles of HIO_3 yield 1.00 mole of I_2O_5, then:

$$\frac{250}{175.9}\ \cancel{\text{moles } HIO_3} \times \frac{1.00 \text{ mole } I_2O_5}{2.00\ \cancel{\text{moles } HIO_3}} = \frac{250}{351.8} \text{ mole } I_2O_5.$$

3. Since there are 333.8 g of I_2O_5 in 1.00 mole, then:

$$\frac{250}{351.8}\ \cancel{\text{mole } I_2O_5} \times 333.8\ \frac{\text{g } I_2O_5}{\cancel{\text{mole } I_2O_5}} = 237 \text{ g } I_2O_5.$$

Combining the above steps gives:

$$\frac{250\ \cancel{\text{g HIO}_3}}{175.9\ \frac{\cancel{\text{g HIO}_3}}{\cancel{\text{mole HIO}_3}}} \times \frac{1.00\ \cancel{\text{mole I}_2\text{O}_5}}{2.00\ \cancel{\text{moles HIO}_3}} \times 333.8\ \frac{\text{g I}_2\text{O}_5}{\cancel{\text{mole I}_2\text{O}_5}} = 237\ \text{g I}_2\text{O}_5.$$

11.4. Problems Involving Volume Relationships among the Reactants and Products. One mole of a gaseous compound at standard conditions occupies a volume of 22.4 liters. Since an equation involves mole quantities of reactants and products, it follows that reactants or products existing in the gaseous state under the given conditions may be interpreted in terms of volumes. For example:

	N_2 +	$3\ H_2$ →	$2\ NH_3$.
Molecules	1	3	2
Moles	1	3	2
Liters (S.C.)	22.4	67.2	44.8

Observe that the ratio 22.4 *l.* N_2 : 67.2 *l.* H_2 : 44.8 *l.* NH_3 is the same as that of the number of molecules entering into the reaction, 1 N_2 : 3 H_2 : 2 NH_3.

Because of the small integer values relating volumes of gases in equations, it is often possible to solve this type of problem by inspection.

Example 11.7. How many liters of ammonia, NH_3, could be prepared from 750 liters of nitrogen, all gases being measured at standard conditions?

Solution. From the above equation we see that, at standard conditions, 1.00 mole of N_2 (22.4 *l*) will yield 2.00 moles of NH_3 (44.8 *l*). That is, the volume of NH_3 formed will be twice the volume of the N_2 used. Therefore:

$$2 \times 750\ l = 1500\ l \text{ of } NH_3 \text{ formed at S.C.}$$

Example 11.8. Iron pyrites, FeS_2, is burned in air to obtain SO_2 according to the equation:

$$4\ FeS_2 + \underset{11.00\text{ moles}}{\underline{11\ O_2}} \rightarrow 2\ Fe_2O_3 + \underset{8.00\text{ moles}}{\underline{8\ SO_2}}.$$

How many liters of SO_2, measured at 300° C and 740 mm of Hg, could be obtained from 100 liters of oxygen at S.C.?

Solution. (Observe step 3 very carefully.)

1. Change 100 *l* of O_2 to moles of O_2.

$$\frac{100\ \cancel{l\ O_2}}{22.4\ \frac{\cancel{l\ O_2}}{\text{mole } O_2}} = \frac{100}{22.4}\text{ mole } O_2.$$

2. Since 11.00 moles of O_2 yield 8.00 moles of SO_2, then:

$$\frac{100}{22.4}\cancel{\text{mole } O_2} \times \frac{8.00 \text{ mole } SO_2}{11.00 \cancel{\text{mole } O_2}} = \frac{800}{246.4} \text{ mole } SO_2.$$

3. We will now use the equation PV = nRT (Sec. 7.8).

$$V = \frac{nRT}{P} = \frac{\frac{800}{246.4}\cancel{\text{mole}} \times 0.0821 \frac{l\text{-}\cancel{\text{atm}}}{\cancel{\text{mole}}\text{-}\cancel{°K}} \times 573° \cancel{K}}{\frac{740}{760}\cancel{\text{atm}}} = 394\ l\ SO_2.$$

11.5. Problems Involving Both Weight and Volume Relationships among the Reactants and Products. This type of problem is a combination of the types discussed in Secs. 11.3 and 11.4.

Example 11.9. What volume of oxygen at standard conditions could be obtained by heating 8.66 grams of potassium chlorate, $KClO_3$?

Solution. The equation for the reaction would be:

$$\begin{array}{lcccl} \underline{2\ KClO_3} & \rightarrow & 2\ KCl & + & \underline{3\ O_2} \\ 2.00 \text{ moles} & & & & 3.00 \text{ moles} \\ 245.2 \text{ g} & & & & 96.0 \text{ g} \end{array}$$

1. Change 8.66 g of $KClO_3$ to moles of $KClO_3$.

$$\frac{8.66 \cancel{\text{g } KClO_3}}{122.6 \frac{\cancel{\text{g } KClO_3}}{\text{mole } KClO_3}} = \frac{8.66}{122.6} \text{ mole } KClO_3.$$

2. Since 2.00 moles of $KClO_3$ yield 3.00 moles of O_2, then:

$$\frac{8.66}{122.6}\cancel{\text{mole } KClO_3} \times \frac{3.00 \text{ mole } O_2}{2.00 \cancel{\text{mole } KClO_3}} = \frac{25.98}{245.2} \text{ mole } O_2.$$

3. Since, at S.C., there are 22.4 l of O_2 in 1.00 mole:

$$\frac{25.98}{245.2}\cancel{\text{mole } O_2} \times 22.4 \frac{l\ O_2}{\cancel{\text{mole } O_2}} = 2.37\ l\ O_2 \text{ at S.C.}$$

Combining the above steps gives:

$$\frac{8.66 \cancel{\text{g } KClO_3}}{122.6 \frac{\cancel{\text{g } KClO_3}}{\cancel{\text{mole } KClO_3}}} \times \frac{3.00 \cancel{\text{moles } O_2}}{2.00 \cancel{\text{moles } KClO_3}} \times 22.4 \frac{l\ O_2}{\cancel{\text{mole } O_2}} = 2.37\ l\ O_2 \text{ at S.C.}$$

Example 11.10. What volume of oxygen, collected at 27° C and 740 mm of Hg, could be obtained by heating 8.66 g of $KClO_3$?

Solution. Note that this problem differs from Example 11.9 in that the oxygen is collected at other than standard conditions. Therefore, the value 2.37 l, obtained in Example 11.9, must be changed to the volume it would occupy under the conditions given in the problem. Then:

$$V_2 = 2.37\ l \times \frac{300°\ \text{K}}{273°\ \text{K}} \times \frac{760\ \text{mm of Hg}}{740\ \text{mm of Hg}}$$

$$= 2.68\ l \text{ of } O_2 \text{ at } 27°\ \text{C and } 740 \text{ mm of Hg.}$$

A second method of solution would be to use the moles of O_2 obtained in step 2 of Example 11.9 and the general gas equation PV = nRT (Sec. 7.8). Then:

$$V = \frac{nRT}{P} = \frac{\frac{25.98}{245.2}\ \text{mole} \times 0.0821\ \frac{l\text{-atm}}{\text{mole-°K}} \times 300°\ \text{K}}{\frac{740}{760}\ \text{atm}} = 2.68\ l\ O_2.$$

Example 11.11. What volume of phosphine, PH_3, could be obtained at 43° C. and 725 mm. of Hg by the action of 5.28 g. of phosphorus on an excess of a strong solution of sodium hydroxide? Assume 100 per cent conversion of phosphorus to PH_3.

Solution. Only phosphorus need be balanced. Therefore:

$$\underset{1.00\ \text{mole}}{\underline{P}} \rightarrow \underset{1.00\ \text{mole}}{\underline{PH_3}}.$$

1. Change 5.28 g of P to moles of P.

$$\frac{5.28\ \text{g P}}{31.0\ \frac{\text{g P}}{\text{mole P}}} = \frac{5.28}{31.0}\ \text{mole P}$$

2. Since 1.00 mole of P yields 1.00 mole of PH_3, then:

$$\frac{5.28}{31.0}\ \text{mole P} \times 1.00\ \frac{\text{mole } PH_3}{\text{mole P}} = \frac{5.28}{31.0}\ \text{mole } PH_3.$$

3. Since PH_3 is a gas, we can use the equation PV = nRT (Sec. 7.8), and:

$$V = \frac{nRT}{P} = \frac{\frac{5.28}{31.0}\ \text{mole} \times 0.0821\ \frac{l\text{-atm}}{\text{mole°K}} \times 316°\ \text{K}}{\frac{725}{760}\ \text{atm}} = 4.64\ l\ PH_3.$$

Example 11.12. Chlorine is prepared by the reaction:

$$\underset{2.00\ \text{moles}}{2\ KMnO_4} + 16\ HCl \rightarrow 2\ KCl + 2\ MnCl_2 + 8\ H_2O + \underset{5.00\ \text{moles}}{5\ Cl_2}.$$

How many grams of $KMnO_4$ would be required to prepare the chlorine to fill a 1500 ml cylinder at 5.00 atm and 20° C?

Solution. The method of solution will be reversed from that of the previous problems.

1. We will use the equation PV = nRT to determine the number of moles of chlorine needed.

$$n = \frac{PV}{RT} = \frac{5.00 \cancel{\text{atm}} \times 1.50 \cancel{l}}{0.0821 \frac{\cancel{l}\text{-}\cancel{\text{atm}}}{\text{mole}°\cancel{K}} \times 293° \cancel{K}} = 0.313 \text{ mole } Cl_2.$$

2. Since 2.00 moles of $KMnO_4$ yield 5.00 moles of Cl_2, then:

$$0.313 \cancel{\text{mole } Cl_2} \times \frac{2.00 \text{ moles } KMnO_4}{5.00 \cancel{\text{moles } Cl_2}} = \frac{0.626}{5.00} \text{ mole } KMnO_4.$$

3. Since there are 158.0 g of $KMnO_4$ in 1.00 mole, then:

$$\frac{0.626}{5.00} \cancel{\text{mole } KMnO_4} \times 158.0 \frac{\text{g } KMnO_4}{\cancel{\text{mole } KMnO_4}} = 19.8 \text{ g } KMnO_4.$$

Combining steps 2 and 3 gives:

$$0.313 \cancel{\text{mole } Cl_2} \times \frac{2.00 \cancel{\text{moles } KMnO_4}}{5.00 \cancel{\text{moles } Cl_2}} \times 158.0 \frac{\text{g } KMnO_4}{\cancel{\text{mole } KMnO_4}} = 19.8 \text{ g } KMnO_4.$$

Problems

Part I

11.1. When limestone, $CaCO_3$, is heated to a sufficiently high temperature it decomposes into lime, CaO, and carbon dioxide. How much lime could be obtained from 150 g of limestone? *Ans.* 84.1 g.

11.2. How many moles of zinc would be required to prepare 12 moles of zinc chloride by the action of zinc on hydrochloric acid? *Ans.* 12 moles.

11.3. How many moles of ferrous sulfate would be formed by the action of 100 g of iron on sulfuric acid? *Ans.* 1.79 moles.

11.4. What weight of aluminum sulfate could be obtained by the action of 25.0 g of aluminum on an excess of sulfuric acid? *Ans.* 158 g.

11.5. Given the equation $2\,Fe + 3\,H_2O \rightarrow Fe_2O_3 + 3\,H_2 \uparrow$:

a. How many grams of water will react with one mole of iron? *Ans.* 27 g.

b. How many grams of Fe_2O_3 would be formed from one mole of water? *Ans.* 53.2 g.

11.6. Given the equation:

$$Cu + 2\,H_2SO_4 \rightarrow CuSO_4 + SO_2 + 2\,H_2O:$$

a. How many grams of copper sulfate would be formed for each mole of copper reacting? *Ans.* 159.6 g.

b. How many grams of copper would be required to produce 100 g of copper sulfate? *Ans.* 39.8 g.

c. How many moles of sulfur dioxide would be found for each mole of acid reacting? *Ans.* 0.50 mole.

d. How many grams of water would be formed for each mole of copper sulfate formed? *Ans.* 36 g.

11.7. How many grams of sulfur will combine with 3.5 g of copper to form CuS? *Ans.* 1.8 g.

11.8. How many grams of iron would be required to replace the copper in 3.96 g of copper sulfate, $CuSO_4$? *Ans.* 1.39 g.

11.9. How many grams of iron would be required to replace one-half gram-atom of copper from a solution of copper sulfate, $CuSO_4$? *Ans.* 27.9 g.

11.10. Calculate the volume of oxygen required to burn 60 liters of propane, C_3H_8, to CO_2 and H_2O, all gases being measured under standard conditions. *Ans.* 300 *l.*

11.11. Given the equation:

$$2\,Al + 3\,H_2SO_4 \rightarrow Al_2(SO_4)_3 + 3\,H_2 \uparrow, \text{ calculate:}$$

a. The number of grams of $Al_2(SO_4)_3$ formed for each gram-atom of aluminum reacting. *Ans.* 171.07 g.

b. The number of grams of $Al_2(SO_4)_3$ formed for each mole of hydrogen formed. *Ans.* 114.05 g.

c. The number of moles of H_2SO_4 required for each mole of $Al_2(SO_4)_3$ formed. *Ans.* 3 moles.

d. The number of grams of aluminum required for each liter of hydrogen formed at standard conditions. *Ans.* 0.803 g.

11.12. Given the equation:

$$6\,FeCl_2 + 14\,HCl + K_2Cr_2O_7 \rightarrow 6\,FeCl_3 + 2\,KCl + 2\,CrCl_3 + 7\,H_2O:$$

a. Indicate the weight relationships expressed in the products.
Ans. 973.3 g $FeCl_3$, 149.2 g KCl, 316.8 g $CrCl_3$, 126.1 g H_2O.

b. How many grams of $CrCl_3$ would have been formed from 0.78 g of $K_2Cr_2O_7$? *Ans.* 0.84 g.

c. How many grams of $FeCl_3$ would be formed from 5.00 g of $FeCl_2$? *Ans.* 6.40 g.

11.13. A portion of a silver coin weighing 2.44 g and containing 90 per cent silver was dissolved in nitric acid. What weight of silver nitrate was formed? *Ans.* 3.46 g.

11.14. How many moles of water will be liberated by the conversion of 500 g of $CuSO_4 \cdot 5\,H_2O$ to the anhydrous salt? *Ans.* 10.0 moles.

11.15. Calculate the gram-equivalent weight of magnesium, given that 1.10 g of magnesium will liberate one liter of hydrogen at standard conditions. *Ans.* 12.3 g.

11.16. Given that the gram-equivalent weight of calcium is 20.04 g :

a. With how many grams of bromine will 1.00 g of calcium combine?
Ans. 3.99 g.

b. How many liters of hydrogen, at 27° C and 725 mm of Hg, could be obtained by the action of 1.00 g of calcium on water?
Ans. 0.645 *l.*

11.17. A mass of 1.00 g of sodium was made to react with water. The hydrogen measured 663 ml collected over water at 27° C. and 640 mm of Hg. Calculate the gram-equivalent weight of sodium.
Ans. 23.0 g.

11.18. In an experiment it was found that 1.365 g of sulfur combined with 9.188 g of silver. The gram-equivalent weight of silver is 107.9 g. Determine the gram-equivalent weight of sulfur.
Ans. 16.03 g.

Part II

11.19. Given the equation $N_2 + 3\ H_2 \rightarrow 2\ NH_3$:

a. How many grams of ammonia would be prepared from 1000 g of nitrogen? *Ans.* 1214 g.

b. How many liters of hydrogen at standard conditions would react with 1000 g of nitrogen? *Ans.* 2400 *l.*

c. Under similar conditions, how many cubic feet of ammonia would be formed from 1000 ft^3 of nitrogen? *Ans.* 2000 $ft.^3$

d. Under similar conditions, how many cubic feet of hydrogen will combine with 1000 ft^3 of nitrogen? *Ans.* 3000 $ft.^3$

11.20. How many grams of $Ca(OH)_2$ would be formed by the reaction of 25.0 g of CaO with water? *Ans.* 33.1 g.

11.21. Given the equation for the burning of acetylene in air:

$$2\ C_2H_2 + 5\ O_2 \rightarrow 4\ CO_2 \uparrow + 2\ H_2O:$$

a. How many grams of water would be formed from 2.5 moles of C_2H_2?
Ans. 45 g.

b. At S.C., how many liters of carbon dioxide would be formed from 1.00 g of C_2H_2? *Ans.* 1.72 *l.*

11.22. Rust may be removed from linen by the action of dilute hydrochloric acid, $Fe_2O_3 + 6\ HCl \rightarrow 2\ FeCl_3 + 3\ H_2O$. The $FeCl_3$ is soluble in water. How many grams of rust could be removed by the action of 100 ml. of acid of density 1.018 g per ml and containing 4.00 per cent acid by weight? *Ans.* 2.97 g.

11.23. How many grams of copper would be precipitated by one gram of iron from a cupric salt in solution? *Ans.* 1.14 g.

11.24. Calculate the number of grams of NaCl required to react with 100 ml of 95 per cent H_2SO_4, density 1.84 g per ml., according to the equation:

$$2\ NaCl + H_2SO_4 \rightarrow 2\ HCl + Na_2SO_4.$$

Ans. 209 g.

11.25. How many pounds of Fe_2O_3 could be obtained from 800 pounds of $Fe(OH)_3$? *Ans.* 598 lb.

11.26. How many grams of $BaSO_4$ would be precipitated by the action of H_2SO_4 on 2.68 g of $BaCl_2$? *Ans.* 3.00 g.

11.27. How many pounds of water would be required to slake 50 pounds of CaO? *Ans.* 16 lb.

11.28. How many pounds of limestone, 80 per cent $CaCO_3$, would be required to make one ton of CaO? *Ans.* 4460 lb.

11.29. How many milliliters of nitric acid, density 1.21 g per ml, and containing 34.3 per cent acid by weight, would be required to dissolve 25.0 g of silver? Note: the products of the reaction are $AgNO_3$, H_2O, and NO_2. *Ans.* 70.4 ml.

11.30. Phosphorus is prepared according to the equation:

$$Ca_3(PO_4)_2 + 3\ SiO_2 + 5\ C \rightarrow 3\ CaSiO_3 + 5\ CO \uparrow + 2\ P.$$

a. How many pounds of phosphorus could be obtained from 2000 lb of phosphate rock containing 70.5 per cent $Ca_3(PO_4)_2$? *Ans.* 282 lb.

b. How many pounds of sand, SiO_2, would be required to react with 500 lb of $Ca_3(PO_4)_2$? *Ans.* 290 lb.

c. How many pounds of carbon would be required for each ton of $Ca_3(PO_4)_2$? *Ans.* 387 lb.

d. What volume of CO, reduced to standard conditions, would be formed for each 100 lb. of carbon used? *Ans.* 8.47×10^4 *l.*

11.31. How many pounds of lime, CaO, could be obtained from 500 lb of limestone containing 95 per cent $CaCO_3$? *Ans.* 266 lb.

11.32. A sample of zinc weighing 10.93 g reacted with 22.5 ml of hydrochloric acid of density 1.18 g per ml and containing 35 per cent HCl by weight. Assuming the impurities in the zinc to be nonreactive toward HCl, what is the per cent of zinc in the sample? *Ans.* 76.2%.

11.33. How much sodium hydroxide could be prepared from 2500 g of calcium hydroxide on the basis of the reaction:

$$Na_2CO_3 + Ca(OH)_2 \rightarrow CaCO_3 + 2\ NaOH?$$ *Ans.* 2700 g.

11.34. A tank holding 25.0 *l* of butane, C_4H_{10}, at 27° C and 5 atmospheres pressure, would require what volume of oxygen, measured at 27° C and 770 mm of Hg, for complete combustion? *Ans.* 802 *l.*

11.35. What weight of oxygen would be required to roast one ton of pyrite, FeS_2, on the basis of the reaction:

$$4\ FeS_2 + 11\ O_2 \rightarrow 2\ Fe_2O_3 + 8\ SO_2?$$ *Ans.* 1467 lb.

11.36. Commercial HCl is prepared according to the equation:

$$NaCl + H_2SO_4 \rightarrow NaHSO_4 + HCl.$$

How many liters of H_2SO_4, density 1.84 g per ml and containing

95 per cent acid by weight, would be required to prepare 25 *l* of hydrochloric acid of density 1.18 g per ml and containing 36 per cent HCl by weight? *Ans*. 16.3 *l*.

11.37. How many moles of NH_4Cl would be required to prepare 0.10 mole of nitrogen, given the reaction:

$$NH_4Cl + NaNO_2 \rightarrow NaCl + 2\ H_2O + N_2?$$

Ans. 0.10 mole.

11.38. Chlorine is to be prepared according to the reaction:

$$MnO_2 + 4\ HCl \rightarrow MnCl_2 + 2\ H_2O + Cl_2.$$

What volume of concentrated hydrochloric acid, density 1.18 g per ml. and containing 36 per cent HCl by weight, and what weight of pyrolusite ore containing 75 per cent by weight MnO_2, would be required to prepare 100 g of chlorine? *Ans*. 484 ml , 164 g.

11.39. How many cubic feet of chlorine will combine with 25.0 ft^3 of hydrogen, all gases measured at S.C.? *Ans*. 25.0 $ft.^3$

11.40. How many tons of 95 per cent H_2SO_4 could be prepared from one ton of pyrite containing 90 per cent FeS_2, assuming all the sulfur to be converted to acid? *Ans*. 1.5 tons.

11.41. How many grams of 68 per cent HNO_3 could be prepared from 25 grams of $NaNO_3$ by the action of H_2SO_4? *Ans*. 27.2 g.

11.42. The CO_2 resulting from the burning of 0.325 g of a compound in oxygen was passed through lime water, resulting in the formation of 1.08 g of $CaCO_3$ as a precipitate. What was the per cent of carbon in the compound? *Ans*. 39.9%.

11.43. What volume of CO_2 would be formed at standard conditions when one cubic foot of CH_4 burns to CO_2 and water? *Ans*. 1 $ft.^3$

11.44. What volume of ozone would be formed from 960 ml of oxygen, temperature and pressure being constant? *Ans*. 640 ml.

11.45. Twelve liters of oxygen were partially converted to ozone. The resultant volume of the oxygen-ozone mixture was 11.8 liters. What volume of ozone was formed? *Ans*. 0.4 *l*.

11.46. What volume of oxygen, measured at 15° C. and 720 mm of Hg, could be obtained from 18 g of HgO? *Ans*. 1.04 *l*.

11.47. How many grams of $KClO_3$ would be required to produce 480 ml of oxygen at 21° C and 640 mm of Hg? *Ans*. 1.37 g.

11.48. How many liters of hydrogen could be obtained by the action of steam on 100 g of iron at 200° C and 760 mm of Hg? *Ans*. 92.7 *l*.

11.49. How many liters of hydrogen at S.C. would be required to reduce 80 g of CuO to free Cu? *Ans*. 22.5 *l*.

11.50. How much zinc would be required to prepare 500 ml of hydrogen at 20° C and 720 mm of Hg? *Ans*. 1.29 g.

11.51. Calculate the weight of 39 per cent HCl required to prepare sufficient chlorine by action with MnO_2 to fill a cylinder of 3500 ml capacity under a pressure of 20 atmospheres at 0° C. *Ans*. 1170 g.

11.52. How many grams of $(NH_4)_2SO_4$ reacting with $Ca(OH)_2$ would be required to prepare sufficient ammonia to react with 10 liters of HCl at S.C.? *Ans.* 29.4 g.

11.53. How many moles of CO_2 would be formed by burning 500 pounds of coal containing 96 per cent carbon? *Ans.* 1.8×10^4 moles.

11.54. What volume of hydrogen at 27° C and 760 mm of Hg will contain the same number of molecules as one liter of oxygen at standard conditions? *Ans.* 1.1 *l*.

11.55. How many moles of hydrogen could be obtained by the action of steam on 1.00 kg of iron? *Ans.* 23.9 moles.

11.56. What weight of nitrogen will contain the same number of molecules as 2500 ml of hydrogen at S.C.? *Ans.* 3.13 g.

11.57. What weight of lead will contain the same number of atoms as 1.00 g of sulfur? *Ans.* 6.46 g.

11.58. How many grams of zinc reacting with H_2SO_4 will liberate the same amount of hydrogen as 2.30 g of sodium reacting with water? *Ans.* 3.27 g.

11.59. At standard conditions, how many cubic feet of hydrogen will unite with 6.0 $ft.^3$ of nitrogen to form ammonia? *Ans.* 18 $ft.^3$

11.60. How many cubic feet of ammonia will be formed in problem 11.59? *Ans.* 12 $ft.^3$

11.61. The following equation represents a commercial method for the preparation of nitric oxide, NO:

$$4\ NH_3 + 5\ O_2 \rightarrow 6\ H_2O + 4\ NO.$$

How many liters each of ammonia and oxygen would be required to produce 80.0 *l* of NO at S.C.? *Ans.* 80 *l*, 100 *l*.

11.62. Sodium chloride weighing 1.225 g was dissolved in 500 ml of water. To this solution was added 1.700 g of $AgNO_3$ in solution. How much AgCl precipitated? *Ans.* 1.43 g.

11.63. How many grams of iron can be oxidized to Fe_2O_3 by one mole of oxygen? *Ans.* 74.5 g.

11.64. What volume of hydrogen at standard conditions would be liberated by the action of 3.22 g of zinc on 50 ml of 40 per cent H_2SO_4, specific gravity 1.30? *Ans.* 1100 ml.

11.65. How many moles of CO_2, measured at 21° C and 740 mm of Hg, would be liberated by the action of 25.0 ml of 20 per cent HCl, density 1.10 g per ml, on 30.0 g of marble containing 90 per cent $CaCO_3$? *Ans.* 0.075 mole.

11.66. How much SO_2 could be obtained from 16.033 g of sulfur? *Ans.* 32.033 g.

11.67. How many moles of SO_2 could be obtained from 25.0 g of sulfur? *Ans.* 0.78 mole.

11.68. How much zinc must react with sulfuric acid in order to obtain 750 ml of hydrogen collected at 33° C and 680 mm of Hg? *Ans.* 1.75 g.

11.69. How many pounds of zinc oxide, ZnO, could be obtained by roasting 100 pounds of zinc blende, ZnS, given the equation:

$$2\ ZnS + 3\ O_2 \rightarrow 2\ ZnO + 2\ SO_2?$$ *Ans.* 83.5 lb.

11.70. How many moles of CO_2 would be formed by burning 120.1 g of carbon? *Ans.* 10 moles.

11.71. How many grams of oxygen must be contained in a flash bulb to oxidize 0.25 g of aluminum? *Ans.* 0.22 g.

11.72. How many gram-atoms of iron would be required to prepare five moles of Fe_2O_3? *Ans.* 10 gram-atoms.

11.73. Given the equation, $2\ Fe + 3\ H_2O \rightarrow Fe_2O_3 + 3\ H_2$, calculate the following:

a. 1.00 g of Fe combines with ____ mole of H_2O. *Ans.* 0.0268 mole.

b. 1.00 mole of Fe combines with ____ g of H_2O. *Ans.* 27.0 g.

c. 1.00 mole of Fe forms ____ g of Fe_2O_3. *Ans.* 79.85 g.

11.74. How many grams of oxygen are required to prepare 100 g of P_2O_5? *Ans.* 56.3 g.

11.75. A sulfuric acid plant uses 2500 tons of SO_2 daily. How many tons of sulfur must be burned to produce this amount of SO_2 gas? *Ans.* 1250 tons.

11.76. How many grams of carbon could be oxidized to CO_2 by the oxygen liberated from 231.76 g of silver oxide, Ag_2O? *Ans.* 6.0 g.

11.77. When 25.0 g of Al was dissolved in HCl, the hydrogen was collected over water at 0° C and 765 mm of Hg. What was the volume of gas collected? *Ans.* 31.1 *l.*

11.78. How many moles of aluminum oxide, Al_2O_3, could be obtained from 100 g of aluminum? *Ans.* 1.85 moles.

11.79. Two moles of hydrochloric acid and 100 g of sodium hydroxide were placed in a beaker containing water. Which reactant was in excess and by how much? *Ans.* NaOH, 0.5 mole.

11.80. Ferric oxide may be reduced with carbon according to the equation:

$$2\ Fe_2O_3 + 6\ C \rightarrow 6\ CO + 4\ Fe.$$

a. How many liters of carbon monoxide will be produced at S.C. for each mole of ferric oxide reduced? *Ans.* 67.2 *l.*

b. How many grams of carbon will be required for each mole of ferric oxide? *Ans.* 36 g.

c. How many pounds of ferric oxide would be required for each 100 lb of iron produced? *Ans.* 143 lb.

11.81. When 0.850 g of calcium reacted with hydrochloric acid, there was liberated 476 ml of hydrogen measured at S.C. From these data calculate the gram-equivalent weight of calcium. *Ans.* 20.0 g.

11.82. When 1.215 g of cadmium reacted with hydrochloric acid, 283 ml of gas was collected over water at 27° C and 740 mm of Hg. Calculate the gram-equivalent weight of cadmium. *Ans.* 56.2 g.

11.83. Water gas is produced from carbon and water according to the equation, $C + H_2O \rightarrow CO + H_2$. A furnace burns 100 ft ^{3} of water gas per hour. What volume of carbon dioxide, at S.C., would be formed each 24 hr period the furnace operates? *Ans.* 1200 ft.^{3}

11.84. How many pounds of water would be produced by the furnace in problem 11.83 for each 24 hr period of operation? *Ans.* 60 lb.

11.85. How many kilograms each of hydrogen and oxygen could be obtained by the electrolysis of one ton of water? *Ans.* 101 kg , 807 kg.

11.86. A grocer bought 100 lb of washing soda, $Na_2CO_3 \cdot 10\ H_2O$, at five cents per pound. During the period of storage the water of hydration was lost. The grocer then sold the washing soda at eight cents per pound. Did the grocer gain or lose in the transaction and how much? *Ans.* Lost $2.04.

11.87. How much weight would be lost by 2.614 g of $BaCl_2 \cdot 2\ H_2O$ if heated until the water of hydration had been liberated? *Ans.* 0.385 g.

11.88. A hydrate of strontium chloride contains 40.54 per cent water. What is the formula for the hydrate? *Ans.* $SrCl_2 \cdot 6\ H_2O$.

11.89. A mass of hydrated sodium phosphate, $Na_3PO_4 \cdot x\ H_2O$, weighing 3.615 g lost 2.055 g when heated to form the anhydrous salt. What is the value of x in the above formula? *Ans.* 12.

11.90. What would be the weight of the residue if 8.375 g of $U(SO_4)_2 \cdot 9\ H_2O$ were heated until the water of hydration had been lost? *Ans.* 6.074 g.

12

Thermochemistry

Every chemical change and every physical change involves an energy change. These energy changes that accompany the transformation of matter help us understand better the nature of chemical and physical changes. Energy changes may be in the form of *heat*, as in burning coal; *electrical energy*, as in a storage battery; *radiant energy*, such as light and X-rays; or *mechanical energy*, as in the gasoline engine.

Energy due to the motion of an object is called *kinetic energy*, and is expressed quantitatively by the formula:

$$\text{Kinetic energy} = \frac{\text{mass} \times (\text{velocity})^2}{2}.$$

When the mass (m) is given in grams and the velocity (v) in centimeters per second, then the unit of kinetic energy is the *erg*. That is, dimensionally, ergs have the value of $\text{g cm}^2 \text{ sec}^{-2}$.

The end product of most forms of energy, when used to do useful work, is heat. For this reason energy frequently is expressed in terms of its equivalence in heat units, such as the *calorie* and the *British thermal unit*. One calorie is equal to 4.185×10^7 ergs.

The equivalence between matter and energy was first postulated by Einstein in 1905. The interconversion of matter and energy, as suggested by Einstein, is expressed in the following mathematical equation:

$$E = (2.2 \times 10^{13})\ g.$$

- E: energy in calories
- 2.2×10^{13}: constant relating matter and energy
- g: grams of matter converted to energy

That is, the conversion of one gram of any form of matter to energy would result in the release of 2.2×10^{13} calories. This quantity of heat would raise the temperature of 250,000 tons of water from 0° C. to 100° C. It is approximately the amount of heat obtained by burning 300 tons of coal. Modern developments in atomic energy support the early postulates of Einstein.

The Energy Involved in Physical Changes

12.1. The Calorie and the British Thermal Unit. Heat is a form of energy which may be measured quantitatively. The unit of heat energy used in the metric system is the *calorie* (cal). For all practical purposes the calorie may be defined as the quantity of heat necessary to raise the temperature of one gram of water one degree centigrade. A larger unit, the *kilogram calorie* (kcal), is equal to 1000 cal. The unit of heat energy used in the English system is the *British thermal unit* (BTU or Btu), which is defined as the amount of heat necessary to raise the temperature of one pound of water one degree Fahrenheit.

Example 12.1. How many calories would be required to heat 125 g of water from 23° C to 100° C ?

Solution. By definition:

$$\text{Calories} = (\text{grams of water})(°\text{C temperature change}).$$

Therefore:

$$\text{Calories} = (125)(100 - 23) = 9625.$$

Example 12.2. How many BTU's would be required to heat one gallon of water from 70° F to 212° F ?

Solution. By definition:

$$\text{BTU} = (\text{lb of water})(°\text{F temperature change}).$$

Since one gallon of water = 8.3 lb, then:

$$\text{BTU} = (8.3)(212 - 70) = 1179.$$

12.2. Thermodynamic Quantities. Chemical thermodynamics is the study of the energy effects accompanying chemical and physical changes. Four thermodynamic quantities will be discussed; internal energy, E, and enthalpy, H, entropy, S, and Gibbs free energy, G.

Internal energy. When energy such as heat or electricity is put into a system such as a beaker of water or a storage battery, part of the energy is used to do work and part is stored within the system.

This stored energy within the system adds to the internal energy of the system. The energy that does work is lost to the system. For example, suppose we vaporize one mole of water (18.01 g) at 100° C and 1.00 atm. As a result of vaporization, the water would now occupy a volume of about 22.4 liters. The energy supplied to the mole of water consists of:

Work done against the atmosphere = 740 cal
Heat retained by the water as internal energy = 8973 cal

When the steam condenses to water there would be a loss of 8973 calories to the internal energy of the water. The above represents a change in internal energy, ΔE. No method has yet been found for determining the absolute internal energy of a system.

Enthalpy. The heat absorbed or released by a system at constant pressure is equal to the change in enthalpy, ΔH. When one mole of water is vaporized, $\Delta H = (8973 + 740)$ cal. That is, enthalpy includes the work done by the system. Most chemical reactions in the laboratory are carried out in an open vessel at constant pressure. Therefore, enthalpy is a more practical quantity to use in the laboratory than is internal energy. When one mole of ice melts at 0° C, 1440 cal of heat are absorbed. When one mole of water freezes at 0° C, 1440 cal of heat are released. These changes can be represented as:

$$H_2O\ (s) \rightarrow H_2O\ (l) \qquad \Delta H = +1440 \text{ cal.}$$
$$H_2O\ (l) \rightarrow H_2O\ (s) \qquad \Delta H = -1440 \text{ cal.}$$

An *endothermic* process is one for which ΔH is positive (heat is absorbed), while an *exothermic* process is one for which ΔH is negative (heat is released).

If the change in enthalpy is included as part of the equation we would write:

$$H_2O\ (s) \rightarrow H_2O\ (l)\ -1440 \text{ cal.}$$
$$H_2O\ (l) \rightarrow H_2O\ (s)\ +1440 \text{ cal.}$$

Entropy. The entropy, S, of a system is a measure of its randomness or, in less technical terms, its disorder. Molecules of H_2O have a more orderly arrangement in the form of ice than in the form of water. That is, a given amount of water will have a greater entropy than will the same amount of ice, both being at the same temperature. The concept of entropy applies only to reversible systems such as the interconversion of ice and water by the gain or loss of heat. Such interconversions involve a change in entropy, designated as ΔS. The change in entropy of a reversible system at a constant temperature is defined as the heat absorbed or released divided by the absolute temperature at which the heat exchange occurs. That is:

$$\Delta S = \frac{\text{cal}}{T} = \text{entropy units (e.u.)}$$

or:

$$\text{cal} = T\Delta S.$$

The heat exchange of a compound is usually expressed as cal/mole. Then:

$$\Delta S = \frac{\text{cal/mole}}{T} = \frac{\text{e.u.}}{\text{mole}}.$$

The absorption of 80 cal by 1.00 g of ice at 0° C and 1.00 atm will yield 1.00 g of water at the same conditions. Therefore, the gain in entropy, ΔS, of 1.00 mole of water (18 g) in changing from ice at 0° C and 1.00 atm to water at 0° C and 1.00 atm would be:

$$\Delta S = \frac{(18 \times 80)\ \text{cal/mole}}{373°\ \text{K}} = 3.9\ \frac{\text{cal}}{\text{mole °K}} = 3.9\ \frac{\text{e.u.}}{\text{mole}}.$$

It is possible to determine experimentally the total entropy, S, of a system. For this purpose it is assumed that a perfect crystal at 0° K has an entropy of zero. The following are the entropies of several substances at 25° C.

Substance	O_2	H_2O (g)	H_2O (l)	CO_2
e.u. at 25° C	49.0	45.1	16.7	51.1

Gibbs Free Energy. When a reversible chemical change occurs, the difference between ΔH and $T\Delta S$ represents the amount of available energy released or absorbed as a result of the chemical change. This energy represents the driving force of the reaction, and is called Gibbs free energy, G. That is:

$$\Delta G = \Delta H - T\Delta S.$$

When ΔG is negative the reaction will be a spontaneous one; when ΔG is positive the reverse of the reaction as written will be the spontaneous one; when ΔG is zero equilibrium exists. The magnitude of ΔG is a measure of the extent to which the reaction will go to completion. Thus, a knowledge of ΔG values enables us to predict the course of a reaction. For example, we can predict that the reaction:

$$Cl_2(1\ \text{atm}) + 2I^-(m) \rightarrow 2Cl^-(m) + I_2(s) \qquad \Delta G = -38\ \text{kcal}$$

will be a spontaneous one since ΔG is negative. Had ΔG been positive, the spontaneous reaction would have been the reverse of the reaction as written.

TABLE 12.1

THE HEAT CAPACITIES OF SOME COMMON SUBSTANCES IN CALORIES PER GRAM

Alcohol	0.581	Iron	0.107
Aluminum	0.214	Lead	0.031
Copper	0.092	Sulfur	0.137
Diamond	0.120	Water	1.000

12.3. The Heat Capacity of Substances. Objects have the ability to change temperature, and thus possess varying amounts of heat energy. At the absolute zero of temperature an object would have no heat energy. The *heat capacity*, sometimes called *specific heat*, of a substance is defined as the number of calories required to raise the temperature of one gram of the substance one degree centigrade.

Example 12.3. How many calories would be required to change the temperature of 125 g of iron from 23° C to 100° C ?

Solution. Note that the mass and temperature changes are the same as for the water in Example 12.1. However, the heat capacity of iron is only 0.107 as compared to 1.000 for water. Taking heat capacity into consideration, we have:

$$\text{Calories} = (\text{grams})(^\circ\text{C temperature change})(\text{heat capacity}).$$

Therefore:

$$\text{Calories} = (125)(100 - 23)(0.107) = 1030.$$

Example 12.4. Calculate the heat capacity of silver, given that 280 g of the metal absorbed 136 cal when changed from 22.13° C to 30.79° C.

Solution. Solving the equation given in Example 12.3 for heat capacity, we have:

$$\begin{aligned}\text{Heat capacity} &= \frac{\text{calories}}{(\text{grams})(^\circ\text{C temperature change})} \\ &= \frac{136}{(280)(30.79 - 22.13)} \\ &= 0.056 \text{ cal per g} = \text{heat capacity of silver.}\end{aligned}$$

Example 12.5. A mass of 350 g of copper pellets at 100.0° C was mixed with 200 g of water at 22.4° C. The resultant temperature of the mixture was 33.2° C. Calculate the heat capacity of copper.

Solution. When mixing the above:

$$\text{Heat lost by copper} = \text{heat gained by water.}$$

Let x = heat capacity of copper. Then:

$$(350)(100.0 - 33.2)(x) = (200)(33.2 - 22.4).$$
$$23{,}380x = 2160.$$
$$x = 0.092 \text{ cal per g} = \text{heat capacity of copper.}$$

A quantity called *molar heat capacity* is defined as the number of calories required to raise the temperature of one mole of a substance one degree centigrade. That is:

$$\text{Molar heat capacity} = \text{heat capacity} \times \text{formula weight.}$$

Example 12.6. The heat capacity of sugar, $C_{12}H_{22}O_{11}$, is 0.299 $\frac{\text{cal}}{\text{g °C}}$. What is the molar heat capacity of sugar?

Solution. Since the formula weight of sugar is 342.3, then:

$$\text{Molar heat capacity of sugar} = 0.299 \frac{\text{cal}}{\cancel{\text{g}}\text{°C}} \times 342.3 \frac{\cancel{\text{g}}}{\text{mole}}$$
$$= 102.3 \frac{\text{cal}}{\text{mole °C}}.$$

12.4. The Heats of Fusion and Vaporization of Substances. The temperature of a crystalline solid remains constant at the melting point until the solid is all melted. This shows that heat is required to melt a crystalline solid without changing its temperature. The *heat of fusion* is defined as the number of calories required to change one gram of crystalline solid to the liquid state without a change in temperature. The heat of fusion of ice is 79.7 calories per gram.

Heat must be applied to a liquid at its boiling point in order to convert it to a vapor without a change in temperature. The *heat of vaporization* is defined as the number of calories required to change one gram of a liquid to vapor without a change in temperature. The heat of vaporization of water is 539.6 calories per gram.

TABLE 12.2

HEATS OF FUSION AND VAPORIZATION IN CALORIES PER GRAM

Substance	*Formula*	*Heat of Fusion*	*Heat of Vaporization*
Ammonia	NH_3	108.1	327.1
Benzene	C_6H_6	30.3	94.3
Carbon dioxide	CO_2	45.3	71.4
Carbon tetrachloride	CCl_4	4.2	46.4
Ethyl alcohol	C_2H_5OH	24.9	204.0
Water	H_2O	79.7	539.6

Example 12.7. How many calories would be required to change 10.0 g of ice at 0° C to steam at 100° C?

Solution. Three steps are involved in the solution of the problem:

(1) The heat required to melt the ice.

$$10.0\,\cancel{g} \times 80\,\frac{\text{cal}}{\cancel{g}} = 800 \text{ calories.}$$

(2) The heat required to raise the temperature to 100° C.

$$10.0\text{ g} \times 100^\circ\text{ C temperature rise} = 1000 \text{ calories.}$$

(3) The heat required to vaporize the water.

$$10.0\,\cancel{g} \times 540\,\frac{\text{cal}}{\cancel{g}} = 5400 \text{ calories.}$$

The total heat required would, therefore, be:

$$800 \text{ cal} + 1000 \text{ cal} + 5400 \text{ cal} = 7200 \text{ cal.}$$

Heats of fusion and vaporization may be explained on the basis that force is required to separate atoms and molecules in the solid or liquid states. That is, heat of fusion represents the energy necessary to overcome the attractive forces holding atoms or molecules in definite positions within a crystalline structure. Heat of vaporization represents the energy required to overcome the attraction between atoms or molecules in the liquid state.

Problems

Part I

12.1. How many calories are there in one BTU? *Ans.* 252 cal.

12.2. How many calories would be required to change the temperature of 750 g of water from 15.0° C to 90.0° C? *Ans.* 5.63×10^4 cal.

12.3. How many calories would be required to change the temperature of 500 g of water from 50° F to 50° C? *Ans.* 20,000 cal.

12.4. How many calories would be required to change the temperature of 250 g of aluminum from 15° C to 75° C? *Ans.* 3210 cal.

12.5. Given 800 g of water at 22° C, calculate the resultant temperature of the water following absorption of 3600 calories. *Ans.* 26.5° C.

12.6. How many calories of heat would be liberated if the temperature of 300 g of iron were changed from 75° C to 17° C? *Ans.* 1.86×10^3 cal.

12.7. The resultant temperature obtained when 150 g of water at 28° C was mixed with 350 g of copper at 100° C was 41° C. Calculate the heat capacity of copper. *Ans.* $0.095 \frac{\text{cal}}{\text{g}}$.

12.8. How many calories would be required to change 10.0 g of ice at 0° C to water at 15° C? *Ans.* 950 cal.

12.9. When one gram-atom of helium is formed from hydrogen in the sun, there is a loss of mass of 0.03 g. What is the energy in calories given off by the sun as a result of this loss of mass?
Ans. 6.6×10^{11} cal.

12.10. How many calories are liberated when one mole of steam at 100° C condenses to water at 100° C? *Ans.* 9710 cal.

12.11. How many calories would be required to vaporize one mole of freon, CCl_2F_2, given that the heat of vaporization is 35.0 cal per gram?
Ans. 4230 cal.

12.12. How many calories of heat would be required to vaporize one mole of ethyl alcohol without a change in temperature? *Ans.* 9384 cal.

12.13. Show that molar heat capacity has the dimensions of calories per mole per degree centigrade.

12.14. What is the molar heat capacity of (a) alcohol, C_2H_5OH, (b) copper, and (c) water? *Ans.* (a) 26.7 cal, (b) 5.85 cal., (c) 18.02 cal.

12.15. What is the kinetic energy of a bullet weighing 5.65 g and traveling at a velocity of 200 m per second? *Ans.* 1.13×10^9 ergs.

Part II

12.16. How many calories would be required to change 25 g of ice at −10° C to steam at 110° C, given that the heat capacity of ice is 0.51 cal per g, and that of steam is 0.48 cal per g?
Ans. 1.82×10^4 cal.

12.17. What temperature would result from mixing 50 g of water at 20° C and 250 g of water at 40° C? *Ans.* 36.7° C.

12.18. Ten grams of ice at 0° C was placed into 100 g of water at 50° C. What was the temperature after the ice had melted? *Ans.* 38° C.

12.19. It is estimated that the sun loses 8000 tons of mass per second due to conversion of matter to energy. What is the caloric output of the sun per second? *Ans.* 1.6×10^{23} cal.

12.20. How many calories would be required to change the temperature of one gallon of water one degree centigrade? *Ans.* 3784 cal.

12.21. How many grams of freon would have to vaporize in a mechanical refrigerator to absorb sufficient heat to freeze 1000 grams of water in the freezing compartment? Note: see problem 12.11.
Ans. 2290 g.

12.22. How many tons of ice at 0° C could be converted to steam at 100° C by the heat liberated when one milligram of matter is converted to energy? *Ans.* 33.6 tons.

12.23. How many BTU units would be required to raise the temperature of one cubic foot of water from a room temperature of 70° F to a boiling point of 212° F? *Ans.* 8.86×10^3 BTU.

12.24. How many BTU units would be required to vaporize one pound of water without a change in temperature? *Ans.* 973 BTU.

12.25. How many BTU units would be required to change 25.0 lb of ice at 32° F to steam at 212° F? *Ans.* 3.24×10^4 BTU.

12.26. What would be the resulting temperature in degrees Fahrenheit if 500 g of aluminum at 25.0° C were dropped into 500 ml of ethyl alcohol, density 0.79 g per ml, at 70° C? *Ans.* 132° F.

12.27 How many grams of water at 100° C could be changed to steam at 100° C by the amount of heat required to vaporize one kilogram of alcohol with no change in temperature? *Ans.* 378 g.

12.28. One kilogram each of copper and aluminum at 100° C is placed in an insulated vessel containing one kilogram of water at 40.0° C. What is the resultant temperature of the mixture? *Ans.* 54.0° C.

12.29. Steam at 100° C was passed through 2000 g of water at 20.0° C in an insulated container. After passing the steam through the water it was found that 10.0 g of steam had condensed, the resulting temperature of the water being 22.7° C. From these data calculate the heat of vaporization of water. *Ans.* 540 cal per g.

12.30. Seventy-five grams of ice at 0° C was placed in 250 g of water at 25.0° C. How much of the ice melted? *Ans.* 78.1 g.

12.31. What is the molar heat of vaporization of water? *Ans.* 9724 cal.

12.32. What is the molar heat of fusion of carbon dioxide? *Ans.* 1994 cal.

12.33. How many calories of heat would be required to convert 5.00 moles of water at 25° C to steam at 100° C? *Ans.* 5.5×10^4 cal.

12.34. The molar heat capacity of iron(III) oxide is 23.63 cal. What is the heat capacity of the oxide? *Ans.* 0.148 cal.

12.35. How many calories of heat would be produced when a bullet weighing 8.75 g, and traveling at a velocity of 1000 ft per sec, hit an impenetrable object? *Ans.* 97.1 cal.

12.36. Calculate the heat capacity of water in ergs per pound per degree Fahrenheit. *Ans.* 1.05×10^{10} ergs.

12.37. Given that the heat of vaporization of water is 540 cal/g at 100° C and 1.00 atm, calculate ΔS when 1.00 mole of water is changed to vapor at 100° C and 1.00 atm. *Ans.* 26 e.u.

12.38. Given that, at 25° C, S = 45.1 e.u. for H_2O (g) and 16.7 e.u. for H_2O (l), calculate the heat of vaporization of water at 25° C in terms of calories per gram. *Ans.* 470 cal/g.

12.39. For the reaction $H_2 + I_2 \rightarrow 2HI$, $\Delta G = +0.30$ e.u./mole.
For the reaction $H_2 + Cl_2 \rightarrow 2HCl$, $\Delta G = -65$ e.u./mole.
Discuss the possibility of preparing HI and HCl by direct union of the elements.

The Energy Involved in Chemical Changes

12.5. Bond Energies. *Bond energy* is a measure of the energy lost by an atom when it forms a bond with another atom. This same amount of energy is required to break the bond. That is, bond energies may be measured as enthalpies, ΔH. Most bond energies are obtained experimentally from calorimetric data. For example:

$$H_2 \rightarrow 2H \quad \Delta H = 104.2 \text{ kcal.}$$

That is, it would require 104.2 kcal of energy to break the bonds in one mole of hydrogen, H_2, to form two moles of atomic hydrogen.

Example 12.8. How many calories of heat would be required to break one hydrogen bond?

Solution. One mole of hydrogen, H_2, contains 6.02×10^{23} hydrogen bonds. Therefore:

$$\frac{104{,}200 \dfrac{\text{cal}}{\cancel{\text{mole}}}}{6.02 \times 10^{23} \dfrac{\text{bonds}}{\cancel{\text{mole}}}} = 1.73 \times 10^{-19} \frac{\text{cal}}{\text{bond}}.$$

Bond energies apply to covalent bonds only. They are sometimes determined by indirect methods.

Example 12.9. Given the following equations in which ΔH has been measured experimentally, calculate ΔH for the CH bond (one mole or 6.02×10^{23} bonds).

$$\begin{array}{llll}
 & C\ (s) + 2\ H_2\ (g) & \rightarrow CH_4\ (g) & \Delta H = -\ \ 18.0 \text{ kcal} \\
 & C\ (g) & \rightarrow C\ (s) & \Delta H = -171.7 \text{ kcal} \\
 & 4\ H\ (g) & \rightarrow 2\ H_2\ (g) & \Delta H = -208.4 \text{ kcal} \\
\hline
\text{Add:} & C\ (g) + 4\ H\ (g) & \rightarrow CH_4\ (g) & \Delta H = -398.1 \text{ kcal}
\end{array}$$

Solution. Adding the three equations algebraically, as shown above, will give us the desired equation:

$$C\ (g) + 4\ H\ (g) \rightarrow CH_4\ (g) \qquad \Delta H = -398.1 \text{ kcal.}$$

That is, to break the bonds in one mole of CH_4 (g) would require 398.1 kcal of energy. However, each molecule of CH_4 contains four CH bonds.

Therefore:

$$\frac{398.1 \frac{\text{kcal}}{\text{mole}}}{4 \text{ bonds}} = 99.5 \text{ kcal to break one mole of CH bonds.}$$

12.6. Thermochemical Equations. Most chemical reactions are accompanied by enthalpy changes, ΔH. Equations in which the ΔH values are given are called *thermochemical equations*.

The *heat of reaction* is defined as the amount of heat released ($-\Delta H$) or absorbed ($+\Delta H$) when *one* mole of a substance reacts. In Sec. 12.5 the heat of reaction for the conversion of molecular hydrogen to atomic hydrogen was given as 104.2 kcal. Two types of enthalpies of reaction will be discussed: (1) heat of combustion and (2) heat of formation.

12.7. Heat of Combustion. Combustion is the rapid reaction of a substance with oxygen, being accompanied by heat and light. *Heat of combustion* is defined as the quantity of heat liberated when one mole of a substance burns in oxygen. All combustion reactions are exothermic. For example:

C (s)	+ O_2 (g)	→ CO_2 (g)	$\Delta H = -94.1$ kcal.
1.00 mole	1.00 mole	1.00 mole	
12.01 g	32.00 g	44.01 g	

The equation could be written:

$$C\ (s) + O_2\ (g) \rightarrow CO_2\ (g) + 94.1 \text{ kcal}$$

indicating that heat is one of the products of the reaction.

The above equation tells us that one mole of carbon, 12.01 g, will react with one mole of oxygen, 32.00 g, to yield one mole of carbon dioxide, 44.01 g, and 94.1 kcal of heat. Therefore, by definition, the heat of combustion of carbon is 94.1 kcal.

Example 12.10. When 1.14 g of sulfur was burned to SO_2, there was liberated 2464 cal. Calculate the heat of combustion of sulfur.

Solution. By definition the heat of combustion of sulfur is the heat liberated when one mole of sulfur, 32.064 g, burns to SO_2. Therefore:

$$\frac{2464 \text{ cal}}{1.14 \text{ g S}} = \text{calories released when 1.00 g of S burns,}$$

and 1.00 mole of S, 32.064 g, would release:

$$32.064 \cancel{\text{g S}} \times \frac{2464 \text{ cal}}{1.14 \cancel{\text{g S}}} = 69{,}300 \text{ cal or 69.3 kcal.}$$

That is: $S + O_2 \rightarrow SO_2 \quad \Delta H = -69.3$ kcal.

Example 12.11. The heat of combustion of heptane, C_7H_{16}, is 1150 kcal. How many kcal would be liberated by the burning of 500 g of heptane?

Solution. One mole, 100.2 g, of C_7H_{16} will liberate 1150 kcal. Since:

$$\frac{500 \; \cancel{\text{g } C_7H_{16}}}{100.2 \; \dfrac{\cancel{\text{g } C_7H_{16}}}{\text{mole } C_7H_{16}}} = \frac{500}{100.2} \text{ mole } C_7H_{16}$$

then: $$\frac{500}{100.2} \; \cancel{\text{mole } C_7H_{16}} \times 1150 \; \frac{\text{kcal}}{\cancel{\text{mole } C_7H_{16}}} = -1150 \text{ kcal.}$$

That is:

$$C_7H_{16} + 11 \; O_2 \rightarrow 7 \; CO_2 + 8 \; H_2O \qquad \Delta H = -5.74 \times 10^3 \text{ kcal.}$$

12.8. Heat of Formation. The *heat of formation* is defined as the quantity of heat liberated or absorbed when one mole of a substance is formed from the elements.

Example 12.12. When 1.14 g of sulfur was burned to SO_2, there was liberated 2464 cal. Calculate the heat of formation of SO_2.

Solution. From the equation:

$$\begin{array}{lcccc} S\,(s) & + & O_2\,(g) & \rightarrow & SO_2\,(g) \\ 1.00 \text{ mole} & & & & 1.00 \text{ mole} \end{array}$$

we see that one mole of SO_2 is formed for each mole of sulfur reacting. Therefore, the heat of combustion of sulfur and the heat of formation of SO_2 are numerically the same, 69,300 cal. This relationship exists in all reactions of direct combination of an element with oxygen.

Example 12.13. When hydrogen was burned in chlorine, 1000 cal was liberated and 1.650 g of HCl formed. Calculate the heat of formation of HCl.

Solution. By definition, the heat of formation of HCl is the heat liberated when one mole, 36.46 g, of HCl is formed from the elements.

Since: $$\frac{1.650 \; \cancel{\text{g HCl}}}{36.46 \; \dfrac{\cancel{\text{g HCl}}}{\text{mole HCl}}} = \frac{1.650}{36.46} \text{ mole HCl,}$$

then: $$\frac{1000 \text{ cal}}{\dfrac{1.650}{36.46} \text{ mole HCl}} = 22{,}100 \; \frac{\text{cal}}{\text{mole HCl}}.$$

That is: $H_2\,(g) + Cl_2\,(g) \rightarrow 2\;HCl\,(g) \quad \Delta H = -44.2$ kcal.
Observe that we found $\Delta H = -22.1$ kcal for one mole of HCl. However, the equation we have written indicates the formation of two moles of HCl. Therefore, ΔH for the equation as it is written is -44.2 kcal.

Example 12.14. Calculate ΔH for the reaction between ethylene, C_2H_4, and oxygen (see Table 12.3).

Solution. First write the equation for the reaction.

$$\begin{array}{llll} C_2H_4\ (g) & +\ 3\ O_2\ (g) & \rightarrow\ 2\ CO_2\ (g) & +\ 2\ H_2O\ (l) \\ -12.6\ \text{kcal} & 0.0\ \text{kcal} & -(2 \times 94.1)\ \text{kcal} & -(2 \times 68.4)\ \text{kcal} \end{array}$$

Then, under each substance, write the heat of formation for the appropriate number of moles indicated in the equation. The ΔH for the reaction would be:

$$\Sigma\ \Delta H \text{ of products} = \Sigma\ \Delta H \text{ of reactants.}$$

Then: $[-(2 \times 94.1)] + [-(2 \times 68.4)] - (-12.6) = -312.0$ kcal is the ΔH for the reaction.

Example 12.15. The heat of combustion of methane, CH_4, is

$$210.8\ \frac{\text{kcal}}{\text{mole}} \left(\Delta H = -210.8\ \frac{\text{kcal}}{\text{mole}} \right).$$

Calculate ΔH for the heat of formation of methane.

Solution. Write the equation and proceed as in Example 12.14.

$$\begin{array}{llll} CH_4\ (g) & +\ 2\ O_2\ (g) & \rightarrow\ CO_2\ (g) & +\ 2\ H_2O\ (l) \\ x\ \text{kcal} & 0.0\ \text{kcal} & -94.1\ \text{kcal} & -(2 \times 68.3)\ \text{kcal} \end{array}$$

Then: $(-94.1) + (-136.6) - (x) = -210.8$ kcal,

or: $x = +19.9\ \frac{\text{kcal}}{\text{mole}}.$

TABLE 12.3

ENTHALPIES OF REACTION IN KILOCALORIES PER MOLE AT 25° C

Substance	*Formula*	ΔH (*formation*)	ΔH (*combustion*)
Ammonia	NH_3	−16.0	—
Benzene	C_6H_6	+11.6	−780.0
Carbon	C	—	−94.4
Ethylene	C_2H_4	−12.6	−312.0
Hydrogen	H_2	—	−68.4
Hydrogen chloride	HCl	−22.2	—
Hydrogen iodide	HI	+6.2	—
Nitric oxide	NO	−21.6	—
Nitrogen dioxide	NO_2	+12.4	—
Sugar	$C_{12}H_{22}O_{11}$	—	−1350.0
Sulfur	S	—	−70.9
Sulfur dioxide	SO_2	−70.9	—
Water	H_2O	−68.4	—

The heat of formation of an element is zero.

12.9. The Measurement of Heat of Reaction. The experimental measurement of heat of reaction may be brought about by *calorimetry.* Many types of calorimeters are in use for such measurements. Fig. 12.1 shows the general construction of a bomb type of instrument.

Example 12.16. A sample of carbon weighing 0.463 g was placed in the cup C in Fig. 12.1. The calorimeter was assembled and 2500 g of water placed in the outer jacket D. As a result of burning the carbon, the temperature of the water rose from 22.54° C to 23.82° C. The metal and glass of the calorimeter were equivalent to 350 g of water in terms of their heat-absorbing ability. Calculate the heat of combustion of carbon.

Solution. The heat from the 0.463 g of burning carbon raised the temperature of an equivalent of 2500 g + 350 g = 2850 g of water 23.82° C − 22.54° C = 1.28° C. Therefore: 2850 g H_2O × 1.28° C = 3648 cal. Since 0.463 g of carbon would be:

$$\frac{0.463\ \cancel{\text{g C}}}{12.01\ \frac{\cancel{\text{g C}}}{\text{mole C}}} = \frac{0.463}{12.01}\ \text{mole C},$$

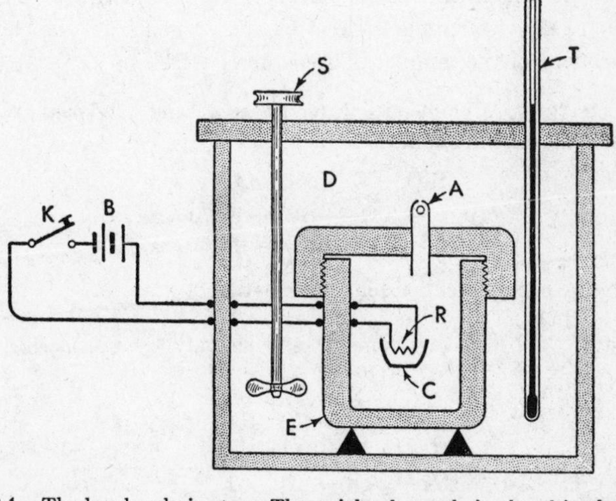

Fig. 12.1. The bomb calorimeter. The weighted sample is placed in the cup C, and the bomb E is filled with oxygen under pressure through the valve A. The outer jacket D is filled with water. S is a stirrer and T a thermometer. When K is closed the wire R becomes red-hot, causing the substance in the cup to burn. The heat is absorbed by the water, metal, and glass of the instrument. By means of the rise in temperature after the ignition, the heat of combustion of the substance in cup C may be calculated.

then 1.00 mole of C (12.01 g) would release:

$$\frac{3648 \text{ cal}}{\frac{0.463}{12.01} \text{ mole C}} = 94{,}600 \frac{\text{cal}}{\text{mole C}}.$$

That is: $C\ (s) + O_2\ (g) \rightarrow CO_2\ (g) \quad \Delta H = -94.6$ kcal.

12.10. The Law of Hess. The law of Hess states that the heat liberated or absorbed as the result of a chemical or physical change depends only on the initial and final substances. The intermediate steps involved do not affect the heat liberated or absorbed. For example, carbon dioxide may be obtained by burning carbon in an excess of oxygen; or it may be prepared by burning carbon to carbon monoxide, then burning the monoxide to the dioxide. The thermochemical equations are given below.

$$(1) \quad C + O_2 \rightarrow CO_2 \qquad \Delta H = -94.1 \text{ kcal}$$

$$(2) \quad \begin{cases} C + \frac{1}{2} O_2 \rightarrow \cancel{CO} + 29.0 \text{ kcal} \\ \cancel{CO} + \frac{1}{2} O_2 \rightarrow CO_2 + 65.1 \text{ kcal} \\ \hline C + O_2 \rightarrow CO_2 + 94.1 \text{ kcal} \end{cases} \quad \text{Add}$$

Recall that chemical equations may be added algebraically. Also, the heat released when one mole of CO_2 is formed from the elements is independent of the number of steps involved.

Example 12.17. Commercial water gas is obtained by passing superheated steam over carbon, in the form of coke.

$$H_2O + C \rightarrow H_2 + CO$$

Given

$$(1) \quad H_2 + \tfrac{1}{2} O_2 \rightarrow H_2O + 58{,}700 \text{ cal.}$$

$$(2) \quad C + \tfrac{1}{2} O_2 \rightarrow CO + 29{,}000 \text{ cal.}$$

Calculate the heat of reaction for the formation of water gas.

Solution. Subtract (1) from (2) and simplify by transposing terms.

Then:

$$\begin{array}{l} C + \tfrac{1}{2} O_2 \rightarrow CO + \quad 29{,}000 \text{ cal} \\ H_2 + \tfrac{1}{2} O_2 \rightarrow H_2O + \quad 58{,}700 \text{ cal} \\ \hline C + \cancel{\tfrac{1}{2} O_2} - H_2 - \cancel{\tfrac{1}{2} O_2} \rightarrow CO - H_2O - 29{,}700 \text{ cal} \end{array}$$

and

$$C + H_2O \rightarrow CO + H_2 - 29{,}700 \text{ cal.}$$

The value $-29{,}700$ cal indicates an endothermic reaction. That is, 29,700 cal are absorbed during the course of the reaction. This may also be stated as:

$$C + H_2O \rightarrow CO + H_2 \quad \Delta H = +29.7 \text{ kcal.}$$

Example 12.18. From the equations:

$$(1)\ H_2\ (\text{gas}) + \tfrac{1}{2}\ O_2\ (\text{gas}) \rightarrow H_2O\ (\text{gas}) \quad + 58{,}700\ \text{cal}$$
$$(2)\ H_2\ (\text{gas}) + \tfrac{1}{2}\ O_2\ (\text{gas}) \rightarrow H_2O\ (\text{liquid}) + 68{,}400\ \text{cal},$$

calculate the heat of vaporization of water.

Solution. Subtract (1) from (2) and transpose.

$$\begin{array}{r} H_2\ (\text{gas}) + \tfrac{1}{2}\ O_2\ (\text{gas}) \rightarrow H_2O\ (\text{liquid}) + 68{,}400\ \text{cal} \\ H_2\ (\text{gas}) + \tfrac{1}{2}\ O_2\ (\text{gas}) \rightarrow H_2O\ (\text{gas}) \quad + 58{,}700\ \text{cal} \\ \hline \cancel{H_2}(g) + \cancel{\tfrac{1}{2}O_2}\ (g) - \cancel{H_2}(g) - \cancel{\tfrac{1}{2}O_2}\ (g) \rightarrow H_2O\ (l) - H_2O\ (g) + 9700\ \text{cal} \end{array}$$

$$\begin{array}{c} H_2O\ (g) \rightarrow H_2O\ (l) + 9700\ \text{cal} \\ 18\ g \qquad 18\ g \end{array}$$

or: $H_2O\ (g) \rightarrow H_2O\ (l) \quad \Delta H = -9.70\ \text{kcal}.$

That is, 18 g of water vapor condenses to liquid with the release of 9700 cal.

Therefore, 1.00 g of water vapor would release $\dfrac{9700\ \text{cal}}{18\ g\ H_2O} = 540\ \dfrac{\text{cal}}{g\ H_2O}.$

That is, the heat of vaporization of water is 540 calories per gram.

Problems

Part I

12.40. When 1.20 g of benzene, C_6H_6, was burned to CO_2 and H_2O, 12.0 kcal was liberated. Calculate the heat of combustion of benzene. *Ans.* 780 kcal.

12.41. When 0.327 g of carbon combined with sulfur to form CS_2, 0.691 kcal was absorbed. Calculate the heat of formation of CS_2. *Ans.* −25.4 kcal.

12.42. How many calories would be liberated by the complete combustion of 250 liters of CH_4 at S.C., given that the heat of combustion of CH_4 is 211 kcal? *Ans.* 2355 kcal.

12.43. The heat of formation of water is 68.4 kcal. How many calories would be liberated by burning 12 g of hydrogen? *Ans.* 410 kcal.

12.44. Magnesium was burned in oxygen until one gram of MgO had formed. The heat liberated was 3.60 kcal. Calculate the heat of formation of MgO. *Ans.* 145 kcal.

12.45. When 0.327 g of carbon was burned to CO_2, 0.257 kcal was liberated. Calculate the heat of formation of CO_2. *Ans.* 94.4 kcal.

12.46. A sample of coal weighing 0.875 g, when burned in a calorimeter, liberated enough heat to raise the temperature of 2500 g of water from 18.50° C to 20.60° C. Calculate the BTU value of the coal per pound. *Ans.* 10,810 BTU.

12.47. The burning of 1.00 g of sulfur to SO_2 resulted in the liberation of 2200 calories. Calculate the heat of formation of SO_2. *Ans.* 70.6 kcal.

12.48. Given the equations:

$$H_2 \text{ (gas)} + \tfrac{1}{2} O_2 \text{ (gas)} \rightarrow H_2O \text{ (liquid)} + 68{,}400 \text{ cal}$$
$$H_2 \text{ (gas)} + \tfrac{1}{2} O_2 \text{ (gas)} \rightarrow H_2O \text{ (ice)} + 69{,}840 \text{ cal.}$$

Calculate the heat of fusion of ice. *Ans.* 80 cal.

12.49. How much heat would be liberated by the combustion of 2.75 moles of carbon? *Ans.* 260 kcal.

12.50. From the following equations calculate the heat of formation of methane:

$$H_2 \text{ (g)} + \tfrac{1}{2} O_2 \text{ (g)} \rightarrow H_2O \text{ (l)} + 68{,}400 \text{ cal}$$
$$C \text{ (s)} + O_2 \text{ (g)} \rightarrow CO_2 \text{ (g)} + 94{,}400 \text{ cal}$$
$$CH_4 \text{ (g)} + 2\, O_2 \text{ (g)} \rightarrow CO_2 \text{ (g)} + 2\, H_2O \text{ (l)} + 210{,}800 \text{ cal.}$$

Ans. 20,400 cal.

12.51. The heat of combustion of acetylene is 312,000 cal per mole. How many liters of carbon dioxide (at S.C.) are released for each kilocalorie produced? *Ans.* 0.144 *l.*

12.52 The burning of sufficient magnesium in oxygen to form 2.00 g of oxide resulted in the liberation of 7200 cal Calculate the heat of combustion of magnesium. *Ans.* 145 kcal.

12.53 Calculate the heat of formation of nitric oxide, NO, from the following equations:

$$N_2 \text{ (g)} + 2\, O_2 \text{ (g)} \rightarrow 2\, NO_2 \text{ (g)} - 15{,}000 \text{ cal}$$
$$2\, NO \text{ (g)} + O_2 \text{ (g)} \rightarrow 2\, NO_2 \text{ (g)} + 28{,}000 \text{ cal.}$$

Ans. −21,500 cal.

Part II

12.54. Assuming methane costs 75 cents per 1000 ft^3, calculate the cost per 1,000,000 BTU. *Ans.* 71 cents.

12.55. Given the equation:

$$C_2H_4 \text{ (g)} + 3\, O_2 \text{ (g)} \rightarrow 2\, CO_2 \text{ (g)} + 2\, H_2O \text{ (l)} + 313{,}000 \text{ cal.}$$

Calculate the heat of formation of ethylene, C_2H_4. *Ans.* −12,600 cal.

12.56. How many liters of water could be changed from 0° C to 100° C by the combustion of one mole of carbon? *Ans.* 0.944 *l.*

12.57. The heat of combustion of acetylene is 311.2 kcal. Calculate the heat of formation of acetylene. *Ans.* −54 kcal.

12.58. Given the equation:

$$N_2 + O_2 \rightarrow 2\, NO - 43.2 \text{ kcal.}$$

Calculate the number of calories required to convert 25 liters of N_2 to NO at standard conditions. *Ans.* 48.2 kcal.

12.59. How much heat would be liberated by 18 g of carbon when burned to CO_2? *Ans.* 141.9 kcal.

12.60. How many grams of magnesium oxide would have to be formed from the elements in order to liberate 18.0 kcal ? *Ans.* 5.0 g.

12.61. How many calories are liberated when 1.00 g of SO_2 is formed from the elements? *Ans.* 1082 cal.

12.62. When 0.500 g of coal was burned in a calorimeter, sufficient heat was formed to raise the temperature of 1500 g. of water from 23.00° C to 25.50° C. Calculate the heat value of the coal in (a) calories per gram, and (b) BTU per pound.
Ans. (a) 7.5 kcal per g ; (b) 13,500 BTU per lb.

12.63. How much heat would be liberated by the combustion of 2.5 pound-moles of carbon? *Ans.* 1.07×10^5 kcal.

12.64. How much heat would be liberated by the formation of one pound-mole of SO_2? *Ans.* 3.15×10^4 kcal.

12.65. Given the thermochemical equations:

$$4\ Fe\ (s) + 3\ O_2\ (g) \rightarrow 2\ Fe_2O_3\ (s) + 398\ kcal$$
$$4\ Al\ (s) + 3\ O_2\ (g) \rightarrow 2\ Al_2O_3\ (s) + 798\ kcal.$$

Calculate the heat of the reaction:

$$Fe_2O_3\ (s) + 2\ Al\ (s) \rightarrow Al_2O_3\ (s) + 2\ Fe\ (s) + x\ kcal.$$

Ans. 200 kcal.

12.66. The heat liberated from the combustion of 1.250 g of coke raised the temperature of 1000 g of water from 22.5° C to 30.1° C. What was the per cent of carbon in the coke, assuming the impurities to be noncombustible? *Ans.* 77.4%.

12.67. A fuel oil has a heat of combustion of 16,500 BTU per lb. The boiler of a furnace holds 100 gallons of water. Assuming 60 per cent efficiency in heat transfer, how many pounds of fuel oil would be required to heat the water in the boiler from 70° F to 180° F ?
Ans. 9.3 lb.

12.68. An anthracite coal containing 95 per cent carbon costs $21 per ton. What is the cost of the fuel in terms of dollars per 1,000,000 BTU?
Ans. $0.78.

12.69. A coal is purchased having a fuel value of 12,630 BTU per lb. Assuming carbon to be the only combustible constituent of the coal, what is the per cent of carbon in the coal? *Ans.* 89.1%.

12.70. Given that $\Delta H = -196$ kcal for breaking the C-C bond, calculate ΔH for breaking all the bonds in ethane, C_2H_6. *Ans.* -793 kcal.

12.71. Given the equation:

$$2\ Cl\ (g) \rightarrow Cl_2\ (g) \quad \Delta H = -57.8\ kcal.$$

How many calories would be required to break one Cl-Cl bond?
Ans. 9.60×10^{-20} cal.

12.72. Calculate ΔH for the decomposition of $CaCO_3$ into CaO and CO_2.
Ans. $42.5 \dfrac{kcal}{mole}$.

12.73. Calculate the heat of combustion of methane, CH_4, given that

$\Delta H_f = 19.9$ kcal for methane. *Ans.* $-210.8 \frac{\text{kcal}}{\text{mole}}$.

12.74. Given that the heat of reaction for the equation

$$N_2\ (g) + 3\ H_2\ (g) \rightarrow 2\ NH_3\ (g)$$

is -22.0 kcal. Calculate ΔH_f for NH_3 *Ans.* $-11.0 \frac{\text{kcal}}{\text{mole}}$.

12.75. How much heat will be evolved when 10.0 lb of quicklime, CaO, is slaked in water to form $Ca(OH)_2$? *Ans.* 1257 kcal.

12.76. One of the fuels for rockets is a mixture consisting of hydrazine, N_2H_4, and hydrogen peroxide, H_2O_2. They react according to the equation:

$$N_2H_4\ (l) + 2\ H_2O_2\ (l) \rightarrow N_2\ (g) + 4\ H_2O\ (g).$$

The mixture contains the reactants in the mole ratio as given in the above equation. Calculate the fuel value of the mixture per gram.

Ans. 1.51 kcal per g.

13

Experimental Determination of Atomic Weights

The use of atomic weights in the quantitative treatment of chemical data is evidence of the need for accurate atomic weights. The atomic weights of the elements are constantly being redetermined as better techniques are developed. The corrections obtained are small but important. It may be of interest to present a few of the experimental methods which have been used to determine atomic weights. The methods may be classified as either *physical* or *chemical*.

Physical Methods for Determining Atomic Weights

13.1. Molar Heat Capacity. The law of Dulong and Petit, proposed about 1818, states that the product of the atomic weight and heat capacity (Sec. 12.3) of an element is essentially a constant equal to 6.3.

That is:

$$\text{Atomic weight} \times \text{heat capacity} = 6.3.$$

The method is principally of historical interest, and is only an approximation, as shown in Table 13.1.

TABLE 13.1

Element	*Heat Capacity*	$\frac{6.3}{\textit{Ht. Cap.}}$	*Accepted Atomic Weight*
Aluminum	0.214	29	26.9815
Copper	0.092	68	63.54
Iron	0.107	59	55.847
Lead	0.031	203	207.19
Magnesium	0.246	26	24.312
Sulfur	0.176	36	32.064

Example 13.1. The experimental value for the heat capacity of silver is 0.058 cal per g. Calculate the approximate atomic weight of silver.

Solution. From the law of Dulong and Petit:

$$\text{At. wt. of silver} = \frac{6.3}{0.058} = 109.$$

Example 13.2. Calculate the approximate heat capacity of gold and compare to the accepted value.

Solution.

$$\text{Heat capacity of gold} = \frac{6.3}{197} = 0.032 \text{ cal per g.}$$

The experimental value for the heat capacity of gold is 0.031 cal per g.

13.2. The Mass Spectrograph. With the discovery of isotopes and methods for producing nuclear fission, more accurate methods were developed for measuring the masses of particles of atomic size. The mass spectrograph is an instrument used for such measurements. The instrument is quite complex. However, the principle upon which its operation is based is relatively simple, as shown in Fig. 13.1.

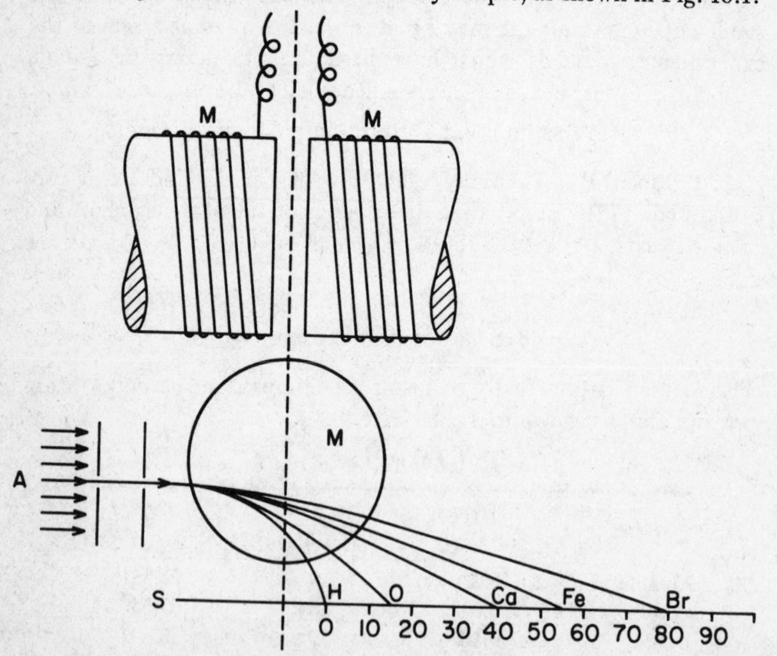

Fig. 13.1. The mass spectrograph. A stream of charged particles, A, the mass of which is to be determined, is passed between the poles of the electromagnet M. The charged particles are deflected in proportion to their masses along the scale S, on a photographic plate. A black spot is produced at the point of impact of the particles when the plate is developed.

Example 13.3. On a photographic plate calibrated from 0 to 100, in terms of a m u , in a mass spectrograph, the black spot for C^{12} was observed at a reading of 26.240. The black spot for aluminum was 59.006. Calculate the weight of the aluminum atoms in a m u.

Solution. The relative weights of the two atoms would be in proportion to the scale readings on the photographic plate. Since C^{12} = 12.000 a m u., then:

$$\frac{12.000 \text{ amu}}{26.240 \text{ scale units}} = \text{amu per scale unit}$$

and: $$59.006 \text{ scale units} \times \frac{12.000 \text{ amu}}{26.240 \text{ scale units}} = 26.984 \text{ amu for Al.}$$

Example 13.4. Analysis with a mass spectrograph showed that naturally occurring chlorine consists of 75.40 per cent of Cl^{35} = 34.969 a m u , and 24.60 per cent Cl^{37} = 36.966 a m u. Calculate the atomic weight for chlorine.

Solution. The atomic weight of chlorine will be between 34.969 a m u. and 36.966 a m u. However, since the lighter isotope predominates, the atomic weight will be closer to the lower value than it will be to the higher value. Then:

$$\begin{aligned} \text{At. wt.} &= (0.7540 \times 34.969) + (0.2460 \times 36.966) \text{ a m u} \\ &= 35.452 \text{ a m u.} \end{aligned}$$

Chemical Methods for Determining Atomic Weights

13.3. Cannizzaro's Method. To determine the atomic weight of an element by this method, the following information must be available for several compounds containing the element: (1) the molecular weight of each compound, and (2) the percentage by weight of the element in each compound. Such data may be obtained by standard laboratory procedures.

Example 13.5. It is desired to determine the atomic weight of chlorine from the data given in the following table.

Compound	*Mol. Wt.*	*% Chlorine*	*Contribution of Chlorine to Mol. Wt.*
Hydrogen chloride	36.5	97.2	35.5
Phosgene	98.9	71.7	70.9
Boron trichloride	117.2	90.8	106.4
Carbon tetrachloride	153.9	92.1	141.7

Solution. First, calculate the contribution of chlorine to the molecular weight of each compound. Then, in hydrogen chloride the chlorine would contribute:

$$36.5 \times 0.972 = 35.5 \text{ units of mol. wt.}$$

Similarly, the contributions of chlorine were calculated and put in column 4 for each of the four compounds.

By inspection, we select the value in column 4 of which all are multiples. This value, in the above example 35.5, is considered to be the atomic weight of the element. That is:

$$\text{Atomic weight of chlorine} = 35.5 \text{ a m u.}$$

13.4. From Percentage Composition. By means of chemical analysis it is possible to determine the composition of a compound to five decimal places in terms of grams (0.00000 g). Atomic weights calculated from such data give correspondingly accurate values.

Example 13.6. Chemical analysis of silver nitrate gave the composition, 63.500 per cent silver, 8.245 per cent nitrogen, and 28.255 per cent oxygen. Calculate the atomic weight of silver with an accuracy consistent with the data. The approximate atomic weight of silver is 109, as determined from heat capacity (Example 13.1).

Solution. Calculate the formula (Sec. 10.3). Then:

$$\text{Silver} = \frac{63.500}{109} = 0.583.$$

$$\text{Nitrogen} = \frac{8.245}{14} = 0.589.$$

$$\text{Oxygen} = \frac{28.255}{16} = 1.766.$$

The simplest integer ratio is 1 : 1 : 3. The formula is therefore $AgNO_3$.

From the formula $AgNO_3$ we see that there are 3 gram-atoms of oxygen associated with 1 gram-atom of silver. Therefore, the atomic weight of silver would be equal numerically to the number of grams of silver combined with $3 \times 15.9994 = 47.9982$ g of oxygen. Then:

$$\frac{63.500 \text{ g Ag}}{28.255 \text{ g } O_2} = \text{g of Ag combined with 1.00 g of } O_2.$$

$$47.9982 \cancel{\text{g } O_2} \times \frac{63.500 \text{ g Ag}}{28.255 \cancel{\text{g } O_2}} = 107.87 \text{ g Ag}$$

That is, the atomic weight of Ag is 107.87 amu.

13.5. From Gram-equivalent Weights. The gram-equivalent weights of the elements may be determined by chemical analysis, and are therefore quite accurate. The atomic weight of an element is a multiple of the gram-equivalent weight, the numerical value of the multiple being termed the *oxidation number*. That is:

$$\text{Oxidation number} = \frac{\text{atomic weight}}{\text{equivalent weight}}.$$

Example 13.7. The heat capacity of gold is 0.031 cal per g, and the gram-equivalent weight is 65.67 g. Calculate the atomic weight of gold.

Solution.

$$\text{Approximate at. wt. of gold} = \frac{6.3}{0.031} = 203.$$

$$\text{Oxidation number of gold} = \frac{203}{65.67} = 3.$$

That is, the accurate atomic weight of gold is 3 times the gram-equivalent weight, or:

$$\text{Atomic weight of gold} = 65.67 \times 3 = 197.0.$$

Example 13.8. The approximate atomic weight of zinc is 65, as determined from its heat capacity. It was found that 1.632 g of zinc replaced 559.7 ml of hydrogen from a solution of H_2SO_4, measured at S.C. Calculate a more accurate atomic weight of zinc.

Solution. One gram-equivalent weight of zinc will replace 1.008 g of hydrogen or 11.2 *l*. However, 11.2 *l* is not so accurate a value as 1.008 g. We will therefore use 1.008 g of hydrogen. Since one liter of hydrogen weighs 0.0899 g, then 559.7 ml will weigh $0.5597 \times 0.0899 = 0.05032$ g. Therefore:

$$\frac{1.632 \text{ g Zn}}{0.05032 \text{ g } H_2} = \text{g of Zn that will replace 1.00 g } H_2.$$

$$1.008\ \cancel{\text{g } H_2} \times \frac{1.632 \text{ g Zn}}{0.05032\ \cancel{\text{g } H_2}} = 32.69 \text{ g Zn}$$

That is, 32.69 g of Zn is 1.00 gram-equivalent of Zn.

Then, $$\text{oxidation number of Zn} = \frac{65}{32.69} = 2,$$

and $$\text{atomic weight of zinc} = 32.69 \times 2 = 65.38.$$

Example 13.9. The following data were recorded in the determination of the atomic weight of tin. The heat capacity is 0.0542 cal per g. It was found that 2.1440 g of tin combined with oxygen to form 2.7219 g of oxide. Calculate the atomic weight of tin, given that its approximate atomic weight is 118.

Solution.

$$\text{Approximate atomic weight} = \frac{6.3}{0.0542} = 116.$$

The gram-equivalent weight would be the amount of tin that will combine with 7.9997 g. of oxygen. Therefore:

$2.7219 \text{ g oxide} - 2.1440 \text{ g Sn} = 0.5779 \text{ g } O_2$ combined with 2.1440 g Sn

and: $$\frac{2.1440 \text{ g Sn}}{0.5779 \text{ g } O_2} = \text{g of Sn combined with 1.00 g of } O_2.$$

Then: $$7.9997 \cancel{\text{g } O_2} \times \frac{2.1440 \text{ g Sn}}{0.5779 \cancel{\text{g } O_2}} = 29.68 \text{ g Sn}.$$

That is, the gram-equivalent weight of Sn is 29.68 g.

The oxidation number of Sn would be $\frac{118}{29.68} = 4$, and the atomic weight of Sn $= 29.68 \times 4 = 118.7$ amu.

As explained earlier, the approximate atomic weight of the element is sufficient to enable us to calculate the oxidation number, which we know is an integer value.

Problems

Part I

13.1. Calculate the approximate heat capacity of uranium. *Ans.* 0.026 cal per g.

13.2. Given the heat capacity of lead as 0.031, calculate the atomic weight of lead and compare to the accepted value. *Ans.* 203.

13.3. The gram-equivalent weight of magnesium was found experimentally to be 12.16 g. The heat capacity of magnesium is 0.24 cal per g. Calculate the accurate atomic weight of magnesium. *Ans.* 24.32.

13.4. An oxide of vanadium was found to contain 2.123 g of vanadium for each 1.000 g of oxygen. Calculate the atomic weight of vanadium, given that the approximate atomic weight is 50. *Ans.* 50.95.

13.5. Sodium oxalate, $Na_2C_2O_4$, contains 34.314 per cent sodium and 47.760 per cent oxygen. Calculate the atomic weight of sodium. *Ans.* 22.991.

13.6. The equivalent weight of zinc is 32.69. Its approximate atomic weight is 65. Calculate a more exact atomic weight. *Ans.* 65.38.

13.7. An oxide of lead contains 13.377 per cent oxygen. Knowing the approximate atomic weight of lead to be 205, calculate a more exact atomic weight for the element. *Ans.* 207.21.

13.8. The approximate atomic weight of aluminum is 25. Calculate a more accurate atomic weight, given that 1.334 g of aluminum displaced 1.778 liters of hydrogen measured at 23° C and 770 mm of Hg. *Ans.* 26.98.

13.9. The ratio, per cent Cl^{35} : per cent Cl^{37}, for naturally occurring chlorine is equal to 3.065. What is the per cent of each of the isotopes in naturally occurring chlorine? *Ans.* 75.4% Cl^{35}, 24.6% Cl^{37}.

13.10. The atomic weight of beryllium was determined on a mass spectrograph by comparison with $Cl^{35} = 34.9689$. On a scale of 0 to 100, calibrated in atomic mass units, beryllium was recorded at 11.650 and Cl^{35} at 42.205. What is the atomic weight of beryllium?

Ans. 9.0122.

13.11. Given the following data in which all compounds are gases:

Compound	*Density* (g l^{-1}) at *S.C.*	*Per Cent*	
		Carbon	*Nitrogen*
Methane	0.716	74.9	
Ethane	1.252	85.6	
Propane	1.969	81.7	
Nitric oxide	1.384		45.2
Nitrogen trioxide	3.393		36.9
Hydrazoic acid	1.920		97.7

Calculate the atomic weight of (a) carbon, and (b) nitrogen.

Ans. Check table of atomic weights.

13.12. Three compounds containing the element X have the following molecular weights, and per cent composition by weight of X. Suggest a value for the atomic weight of X.

Compound	*Molecular Weight*	*Per Cent X*
A	64.06	50.05
B	135.03	47.49
C	194.20	49.53

Ans. 32.06.

Part II

13.13. One hundred grams of a metal at 100° C, when added to 300 g of water at 28.3° C, gave a mixture with a temperature of 31.6° C. Calculate the approximate atomic weight of the metal. *Ans.* 43.

13.14. The heat capacity of an element was found to be 0.100 cal per g. It was found that 0.749 g of the element formed 0.937 g of oxide. Calculate the atomic weight of the element. *Ans.* 63.6.

13.15. It was found that 144 cal was required to change the temperature of 260 g of aluminum from 20.00° C to 22.50° C. Calculate the approximate atomic weight for aluminum. *Ans.* 29.

13.16. An oxide of iron was found to contain 30.06 per cent oxygen. When 400 g of iron at 100° C was mixed with 220 g of water at 20° C, the resulting temperature was 33.0° C. Calculate the atomic weight of iron. *Ans.* 55.83.

13.17. Calculate the number of calories required to change the temperature of one atom of a metal one degree centigrade.

Ans. 1.06×10^{-23} cal.

13.18. When 350 g of platinum at 100° C was added to 250 g of water at 16° C, the temperature of the resulting mixture was 19.5° C. Calculate the atomic weight of platinum. *Ans.* 206.

13.19. The gram-equivalent weight of chlorine was found to be 8.865 as calculated from an oxide of the element. Calculate (a) the atomic weight of chlorine given that the heat capacity is 0.19 cal per g, and (b) the formula of the oxide. *Ans.* (a) 35.46; (b) ClO_2.

13.20. Analysis showed that magnesium pyrophosphate, $Mg_2P_2O_7$, contains 0.1086 g of magnesium to 0.2501 g of oxygen. Calculate the atomic weight of magnesium as accurately as the data will permit.

Ans. 24.32.

13.21. The oxide of a metal contains 21.35 per cent oxygen. The heat capacity of the metal is 0.100 cal per g. What is the metal?

Ans. Co.

13.22. It was found that 0.709 g of iron combined with 1.350 g of chlorine. Determine the atomic weight of iron. *Ans.* 55.9.

13.23. A metal X forms two oxides containing 36.81 per cent oxygen and 46.63 per cent oxygen, respectively. The heat capacity of the metal is 0.121 cal per g. What are the formulas for the oxides?

Ans. MnO_2 and MnO_3.

13.24. A mass of metal at 100° C and weighing 150 g was added to 246 g of water at 25.0° C. The resulting temperature of the mixture was 27.4° C. When 1.395 g of the metal was added to a sulfuric acid solution there was collected over water 358.9 ml of gas at 23° C and 660 mm of Hg.

a. Determine the atomic weight of the metal. *Ans.* 112.4.

b. What was the metal? *Ans.* Cd.

13.25. Naturally occurring carbon contains 98.86 per cent C^{12} and 1.14 per cent C^{13}. What is the isotopic weight of C^{13}? *Ans.* 12.98 a m u.

13.26. The ratio for the masses of Cl : Ag is equal to 0.3287, as measured on a mass spectrograph. Calculate the atomic weight of silver.

Ans. 107.9

14

Oxidation-Reduction Equations

In this chapter we shall discuss the more complex types of oxidation-reduction reactions. In Sec. 8.1 oxidation was defined as any process involving the loss of electrons by atoms, and reduction as any process involving the gain of electrons by atoms. It is evident that the processes of oxidation and reduction occur simultaneously in a chemical reaction. Such reactions are called *oxidation-reduction reactions* (sometimes abbreviated *redox reactions*).

In an oxidation-reduction reaction the element that is effective in bringing about oxidation is called the *oxidant* (or oxidizing agent). The element that is effective in bringing about reduction is called the *reductant* (or reducing agent). As an example we will use the following equation.

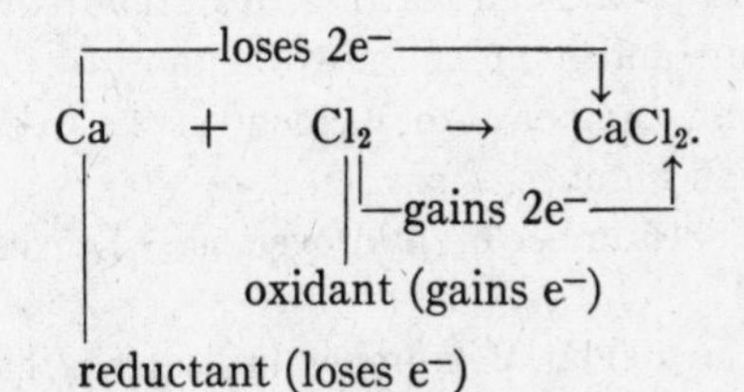

In the above reaction each calcium atom loses 2 e^-, and is therefore oxidized; each of the two atoms in the chlorine molecule gains 1 e^-, and is therefore reduced. The processes of oxidation and reduction may each be represented as a *half-reaction*. For the above equation the half-reactions are:

$$Ca \rightarrow Ca^{+2} + 2e^-. \qquad \text{(oxidation half-reaction)}$$
$$Cl_2 + 2e^- \rightarrow 2\,Cl^-. \qquad \text{(reduction half-reaction)}$$

Oxidation-Reduction Reactions

14.1. Change in Oxidation Number. In oxidation-reduction reactions some of the atoms undergo a change in oxidation number. For example, in the reaction:

$$\overset{(0)}{Ca} + \overset{(0)}{Cl_2} \rightarrow \overset{(+2)(-1)}{CaCl_2}$$

the calcium has changed its oxidation number from 0 to +2, and the chlorine from 0 to −1.

In more complex types of reactions involving compounds or complex ions, the change in oxidation number is somewhat more difficult to determine. For example, in the reaction:

$$\overset{(+1)(-2)}{H_2S} + \overset{(0)}{Cl_2} \rightarrow 2\,\overset{(+1)(-1)}{HCl} + \overset{(0)}{S}$$

the oxidation number of sulfur increased from −2 to 0, and the oxidation number of chlorine decreased from 0 to −1. The oxidation number of hydrogen was unchanged.

In the reaction:

$$\overset{(+4)(-2)}{MnO_2} + 4\,\overset{(+1)(-1)}{HCl} \rightarrow \overset{(+2)(-1)}{MnCl_2} + 2\,\overset{(+1)(-2)}{H_2O} + \overset{(0)}{Cl_2}$$

we see that the oxidation number of manganese decreased from +4 to +2, and the oxidation number of chlorine increased from −1 to 0. The oxidation numbers of hydrogen and oxygen remained unchanged.

14.2. Some Generalities Involving Redox Reactions. The following information is essential in writing oxidation-reduction reactions.

1. The reactants and products must be known.
2. The oxidation number of an element in the free state is zero.
3. The oxidation number of oxygen is −2.
4. The oxidation number of hydrogen is +1, except for hydrides in which it is −1.
5. A decrease in oxidation number indicates a gain of electrons, and an increase in oxidation number a loss of electrons. For example, in the previous equations given:

$$\overset{(0)}{Ca} + \overset{(0)}{Cl_2} \rightarrow \overset{(+2)(-1)}{CaCl_2}.$$

gains 1e⁻ per atom (Cl)

loses 2e⁻ (Ca)

$$\overset{(-2)}{H_2S} + \overset{(0)}{Cl_2} \rightarrow 2\,\overset{(-1)}{HCl} + \overset{(0)}{S}.$$

gains 1e⁻ per atom (Cl)

loses 2e⁻ (S)

$$\overset{(+4)}{MnO_2} + 4\ \overset{(-1)}{HCl} \rightarrow \overset{(+2)}{MnCl_2} + 2\ H_2O + \overset{(0)}{Cl_2}.$$

loses $1e^-$

gains $2e^-$

6. The algebraic sum of the oxidation numbers in a formula is equal to zero. For example:

$$\overset{(+1)(-1)}{HCl} \qquad \overset{(+3)(-2)}{Al_2O_3}$$

$$(+1) + (-1) = 0. \qquad [2 \times (+3)] + [3 \times (-2)] = 0.$$

$$\overset{(+1)(+7)(-2)}{KMnO_4} \qquad \overset{(+1)(+6)(-2)}{K_2Cr_2O_7}$$

$$(+1)+(+7)+[4\times(-2)]=0. \qquad [2\times(+1)]+[2\times(+6)]+[7\times(-2)]=0.$$

Example 14.1. What is the oxidation number of:
(a) Sulfur in H_2SO_4?
(b) Arsenic in H_3AsO_4?
(c) Phosphorus in $Ca_3(PO_4)_2$?

Solution. In each compound let x be the oxidation number of the unknown element.

(a) $\overset{(+1)(x)(-2)}{H_2SO_4}$

$$[2 \times (+1)] + (x) + [4 \times (-2)] = 0$$

or $x = +6 =$ oxidation number of sulfur.

(b) $\overset{(+1)(x)(-2)}{H_3AsO_4}$

$$[3 \times (+1)] + (x) + [4 \times (-2)] = 0$$

or $x = +5 =$ oxidation number of arsenic.

(c) $\overset{(+2)(x)(-2)}{Ca_3(PO_4)_2}$

$$[3 \times (+2)] + (2x) + [8 \times (-2)] = 0$$

or $x = +5 =$ oxidation number of phosphorus.

Two methods will be given for writing oxidation-reduction equations: (1) the *electron transfer method*, and (2) the *ion-electron method*.

Writing Oxidation-Reduction Equations

14.3. By the Electron Transfer Method. This method is based upon the premise that the number of electrons lost by atoms in an equation must equal the number of electrons gained. Change in oxidation number may be interpreted in terms of gain and loss of electrons. Therefore, the electron transfer method is sometimes described as a method based on change in oxidation state. A number of equations will now be written using the electron transfer method.

Example 14.2. Write the equation for the reaction between HNO_3 and H_2S.

Solution. Write down the reactants and products. Above each atom indicate its oxidation number.

loses $2e^-$ (from S in H_2S to S)

$$\begin{array}{ccccccccc} (+1)(+5)(-2) & & (+1)(-2) & & (+1)(-2) & & (+2)(-2) & & (0) \\ HNO_3 & + & H_2S & \rightarrow & H_2O & + & NO & + & S. \end{array}$$

gains $3e^-$ (from N in HNO_3 to NO)

In changing from S^{-2} to S^0 there is a loss of $2e^-$, and in changing from N^{+5} to N^{+2} there is a gain of $3e^-$. Evidently, in any oxidation-reduction equation:

Number of e^- lost = number of e^- gained.

To meet this requirement there must be:

$$\underline{3}\ S^{-2} \rightarrow \underline{3}\ S^0 + 6e^-,$$

and

$$\underline{2}\ N^{+5} + \underline{6}\ e^- \rightarrow \underline{2}\ N^{+2}. \text{ Therefore:}$$

$$\underline{2}\ HNO_3 + \underline{3}\ H_2S \rightarrow \overline{H}_2O + \underline{2}\ NO + \underline{3}\ S.$$

The last step is indicating the number of molecules of water. Since there are 8 H^+ there must be 4 H_2O. The complete equation will be:

$$2\ HNO_3 + 3\ H_2S \rightarrow 4\ H_2O + 2\ NO + 3\ S.$$

or

$$2\ H^+ + 2\ NO_3^- + 3\ H_2S \rightarrow 4\ H_2O + 2\ NO + 3\ S.$$

Example 14.3. Balance the reaction.

$$KMnO_4 + FeSO_4 + H_2SO_4 \rightarrow K_2SO_4 + MnSO_4 + Fe_2(SO_4)_3 + H_2O.$$

Solution. Proceed as in Example 14.2. Then:

gains $5e^-$ (from Mn in $KMnO_4$ to Mn in $MnSO_4$)

$$\begin{array}{ccccccccccccc} (+1)(+7)(-2) & & (+2)(+6)(-2) & & (+1)(+6)(-2) & & (+1)(+6)(-2) & & (+2)(+6)(-2) & & (+3)(+6)(-2) & & (+1)(-2) \\ KMnO_4 & + & FeSO_4 & + & H_2SO_4 & \rightarrow & K_2SO_4 & + & MnSO_4 & + & Fe_2(SO_4)_3 & + & H_2O\ . \end{array}$$

loses $1e^-$ (from Fe in $FeSO_4$ to Fe in $Fe_2(SO_4)_3$)

Then

$$5\ Fe^{+2} \rightarrow 5\ Fe^{+3} + 5e^-,$$

or, writing as an ionic equation, and canceling like entities, we have:

$$\cancel{2\,K^+} + 2\,MnO_4^- + 10\,Fe^{+2} + \cancel{10\,SO_4^-} + 16\,H^+ + \cancel{8\,SO_4^{-2}} \rightarrow$$
$$\cancel{2\,K^+} + \cancel{SO_4^{-2}} + 2\,Mn^{+2} + \cancel{2\,SO_4^{-2}} + 10\,Fe^{+3} + \cancel{15\,SO_4^{-2}} + 8\,H_2O$$

or $\quad 2\,MnO_4^- + 10\,Fe^{+2} + 16\,H^+ \rightarrow 2\,Mn^{+2} + 10\,Fe^{+3} + 8\,H_2O.$

Reducing to the simplest ratio of coefficients we have:

$$MnO_4^- + 5\,Fe^{+2} + 8\,H^+ \rightarrow Mn^{+2} + 5\,Fe^{+3} + 4\,H_2O.$$

Example 14.4. Balance the reaction:

$$K_2Cr_2O_7 + HCl \rightarrow KCl + CrCl_3 + H_2O + Cl_2.$$

Solution. Proceeding as in the previous examples:

2 atoms gain $6e^-$

$$\overset{(+6)}{K_2Cr_2O_7} + \overset{(-1)}{HCl} \rightarrow \overset{(-1)}{KCl} + \overset{(+3)(-1)}{CrCl_3} + H_2O + \overset{(0)}{Cl_2}.$$

loses $1e^-$

Observe: (1) The gain or loss of electrons is given in terms of the expressed formula. The two Cr atoms gain $3e^-$ each, or a total of $6e^-$. (2) Some of the Cl^- ions from HCl form free Cl_2, while the Cl^- ions forming KCl and $CrCl_3$ have not changed their oxidation state. Then:

$$K_2Cr_2O_7 + 6\,HCl + x\,HCl \rightarrow 2\,KCl + 2\,CrCl_3 + y\,H_2O + 3\,Cl_2.$$

The arrows indicate the origin of the Cl^- and Cl_2 in the products. Evidently $x = 8$, and $y = 7$. Collecting the HCl molecules gives:

$$K_2Cr_2O_7 + 14\,HCl \rightarrow 2\,KCl + 2\,CrCl_3 + 7\,H_2O + 3\,Cl_2.$$

Again, writing in the ionic form and canceling, we have:

$$\cancel{2\,K^+} + Cr_2O_7^{-2} + 14\,H^+ + \overset{6}{\cancel{14}}\,Cl^- \rightarrow$$
$$\cancel{2\,K^+} + \cancel{2\,Cl^-} + 2\,Cr^{+3} + \cancel{6\,Cl^-} + 7\,H_2O + 3\,Cl_2$$

or $\qquad Cr_2O_7^{-2} + 14\,H^+ + 6\,Cl^- \rightarrow 2\,Cr^{+3} + 7\,H_2O + 3\,Cl_2.$

Example 14.5. Balance the reaction:

$$Na_2TeO_3 + NaI + HCl \rightarrow NaCl + H_2O + Te + I_2.$$

Solution.

gains $4e^-$

$$\overset{(+4)}{Na_2TeO_3} + NaI + HCl \rightarrow NaCl + H_2O + \overset{(0)}{Te} + I_2.$$

loses $1e^-$

$$Na_2TeO_3 + 4\,NaI + x\,HCl \rightarrow 6\,NaCl + y\,H_2O + Te + 2\,I_2.$$

$x = 6$ and $y = 3$.

$$Na_2TeO_3 + 4\ NaI + 6\ HCl \rightarrow 6\ NaCl + 3\ H_2O + Te + 2\ I_2.$$

And

$$\cancel{2\ Na^+} + TeO_3^{-2} + \cancel{4\ Na^+} + 4\ I^- + 6\ H^+ + \cancel{6\ Cl^-} \rightarrow \cancel{6\ Na^+} + \cancel{6\ Cl^-} + 3\ H_2O + Te + 2\ I_2.$$

or $$TeO_3^{-2} + 4\ I^- + 6\ H^+ \rightarrow 3\ H_2O + Te + 2\ I_2.$$

14.4. By the Ion-Electron Method. In the ion-electron method only the molecules and ions which participate in the chemical change are shown. There are four steps to follow in writing oxidation-reduction equations by the ion-electron method.

Step 1. Pick out the oxidant and the reductant molecule or ion and their products.

Step 2. Write the partial equation for each, balancing in terms of both atoms and electrons.

Step 3. Multiply each partial equation by a number such that there are the same number of electrons in each partial equation.

Step 4. Add the two partial equations and simplify by canceling molecules and ions. The four equations given in the preceding section will now be written using the ion-electron method.

TABLE 13.1

SOME OXIDANTS AND REDUCTANTS AND THEIR PRODUCTS

Oxidants	*Reductants*	*Oxidant* ⇆ *Reductant*
$PbO_2 \rightarrow Pb^{+2}$	$NO_2^- \rightarrow NO_3^-$	$I_2 \leftrightarrows I^-$
$MnO_2 \rightarrow Mn^{+2}$	$AsO_3^{-3} \rightarrow AsO_4^{-3}$	$Cl_2 \leftrightarrows Cl^-$
$MnO_4^- \rightarrow Mn^{+2}$	$H_2O_2 \rightarrow H^+ + O_2$	$Br_2 \leftrightarrows Br^-$
$NO_3^- \rightarrow NO$	$SO_3^{-2} \rightarrow SO_4^{-2}$	$Fe^{+3} \leftrightarrows Fe^{+2}$
$NO_3^- \rightarrow NO_2$	$H_2S \rightarrow S$	$Sn^{+4} \leftrightarrows Sn^{+2}$
$Cr_2O_7^{-2} \rightarrow Cr^{+3}$	$S^{-2} \rightarrow SO_4^{-2}$	$Hg^{+2} \leftrightarrows Hg^+$
$ClO_3^- \rightarrow Cl^-$	$TeO_3^- \rightarrow Te$	M (*-ic*) ⇆ M (*-ous*)
$ClO_2^- \rightarrow Cl^-$	$AsO_2^- \rightarrow AsO_4^{-3}$	

Table 13.1 lists a number of oxidants and reductants and their products. Theoretically, any oxidant and any reductant could be combined to give an oxidation-reduction equation. Refer to Table 13.1 when studying the examples that follow. Also, the table should be used as a reference when working the problems at the end of the chapter.

Example 14.6. Write the equation for the reaction between HNO_3 and H_2S.

Solution.

Step 1. Oxidant: $NO_3^- \rightarrow NO$.
Reductant: $H_2S \rightarrow S$.

Since H_2S exists predominantly in the molecular state, the molecule H_2S is used as the reductant rather than the sulfide ion, S^{-2}.

Step 2. For the oxidant, $NO_3^- \rightarrow NO$, the 2 remaining oxygen atoms require 4 H^+ to form H_2O. Then, $NO_3^- + 4\ H^+ \rightarrow NO + 2\ H_2O$. To balance in terms of electrons, $3e^-$ are needed on the reactant side. Then:

$$3e^- + NO_3^- + 4\ H^+ \rightarrow NO + 2\ H_2O. \qquad (1)$$

For the reductant, $H_2S \rightarrow S$:

$$H_2S \rightarrow S + 2\ H^+.$$

To balance in terms of electrons:

$$H_2S \rightarrow S + 2\ H^+ + 2e^-. \qquad (2)$$

Step 3. Multiply equation (1) by 2, and equation (2) by 3. Then:

$$6e^- + 2\ NO_3^- + 8\ H^+ \rightarrow 2\ NO + 2\ H_2O \qquad (3)$$

and

$$3\ H_2S \rightarrow 3\ S + 6\ H^+ + 6e^-. \qquad (4)$$

Step 4. Add equations (3) and (4) and cancel terms.

$$\cancel{6e^-} + 2\ NO_3^- + \cancelto{2}{8}\ H^+ + 3\ H_2S \rightarrow 2\ NO + 2\ H_2O + 3\ S + \cancel{6\ H^+} + \cancel{6e^-}$$

or

$$2\ NO_3^- + 2\ H^+ + 3\ H_2S \rightarrow 2\ NO + 2\ H_2O + 3\ S.$$

Compare the above equation to the one obtained in Example 14.2.

Example 14.7. Balance the reaction:

$$KMnO_4 + FeSO_4 + H_2SO_4 \rightarrow K_2SO_4 + MnSO_4 + Fe_2(SO_4)_3 + H_2O.$$

Solution.

Step 1. Oxidant: $MnO_4^- \rightarrow Mn^{+2}$.
Reductant: $Fe^{+2} \rightarrow Fe^{+3}$.

Step 2.

$$MnO_4^- + 8\ H^+ + 5e^- \rightarrow Mn^{+2} + 4\ H_2O. \qquad (5)$$

Observe that, in terms of electrons:

$$(1-) + (8+) + (5-) \rightarrow (2+)$$

or

$$(2+) \rightarrow (2+).$$

$$Fe^{+2} \rightarrow Fe^{+3} + 1e^-. \qquad (5a)$$

Step 3. Multiply equation (5a) by 5. Then:

$$5\ Fe^{+2} \rightarrow 5\ Fe^{+3} + 5e^-. \quad (6)$$

Step 4. Add equations (5) and (6). Then:

$$MnO_4^- + 8\ H^+ + \cancel{5e^-} \rightarrow Mn^{+2} + 4\ H_2O$$
$$5\ Fe^{+2} \rightarrow 5\ Fe^{+3} + \cancel{5e^-}$$

Add: $$MnO_4^- + 8\ H^+ + 5\ Fe^{+2} \rightarrow Mn^{+2} + 4\ H_2O + 5\ Fe^{+3}.$$

Again, the equation is correct in terms of electrons.

$$(1-) + (8+) + (10+) \rightarrow (2+) + (15+)$$

or

$$(17+) \rightarrow (17+).$$

Compare the above equation to the one obtained in Example **14.3.**

Example 14.8. Balance the reaction:

$$K_2Cr_2O_7 + HCl \rightarrow KCl + CrCl_3 + H_2O + Cl_2.$$

Solution.

Step 1. $Cr_2O_7^{-2} \rightarrow Cr^{+3}$.

$Cl^- \rightarrow Cl_2$.

Step 2. $Cr_2O_7^{-2} + 14\ H^+ + 6e^- \rightarrow 2\ Cr^{+3} + 7\ H_2O$.

$2\ Cl^- \rightarrow Cl_2 + 2e^-$.

Step 3. $6\ Cl^- \rightarrow 3\ Cl_2 + 6e^-$.

Step 4.

$$Cr_2O_7^{-2} + 14\ H^+ + \cancel{6e^-} \rightarrow 2\ Cr^{+3} + 7\ H_2O$$
$$6\ Cl^- \rightarrow 3\ Cl_2 + \cancel{6e^-}$$
$$Cr_2O_7^{-2} + 14\ H^+ + 6\ Cl^- \rightarrow 2\ Cr^{+3} + 7\ H_2O + 3\ Cl_2.$$

The equation obtained in Example 14.4 is identical to the above. This may be shown by writing the equation from Example 14.4 in ionic form and canceling like terms. Then:

$$K_2Cr_2O_7 + 14\ HCl \rightarrow 2\ KCl + 2\ CrCl_3 + 7\ H_2O + 3\ Cl_2.$$

$$\cancel{2K^+} + Cr_2O_7^{-2} + 14\ H^+ + \overset{6}{\cancel{14}}\ Cl^- \rightarrow \cancel{2K^+} + \cancel{2Cl^-} + 2\ Cr^{+3} + \cancel{6Cl^-} + 7\ H_2O + 3\ Cl_2.$$

or $$Cr_2O_7^{-2} + 14\ H^+ + 6\ Cl^- \rightarrow 2\ Cr^{+3} + 7\ H_2O + 3\ Cl_2.$$

The ions canceled served no function in the oxidation-reduction process.

Example 14.9. Balance the reaction:

$$Na_2TeO_3 + NaI + HCl \rightarrow NaCl + H_2O + Te + I_2.$$

Solution.

Step 1. $TeO_3^{-2} \rightarrow Te$.

$I^- \rightarrow I_2$.

Step 2. $TeO_3^{-2} + 6\ H^+ + 4e^- \rightarrow Te + 3\ H_2O$.

$2\ I^- \rightarrow I_2 + 2e^-$.

Step 3. $4\ I^- \rightarrow 2\ I_2 + 4e^-$.

Step 4. $TeO_3^{-2} + 6\ H^+ + \cancel{4e^-} + 4\ I^- \rightarrow Te + 3\ H_2O + 2\ I_2 + \cancel{4e^-}$

or $TeO_3^{-2} + 6\ H^+ + 4\ I^- \rightarrow Te + 3\ H_2O + 2\ I_2$.

Problems

Part I

14.1. In the rection

$$KClO_3 + Na_2SnO_2 \rightarrow KCl + Na_2SnO_3$$

a. What atoms undergo a change in oxidation number?
b. What ions undergo chemical change?
c. What atoms lose electrons?
d. What atoms gain electrons?
e. What is oxidized? What is reduced?
f. What is the oxidant? What is the reductant?
g. Write the equation using (1) the electron transfer method, and (b) the ion-electron method.

14.2. Given:

$$KMnO_4 + HCl \rightarrow KCl + MnCl_2 + H_2O + Cl_2.$$

a. Write the equation using the electron transfer method.
b. Write the equation using the ion-electron method.
c. Show that the two are identical equations.

14.3. Balance the following in terms of atoms and electrons:
a. $Sn^{+2} \rightarrow Sn^{+4}$.
b. $I^- \rightarrow I_2$.
c. $H_2O_2 \rightarrow O_2 + H^+$.
d. $H_2O + Mn^{+2} \rightarrow MnO_2 + H^+$.
e. $ClO_3^- + H^+ \rightarrow Cl_2 + H_2O$.
f. $ClO_3^- + H^+ \rightarrow Cl^- + H_2O$.

14.4. Balance the following reactions using the electron transfer method:
a. $PbO_2 + HCl \rightarrow PbCl_2 + Cl_2 + H_2O$.
b. $CuS + HNO_3 \rightarrow Cu(NO_3)_2 + S + H_2O + NO$.
c. $KMnO_4 + Na_2SO_4 + H_2O \rightarrow MnO_2 + Na_2SO_4 + KOH$.
d. $K_2Cr_2O_7 + NO_2 + HNO_3 \rightarrow KNO_3 + Cr(NO_3)_3 + H_2O$.

14.5. Balance the following reactions using the ion-electron method:
a. $Cu + HNO_3 \rightarrow Cu(NO_3)_2 + H_2O + NO$.
b. $Cu + HNO_3 \rightarrow Cu(NO_3)_2 + H_2O + NO_2$.
c. $KMnO_4 + H_2SO_4 + H_2S \rightarrow K_2SO_4 + MnSO_4 + H_2O + S$.
d. $Zn + NaOH + NaNO_3 \rightarrow Na_2ZnO_2 + NH_3 + H_2O$.

14.6. What is the oxidation number of the underscored atoms in each of the following formulas?
a. $Cr_2(\underline{S}O_4)_3$.
b. $Ca(\underline{Cl}O)_2$.

c. PO_3^{-3}.
d. PO_4^{-3}.
e. $K_2Cr_2O_7$.

Part II

14.7. Balance the reactions in problem 14.4 using the ion-electron method.
14.8. Balance the reactions in problem 14.5 using the electron transfer method.
14.9. Balance the following in terms of atoms and electrons:
a. $Cr^{+6} \rightarrow Cr^{+3}$.
b. $CrO_4^{-2} + H^+ \rightarrow Cr^{+3} + H_2O$.
c. $BrO_3^- + H^+ \rightarrow Br_2 + H_2O$.
d. $MnO_2 + H^+ \rightarrow Mn^{+2} + H_2O$.
e. $NO_3^- + H^+ \rightarrow NO + H_2O$.
f. $NO_3^- + H^+ \rightarrow NO_2 + H_2O$.
g. $N^{+5} \rightarrow N^{-3}$.
h. $S_2O_3^{-2} \rightarrow S_4O_6^{-2}$.
i. $N^{+5} \rightarrow N^{+3}$.
j. $I_2 + H_2O \rightarrow IO_3^- + H^+$.
14.10. Balance the following reactions (1) by the electron transfer method, (2) by the ion-electron method, and (3) show that in each case the two equations are identical:
a. $As_2S_5 + HNO_3 \rightarrow H_3AsO_4 + H_2SO_4 + H_2O + NO_2$.
b. $CdS + I_2 + HCl \rightarrow CdCl_2 + HI + S$.
c. $CrI_3 + KOH + Cl_2 \rightarrow K_2CrO_4 + KIO_4 + KCl + H_2O$.
d. $HNO_3 + I_2 \rightarrow HIO_3 + NO_2 + H_2O$.
14.11. Convert the following to the molecular form; then write the equation by the electron transfer method:
a. $Cr_2O_7^{-2} + H^+ + H_2SO_3 \rightarrow Cr^{+3} + H_2SO_4 + H_2O$.
b. $NO_3^- + H^+ + Cl^- \rightarrow NO + H_2O + Cl_2$.
c. $Zn + NO_3^- + H^+ \rightarrow Zn^{+2} + NH_3 + H_2O$.
d. $Cu + NO_3^- + H^+ \rightarrow Cu^{+2} + NO + H_2O$.
14.12. Balance the following reactions by the ion-electron method:
a. $Zn + HNO_3 \rightarrow Zn(NO_3)_2 + NH_4NO_3 + H_2O$.
b. $KMnO_4 + H_2O_2 + H_2SO_4 \rightarrow KHSO_4 + MnSO_4 + H_2O + O_2$.
c. $Ag + HNO_3 \rightarrow AgNO_3 + NO + H_2O$.
d. $Ag + HNO_3 \rightarrow AgNO_3 + NO_2 + H_2O$.
14.13. Balance the following reactions by the ion-electron method:
a. $MnO_4^- + H_2O_2 + H^+ \rightarrow Mn^{+2} + H_2O + O_2$.
b. $S^{-2} + H_2O_2 \rightarrow SO_4^{-2} + H_2O$.
c. $MnO_2 + H^+ \rightarrow Mn^{+2} + H_2O + Cl_2$.
d. $S_2O_3^{-2} + I_2 \rightarrow S_4O_6^{-2} + I^-$.

14.14. Balance the reactions in problem 14.12 using the electron transfer method.

14.15. Balance the reactions in problem 14.13 using the electron transfer method.

14.16. Show how the equation in Example 14.6 may be obtained from the equation in Example 14.2.

14.17. Show how the equation in Example 14.7 may be obtained from the equation in Example 14.3.

14.18. Write the partial equations (half-reactions) for each of the following.

a. $Al + CuSO_4 \rightarrow Al_2(SO_4)_3 + Cu$.
b. $I + H_2S \rightarrow HI + S$.
c. $FeCl_3 + SnCl_2 \rightarrow FeCl_2 + SnCl_4$.
d. $FeCl_3 + H_2S \rightarrow FeCl_2 + HCl + S$.

14.19. Balance the following reactions:

a. $Bi(OH)_3 + Na_2SnO_2 \rightarrow Na_2SnO_3 + H_2O + Bi$.
b. $Br_2 + SO_2 + H_2O \rightarrow H_2SO_4 + HBr$.
c. $H_2SO_3 + HNO_3 \rightarrow H_2SO_4 + H_2O + NO$.
d. $KClO_3 + Na_2SnO_2 \rightarrow KCl + Na_2SnO_3$.

14.20. Balance the following reactions:

a. $KClO_3 + H_2SO_4 + FeSO_4 \rightarrow Fe_2(SO_4)_3 + KCl + H_2O$.
b. $KClO_3 + HCl + SnCl_2 \rightarrow SnCl_4 + KCl + H_2O$.
c. $Na_2SO_3 + Br_2 + H_2O \rightarrow Na_2SO_4 + HBr$.
d. $KMnO_4 + HCl + FeCl_2 \rightarrow FeCl_3 + MnCl_2 + KCl + H_2O$.

14.21. Balance the following reactions:

a. $Cr_2O_7^{-2} + I^- + H^+ \rightarrow Cr^{+3} + H_2O + I_2$.
b. $CrO_4^{-2} + S^{-2} + H^+ \rightarrow Cr^{+3} + H_2O + S$.
c. $IO_3^- + I^- + H^+ \rightarrow H_2O + I_2$.
d. $MnO_4^- + H_2O_2 + H^+ \rightarrow Mn^{+2} + H_2O + O_2$.

14.22. Given the following half-reactions, of which I and II are oxidation half-reactions and III and IV are reduction half-reactions:

I. $H_2S \rightarrow S + 2\,H^+ + 2\,e^-$.
II. $SO_3^{-2} + H_2O \rightarrow SO_4^{-2} + 2\,H^+ + 2\,e^-$.
III. $CrO_4^{-2} + 4\,H_2O + 3\,e^- \rightarrow Cr(OH)_4^- + 4\,OH^-$.
IV. $ClO_3^- + 6\,H^+ + 6\,e^- \rightarrow Cl^- + 3\,H_2O$.

Write the oxidation-reduction reactions involving the following pairs of the above half-reactions:

a. I and III.
b. I and IV.
c. II and III.
d. II and IV.

15

Solutions

Many chemical reactions are carried out in water solution, or under conditions in which the reactants are in the liquid state. In solution the interacting particles mix readily and can easily be handled. In order to treat solutions quantitatively it is necessary to have standards for representing the strengths of solutions. Any solution in which the amounts of solute and solvent are known is a *standard solution.* A number of methods for expressing the strengths of solutions have been discussed, such as percentage by *weight* and by *volume.* Four other methods for expressing the strengths of solutions will be discussed. They are (1) *mole fraction*, (2) *molarity*, (3) *molality*, and (4) *normality*.

Standard Solutions in Terms of Mole Quantities

15.1. Mole Fraction. *Mole fraction*, N_x, is defined as the fractional part of the total number of moles in a solution contributed by the component x of the solution.

Example 15.1. A solution was prepared by mixing 10 moles of alcohol, C_2H_5OH, and 10 moles of water. What is the mole fraction composition of the solution? *Note:* 10 moles of alcohol would be $10 \times 46 = 460$ g., and 10 moles of water would be $10 \times 18 = 180$ g.

Solution. By definition:

$$\text{Mole fraction of alcohol} = N_{C_2H_5OH} = \frac{10}{10 + 10} = 0.50$$

and

$$\text{Mole fraction of water} = N_{H_2O} = \frac{10}{10 + 10} = 0.50.$$

That is, in terms of mole fraction composition the solution is 0.50 alcohol and 0.50 water.

Observe that the sum of the mole fractions representing the components of a given solution equals one. In Example 15.1:

$$N_{C_2H_5OH} + N_{H_2O} = 0.50 + 0.50 = 1.00.$$

Mole fraction is an important method of expressing the composition of a solution since it represents the ratio of particles of atomic or molecular dimensions in the solution. In the solution given in Example 15.1, the ratio of alcohol to water molecules is 1:1.

Example 15.2. A given solution contains 100 g of salt, NaCl, and 900 g of water. What are the mole fractions of the components of the solution?

Solution. Since the unit of concentration must be the mole, each of the given quantities of the components must be changed to moles. Then:

$$100 \text{ g. NaCl} = 100 \not{g} \div 58.45 \frac{\not{g}}{\text{mole}} = \textbf{1.71 moles of NaCl.}$$

$$900 \text{ g. } H_2O = 900 \not{g} \div 18.02 \frac{\not{g}}{\text{mole}} = \textbf{49.94 moles of } \mathbf{H_2O.}$$

$$\text{Total} = \overline{\textbf{51.65 moles.}}$$

Therefore, by definition:

$$\text{Mole fraction of NaCl} = N_{NaCl} = \frac{1.71}{51.65} = 0.0331.$$

$$\text{Mole fraction of } H_2O = N_{H_2O} = \frac{49.94}{51.65} = 0.9669.$$

$$N_{NaCl} + N_{H_2O} = 1.0000.$$

Example 15.3. Given that $N_{KOH} = 0.100$ for a water solution of potassium hydroxide, how many grams of KOH are there in 25.0 g of the solution?

Solution.

Since $N_{KOH} + N_{H_2O} = 1.000,$

then $N_{H_2O} = 1.000 - 0.100 = 0.900.$

Let $x =$ g KOH in 25.0 g of solution.

Then $25.0 - x =$ g H_2O in 25.0 g of solution.

We can now set up the expression for the mole fraction of either KOH or H_2O. For KOH:

$$\frac{\frac{x}{56.1} \cancel{\text{mole KOH}}}{\frac{x}{56.1} \cancel{\text{mole KOH}} + \frac{25.0 - x}{18.0} \text{ mole } H_2O} = 0.100 \text{ mole fraction of KOH}$$

and: $x = 6.44$ g KOH in 25.0 g of solution.

15.2. Molarity. *Molarity*, M, is defined as the number of moles of solute per liter of solution. That is:

$$\text{Molarity} = \frac{\text{moles of solute}}{\text{liters of solution}}.$$

As with "mole fraction," the molarity of a solution may be used to determine the ratio of ions or molecules of solute to molecules of water in any given solution. The solvent need not be water. However, unless otherwise stated, the discussion will be limited to water solutions.

Example 15.4. Calculate the molarity of a solution containing 441 g of HCl dissolved in sufficient water to make 1500 ml of solution.

Solution. As shown in the above formula, the amount of solute must be given in moles, and the volume of solution in liters.

Then:

$$\frac{441\ \cancel{\text{g HCl}}}{36.46\ \dfrac{\cancel{\text{g HCl}}}{\text{mole HCl}}} = 12.1 \text{ mole HCl (solute)}$$

and 1500 ml of solution = 1.50 l of solution.

Therefore, the molarity is:

$$\frac{12.1 \text{ moles HCl}}{1.50\ l \text{ solution}} = 8.07\ \frac{\text{moles HCl}}{l \text{ solution}} = 8.07\text{M HCl}.$$

Example 15.5. How many grams of NaOH would be required to prepare 5 liters of 0.100M solution?

Solution. In the above equation there are three variables—molarity, moles of solute, and liters of solution. In this problem the unknown variable is "moles of solute." Moles and grams are readily interchangeable. Solving the above formula for "moles of solute" we have:

$$\text{Moles of solute} = (\text{molarity})(\text{liters of solution}).$$

Since the dimensions of molarity are $\dfrac{\text{moles solute}}{l \text{ solution}}$, then:

$$0.100\ \frac{\text{mole NaOH}}{\cancel{l \text{ solution}}} \times 5.00\ \cancel{l \text{ solution}} = 0.500 \text{ mole NaOH (solute)}.$$

Since there are 40.0 g of NaOH in 1.00 mole, then:

$$0.500\ \cancel{\text{mole NaOH}} \times 40.0\ \frac{\text{g NaOH}}{\cancel{\text{mole NaOH}}} = 20.0 \text{ g NaOH}.$$

Example 15.6. How many grams of $Al_2(SO_4)_3$ are there in 300 ml of 1.50M solution?

Solution. As in Example 15.5, we will determine moles of solute, then change moles to grams.

$$1.50 \frac{\text{moles } Al_2(SO_4)_3}{\cancel{l \text{ solution}}} \times 0.300 \cancel{l \text{ solution}} = 0.450 \text{ mole } Al_2(SO_4)_3.$$

Since there are 342.1 g of $Al_2(SO_4)_3$ in 1.00 mole, then:

$$0.450 \cancel{\text{mole } Al_2(SO_4)_3} \times 341.2 \frac{\text{g } Al_2(SO_4)_3}{\cancel{\text{mole } Al_2(SO_4)_3}} = 154 \text{ g } Al_2(SO_4)_3.$$

15.3. Molality. *Molality*, m, is defined as the number of moles of solute per 1000 grams of solvent. That is:

$$\text{Molality} = \frac{\text{moles of solute}}{\text{kilograms of solvent}}.$$

Example 15.7. Calculate the molality of a solution containing 441 g of HCl dissolved in 1500 g of water. Compare with Example 15.4.

Solution. The amount of solute must be given in moles, and the amount of solvent in kilograms. (See the solution to Example 15.4.)

$$441 \text{ g HCl} = 12.1 \text{ moles HCl}$$

and:

$$1500 \text{ g } H_2O = 1.50 \text{ kg } H_2O.$$

The molality would be:

$$\frac{12.1 \text{ moles HCl}}{1.50 \text{ kg } H_2O} = 8.07 \frac{\text{moles HCl}}{\text{kg } H_2O} = 8.07\text{m HCl}.$$

In Examples 15.4 and 15.7 the molarity (M) and molality (m) are numerically the same. However, note the difference. In Example 15.4 the solution was prepared by dissolving 441 g. of HCl in sufficient water to make a total volume of 1500 ml. In Example 15.7, the 441 g. of HCl were dissolved in 1500 g. of water, resulting in a volume of more than 1500 ml.

Example 15.8. How many grams of NaOH would have to be added to 5000 g. of water in order to prepare a 0.100m solution of NaOH?

Solution. Solving the above formula for the unknown, moles of solute, gives:

$$\text{Moles of solute} = (\text{molality})(\text{kilograms of solvent}).$$

or

$$= (0.100)(5) = 0.500.$$

That is, 0.500 mole of NaOH would be required, and 0.500 mole NaOH $= 0.500 \times 40.0 = 20.0$ g. of NaOH. That is, 20.0 g. of NaOH dissolved in 5000 g. of water will give a 0.100m solution of NaOH.

$$5000 \text{ g } H_2O = 5.00 \text{ kg } H_2O \text{ (solvent)}.$$

The dimensions of molality are $\frac{\text{moles}}{\text{kg solvent}}$. Then:

$$0.100 \frac{\text{mole NaOH}}{\cancel{\text{kg H}_2\text{O}}} \times 5.00 \cancel{\text{kg H}_2\text{O}} = 0.500 \text{ mole NaOH (solute).}$$

Since there are 40.0 g of NaOH in 1.00 mole, then:

$$0.500 \cancel{\text{mole NaOH}} \times 40.0 \frac{\text{g NaOH}}{\cancel{\text{mole NaOH}}} = 20.0 \text{ g NaOH.}$$

Standard Solutions in Terms of Gram-equivalent Quantities

15.4. Gram-equivalent Weights of Acids, Bases, and Salts. The gram-equivalent weights of elements were discussed in Chapter 9. We will now apply the concept of gram-equivalent weight to acids, bases, and salts. As in Chapter 9, one gram-equivalent weight of an acid, base, or salt represents the number of grams of the compound that will involve an exchange of the Avogadro number of electrons (6.02×10^{23}). This statement can be made more practical by stating that one gram-equivalent weight of an acid, base, or salt is equal numerically to the formula weight of the compound divided by either the net positive or the net negative charges represented by the formula (Table 15.1). That is:

$$1.00 \text{ g -equivalent weight} = \frac{\text{formula weight}}{\text{charge}}.$$

TABLE 15.1

THE GRAM-EQUIVALENT WEIGHTS OF A NUMBER OF SUBSTANCES

Compound	*Ions*	*Ionic Charge*	1.00 *Mole* (g)	1.00 *Gram-equivalent Weight*
HCl	$H^+ + Cl^-$	1	36.46	36.46/1 = 36.46 g.
H_2SO_4	$2\ H^+ + SO_4^{-2}$	2	98.08	98.08/2 = 49.04 g.
$NaOH$	$Na^+ + OH^-$	1	40.00	40.00/1 = 40.00 g.
$Ca(OH)_2$	$Ca^{+2} + 2\ OH^-$	2	74.10	74.10/2 = 37.05 g.
K_3PO_4	$3\ K^+ + PO_4^{-3}$	3	212.3	212.3/3 = 70.77 g.
$Al_2(SO_4)_3$	$2\ Al^{+3} + 3\ SO_4^{-2}$	6	342.1	342.1/6 = 57.02 g.

Example 15.9. How many grams are there in 1.65 gram-equivalent weights of $Fe(NO_3)_3$?

Solution. Since $Fe(NO_3)_3$ ionizes to give Fe^{+3} and $3\ NO_3^-$, then:

$$\frac{241.86 \frac{\text{g Fe(NO}_3)_3}{\cancel{\text{mole Fe(NO}_3)_3}}}{3 \frac{\text{g-eq. wt. Fe(NO}_3)_3}{\cancel{\text{mole Fe(NO}_3)_3}}} = 80.62 \frac{\text{g Fe(NO}_3)_3}{\text{g-eq. wt. Fe(NO}_3)_3}.$$

That is, 1.00 g-eq. wt. $Fe(NO_3)_3$ = 80.62 g and 1.65 g-eq. wt. of $Fe(NO_3)_3$ would be:

$$1.65\ \cancel{\text{g-eq. wt. Fe(NO}_3)_3} \times 80.62\ \frac{\text{g Fe(NO}_3)_3}{\cancel{\text{g-eq. wt. Fe(NO}_3)_3}} = 133\ \text{g Fe(NO}_3)_3.$$

It is sometimes more convenient to use smaller units than gram-equivalent weight units. For this reason ***milligram-equivalent weight*** units are sometimes used. For convenience, this smaller unit is frequently abbreviated *milliequivalent.* A milliequivalent is 0.001 part of a gram-equivalent weight. For convenience, the word milliequivalent will be abbreviated meq.

Example 15.10. How many milliequivalents are there in 1.36 g of $Ca(OH)_2$?

Solution. 1.00 g-eq. wt. of $Ca(OH)_2$ would be:

$$\frac{74.10\ \dfrac{\text{g Ca(OH)}_2}{\cancel{\text{mole Ca(OH)}_2}}}{2\ \dfrac{\text{g-eq. wt. Ca(OH)}_2}{\cancel{\text{mole Ca(OH)}_2}}} = 37.05\ \frac{\text{g Ca(OH)}_2}{\text{g-eq. wt. Ca(OH)}_2}$$

and 1.36 g of $Ca(OH)_2$ would be:

$$\frac{1.36\ \cancel{\text{g Ca(OH)}_2}}{37.05\ \dfrac{\cancel{\text{g Ca(OH)}_2}}{\text{g-eq. wt. Ca(OH)}_2}} = 0.0367\ \text{g-eq. wt. Ca(OH)}_2.$$

Since 1.00 g-eq. wt. = 1000 meq., then:

0.0367 g-eq. wt. $Ca(OH)_2$ = 36.7 meq. $Ca(OH)_2$.

Observe that grams per gram-equivalent weight is equal numerically to milligrams per milliequivalent. For example:

1.00 gram-equivalent weight of $Ca(OH)_2$ = 37.05 g

and 1.00 milliequivalent of $Ca(OH)_2$ = 37.05 mg.

Ions in solution may be considered in terms of gram-equivalent quantities. For example:

$$1.00 \text{ gram-equivalent weight of } SO_4^{-2} = \frac{96.06}{2} = 48.03 \text{ g}$$

and

$$1.00 \text{ gram-equivalent weight of } Al^{+3} = \frac{26.98}{3} = 8.99 \text{ g.}$$

In any solution prepared by adding an acid, base, or salt to water there are equal gram-equivalent quantities of + and − ions.

Example 15.11. How many gram-equivalent weights each of Mg^{+2} and Cl^- ions would there be in a solution prepared by dissolving 23.81 g of $MgCl_2$ in water, assuming complete ionization of the salt?

Solution. Following is an interpretation of the ionization equation for $MgCl_2$ in terms of gram-equivalent quantities.

	$MgCl_2$	$\rightarrow$	Mg^{+2}	$+$	$2\ Cl^-$
	95.22 g		24.31 g		70.91 g
or	2.00 g -eq.wt.		2.00 g -eq.wt.		2.00 g -eq.wt.

When a salt ionizes completely in water, equal gram-equivalent quantities of ions and salt are involved. Since 23.81 g. of $MgCl_2$ is 0.500 gram-equivalent weight of $MgCl_2$, there will be formed 0.500 gram-equivalent weight each of Mg^{+2} and Cl^- ions.

15.5. Gram-equivalent Weights of Oxidants and Reductants. The gram-equivalent weight value of an acid, base, or salt depends upon its role as a reactant in a particular reaction. One gram-equivalent weight of an oxidant or reductant is equal to the formula weight divided by the sum of the changes in oxidation number of the atoms in the formula of the oxidant or reductant. That is:

$$1.00 \text{ gram-equivalent weight} = \frac{\text{formula weight}}{\text{change in oxidation number}}.$$

Example 15.12. How many grams are there in 1.00 gram-equivalent weight of both oxidant and reductant in each of the following reactions?

(a) $HNO_3 + H_2S \rightarrow H_2O + NO + S$.

(b) $HNO_3 + Cu \rightarrow Cu(NO_3)_2 + NO_2 + O_2$.

(c) $KMnO_4 + FeSO_4 + H_2SO_4 \rightarrow K_2SO_4 + MnSO_4 + Fe_2(SO_4)_3 + H_2O$.

(d) $K_2Cr_2O_7 + HCl \rightarrow KCl + CrCl_3 + H_2O + Cl_2$.

Solution. Only the reactants and products need be known. It is not necessary to write the equation. Also, the answers may be obtained using either the molecular or ionic expressions as given in Chapter 14.

(a) Oxidant: $\overset{(+5)}{HNO_3} \rightarrow \overset{(+2)}{NO}$ or $\overset{(+5)}{NO_3^-} \rightarrow \overset{(+2)}{NO}$. Therefore:

$$1.00 \text{ g -eq. wt. of } HNO_3 = \frac{63.02}{3} = 21.01 \text{ g.}$$

Reductant: $\overset{(-2)}{H_2S} \rightarrow \overset{(0)}{S}$. Therefore:

$$1.00 \text{ g -eq. wt. of } H_2S = \frac{34.09}{2} = 17.05 \text{ g.}$$

(b) Oxidant: $\overset{(+5)}{HNO_3} \rightarrow \overset{(+4)}{NO_2}$ or $\overset{(+5)}{NO_3^-} \rightarrow \overset{(+4)}{NO_2}$. Therefore:

$$1.00 \text{ g-eq. wt. of } HNO_3 = \frac{63.02}{1} = 63.02 \text{ g.}$$

Reductant: $\overset{(0)}{Cu} \rightarrow \overset{(+2)}{Cu}(NO_3)_2$ or $\overset{(0)}{Cu} \rightarrow \overset{(+2)}{Cu}$. Therefore:

$$1.00 \text{ g-eq. wt. of Cu} = \frac{63.54}{2} = 31.77 \text{ g.}$$

(c) Oxidant: $K\overset{(+7)}{Mn}O_4 \rightarrow \overset{(+2)}{Mn}SO_4$ or $\overset{(+7)}{Mn}O_4^- \rightarrow \overset{(+2)}{Mn}$. Therefore:

$$1.00 \text{ g-eq. wt. of } KMnO_4 = \frac{158.04}{5} = 31.61 \text{ g.}$$

Reductant: $\overset{(+2)}{Fe}SO_4 \rightarrow \overset{(+3)}{Fe_2}(SO_4)_3$ or $\overset{(+2)}{Fe} \rightarrow \overset{(+3)}{Fe}$. Therefore:

$$1.00 \text{ g-eq. wt. of } FeSO_4 = \frac{151.92}{1} = 151.92 \text{ g.}$$

(d) Oxidant: $K_2\overset{2(+6)}{Cr_2}O_7 \rightarrow \overset{2(+3)}{Cr}Cl_3$ or $\overset{2(+6)}{Cr_2}O_7^{-2} \rightarrow \overset{2(+3)}{Cr}$. Therefore:

$$1.00 \text{ g-eq. wt. of } K_2Cr_2O_7 = \frac{294.21}{6} = 49.04 \text{ g.}$$

Reductant: $H\overset{(-1)}{Cl} \rightarrow \overset{(0)}{Cl_2}$. or $\overset{(-1)}{Cl^-} \rightarrow \overset{(0)}{Cl_2}$. Therefore:

$$1.00 \text{ g-eq. wt. of HCl} = \frac{36.47}{1} = 36.47 \text{ g.}$$

15.6. Normality. *Normality*, N, is defined as the number of gram-equivalent weights of the solute per liter of solution. That is:

$$\text{Normality} = \frac{\text{gram-equivalent weights of solute}}{\text{liters of solution}}.$$

Compare the above formula with the one given for molarity in Sec. 15.2.

If "gram-equivalent weights of solute" is represented by E, and "liters of solution" by V, the above equation becomes:

$$N = \frac{E}{V}.$$

Example 15.13. Sufficient water was added to 100 g. of NaOH to make one liter of solution. What was the normality of the solution?

Solution. In the formula $N = \frac{E}{V}$:

$$E = \frac{100\ \cancel{\text{g NaOH}}}{40\ \dfrac{\cancel{\text{g NaOH}}}{\text{g-eq. wt. NaOH}}} = 2.5\ \text{g-eq. wt. NaOH}$$

and V = 1.0 liter.

Therefore, the normality of the solution would be:

$$\frac{2.5\ \text{g-eq. wt. NaOH}}{1.0\ l\ \text{solution}} = 2.5N\ \text{NaOH}.$$

Example 15.14. How much $Ca(OH)_2$ must be dissolved in water to prepare five liters of 0.050N solution?

Solution. From the formula for normality we see that E = NV. Therefore the number of g-eq. wt. of $Ca(OH)_2$ would be:

$$0.050\ \frac{\text{g-eq. wt. Ca(OH)}_2}{\cancel{l\ \text{solution}}} \times 5.00\ \cancel{l\ \text{solution}} = 0.25\ \text{g-eq. wt. Ca(OH)}_2$$

and: $$0.25\ \cancel{\text{g-eq. wt. Ca(OH)}_2} \times 37.05\ \frac{\text{g Ca(OH)}_2}{\cancel{\text{g-eq. wt. Ca(OH)}_2}} = 9.26\ \text{g Ca(OH)}_2.$$

Example 15.15. What is the normality of a solution containing 35.0 g of KOH dissolved in sufficient water to make 400 ml of solution?

Solution. The g-eq. wt. in 35.0 g of KOH are:

$$\frac{35.0\ \cancel{\text{g KOH}}}{56.1\ \dfrac{\cancel{\text{g KOH}}}{\text{g-eq. wt. KOH}}} = 0.624\ \text{g-eq. wt. KOH}$$

and: $$\frac{0.624\ \text{g-eq. wt. KOH}}{0.400\ l\ \text{solution}} = 1.56N\ \text{KOH}.$$

Example 15.16. How many grams of $KMnO_4$ would be required to prepare 2500 ml of 0.010N solution, based on the reaction:

$$\overset{(+7)}{MnO_4^-} + Fe^{+2} \rightarrow Mn^{+2} + Fe^{+3}?$$

Solution. In the above equation manganese changes from +7 to +2 in oxidation number.

The number of g-eq. wt. of $KMnO_4$ required would be:

$$0.010\ \frac{\text{g-eq. wt. KMnO}_4}{\cancel{l\ \text{solution}}} \times 2.5\ \cancel{l\ \text{solution}} = 0.025\ \text{g-eq. wt. KMnO}_4.$$

Since 1.00 g-eq. wt. of $KMnO_4 = \frac{158}{5} = 31.6$ g,

then: $$0.025\ \cancel{\text{g-eq. wt. KMnO}_4} \times 31.6\ \frac{\text{g KMnO}_4}{\cancel{\text{g-eq. wt. KMnO}_4}} = 0.790\ \text{g KMnO}_4.$$

Example 15.17. How many milliequivalents of Na^+ ion are there in 25.0 ml of a solution containing 5.00 g of NaCl per liter?

Solution. Since 1.00 gram-equivalent weight = 1000 milliequivalents, we will first calculate the number of gram-equivalent weights of NaCl in 25.0 ml of the solution, then convert to milliequivalents.

The number of g of NaCl in 25.0 ml of the solution would be:

$$25.0\ \cancel{\text{ml solution}} \times \frac{5.00\text{ g NaCl}}{1000\ \cancel{\text{ml solution}}} = 0.125\text{ g NaCl.}$$

Since there are 58.5 g NaCl in 1.00 g-eq. wt., then:

$$\frac{0.125\ \cancel{\text{g NaCl}}}{58.5\ \dfrac{\cancel{\text{g NaCl}}}{\text{g-eq. wt. NaCl}}} = 0.00214\text{ g-eq. wt. NaCl}$$

and 0.00214 g-eq. wt. NaCl = 2.14 meq. of NaCl in 25.0 ml of solution.

Since 1.00 meq. of NaCl yields 1.00 meq. each of Na^+ and Cl^- ions, then 2.14 meq. of NaCl would yield 2.14 meq. of Na^+ ions.

Problems

Part I

15.1. What is the percentage composition by weight of the solution given in Example 15.1? *Ans.* 71.9% alcohol, 28.1% water.

15.2. What is the ratio of Na^+ ions to water molecules in Example 15.2? Assume that the NaCl is completely ionized. *Ans.* 1.71 : 49.94.

15.3. What is the mole fraction composition of a solution consisting of 250 g of sugar, $C_{12}H_{22}O_{11}$, and 400 g of water? *Ans.* Sugar = 0.032, water = 0.968.

15.4. What is the composition of the solution in Example 15.3 in terms of percentage by weight? *Ans.* 25.8% KOH, 74.2% H_2O.

15.5. How many milligrams are there in 2.50 milliequivalents of (a) $CaCl_2$, and (b) $Al_2(SO_4)_3$? *Ans.* (a) 139 mg, (b) 143 mg.

15.6. How many milligrams are there in 3.50 milliequivalents of (a) Cl^- ions, and (b) PO_4^{-3} ions? *Ans.* (a) 124 mg, (b) 111 mg.

15.7. How many milliequivalents each are there of Mg^{+2} and Cl^- ions in 0.760 g of $MgCl_2$? *Ans.* 16.0 milliequivalents of each ion.

15.8. A water solution contains 8.00 per cent sugar by weight and has a density of 1.03 g per ml. How many grams of sugar are there in 400 ml of the solution? *Ans.* 33.0 g.

15.9. Concentrated hydrochloric acid has a density of 1.20 g per ml and is 35.0 per cent HCl by weight. What is the mole fraction composition of concentrated hydrochloric acid? *Ans.* $N_{HCl} = 0.210$.

15.10. How many grams of NaCl are there in 250 ml of a 2.50M solution? *Ans.* 36.53 g.

15.11. What volume of 0.75M solution could be prepared from 500 g of Na_2SO_4? *Ans.* 4.70 *l*.

15.12. What is the molarity of a solution containing 250 g of $CaCl_2$ in 1500 ml of solution? *Ans.* 1.50M.

15.13. One liter of 12M HCl is diluted to 20 liters. What is the molarity of the diluted solution? *Ans.* 0.6M.

15.14. A solution contains 17.0 g of sodium nitrate in 100 ml of solution, the density of which is 1.10 g per ml. Calculate:

a. The per cent by weight of sodium nitrate. *Ans.* 15.5%.

b. The molarity of the solution. *Ans.* 2.00M.

15.15. How many milligrams each of Na^+ ion and Cl^- ion are there in 1.00 ml of 1.00M NaCl solution?

Ans. 23.0 mg Na^+, 35.5 mg Cl^-.

15.16. What is the molarity of a solution obtained by diluting 250 ml. of 6.0M HCl to one liter volume? *Ans.* 1.5M.

15.17. What is the molality of a solution in which 250 g of $CaCl_2$ is dissolved in 1500 g of water? *Ans.* 1.50m.

15.18. How many grams of water would have to be added to 1000 g of sugar, $C_{12}H_{22}O_{11}$, in order to prepare a 1m solution? *Ans.* 2924 g.

15.19. To what volume would 10.0 ml of 2.0M $Pb(NO_3)_2$ have to be diluted in order that the resulting solution will contain 10.0 mg of Pb^{+2} ion per milliliter? *Ans.* 414 ml.

15.20. How many grams each of H_3PO_4 and $Ca(OH)_2$ would be required to prepare 250 ml of 0.100N solution? *Ans.* 0.82 g, 0.93 g.

15.21. Calculate in terms of normality and molarity the strength of a solution containing 275 g of KOH in 800 ml of solution.

Ans. 6.13N, 6.13M.

15.22. How many milliliters of 0.50N solution could be prepared from 50.0 g of NaOH? *Ans.* 2500 ml.

15.23. What is the strength of each of the following solutions in terms of normality?

a. 6.00M HCl. *Ans.* 6.00N.

b. 0.75M $CaCl_2$. *Ans.* 1.50N.

c. 0.20M H_2S. *Ans.* 0.40N.

15.24. What is the gram-equivalent weight of the oxidant and reductant in the reaction:

$$Na_2SO_3 + Br_2 + H_2O \rightarrow Na_2SO_4 + HBr?$$

Ans. 79.92 g Br_2, 63.1 g Na_2SO_3.

15.25. What is the gram-equivalent weight of the oxidant and reductant in the reaction: $MnO_4^- + Fe^{+2} \rightarrow Mn^{+2} + Fe^{+3}$? Assume that $KMnO_4$ is the source of the MnO_4^- ion and $FeSO_4$ the source of the Fe^{+2} ion. *Ans.* 31.6 g $KMnO_4$, 151.9 g $FeSO_4$.

15.26. How many gram-equivalent weights of H_2SO_4 are there in 250 ml of 0.15M H_2SO_4? *Ans.* 0.075.

15.27. How many gram-equivalent weights of Ba^{+2} ion are there in 100 g of $BaSO_4$? *Ans.* 0.857.

15.28. A solution containing 1.20 millimoles per milliliter of Na_2SO_4 would contain how many grams per liter? *Ans.* 170 g. per liter.

15.29. In Example 15.7 the volume of the solution, as determined experimentally, was 1720 ml. What was the density of the solution? *Ans.* 1.13 g $ml.^{-1}$.

15.30. How many milligrams of Cu^{+2} ions are there in 1.00 ml of 1.00N $CuSO_4$ solution? *Ans.* 31.8 mg.

15.31. How many milliequivalents of Na^+ ions are there in 1.00 ml of a solution containing 50.0 g of NaCl per liter? *Ans.* 855 milliequivalents.

15.32. How many grams of $Al_2(SO_4)_3$ are there in 300 ml of 1.50m solution, the density of which is 1.40 g per ml.? *Ans.* 142 g.

15.33. How many gram-equivalent weights each of Fe^{+2} and SO_4^{-2} ions are there in 50.0 g of $FeSO_4$? *Ans.* 0.658 gram-equivalent weight of each.

Part II

15.34. Calculate the mole fraction composition of a solution containing 500 g of C_2H_5OH and 500 g of H_2O. *Ans.* C_2H_5OH = 0.28, H_2O = 0.72.

15.35. Calculate the mole fraction composition of a saturated solution of NaCl at 0° C. At 0° C, 35.7 g of NaCl will dissolve in 100 g of water. *Ans.* NaCl = 0.099, H_2O = 0.901.

15.36. How many molecules of solute are there in 100 ml of a 0.100M solution? *Ans.* 6.0×10^{21}.

15.37. A solution of NaCl contained 30 g of NaCl in 1500 ml of solution. What was the molarity of the solution? *Ans.* 0.34M.

15.38. How many grams of KBr could be obtained by evaporating 50 ml of a 0.50M solution of the salt? *Ans.* 2.98 g.

15.39. What would be the molarity of a solution in which 50 liters of HCl gas measured at standard conditions is dissolved in 2.0 liters of water? Assume no change in volume. *Ans.* 1.12M.

15.40. Calculate the molarity of a solution prepared by adding 500 ml of water to 100 ml of 0.60M solution. *Ans.* 0.10M.

15.41. Which solution would have the greater concentration of NaCl, a 1.00M solution of NaCl or one containing 10 per cent by weight NaCl (density 1.07 g per ml)? *Ans.* 10% solution.

15.42. Which of the following solutions would contain the greater number of molecules of solute, 100 ml of 3M sugar solution or 200 ml of 2M sugar solution? *Ans.* 2M solution.

15.43. How many grams of HCl are there in 750 ml of 0.50M HCl? *Ans.* 13.7 g.

15.44. What is the molarity of a solution containing 75 g of KOH dissolved in water to make one liter of solution? *Ans.* 1.34M.

15.45. What weight of $BaCl_2$ would be required to prepare 1500 ml of 1.50M solution? *Ans.* 469 g.

15.46. How many grams each of Na^+ ion and of Cl^- ion are there in 500 ml of 1.5M NaCl solution? *Ans.* 17.3 g Na^+, 26.6 g Cl^-.

15.47. How many milligrams each of Na^+ ion and Cl^- ion are there in 5.00 ml of 0.100M solution of NaCl?

Ans. 11.5 mg Na^+, 17.8 mg Cl^-.

15.48. To what volume must 100 ml of 6.0M HCl be diluted in order that the resulting solution be 1.00M? *Ans.* 600 ml.

15.49. How many grams of $CuSO_4 \cdot 5\ H_2O$ would be required to prepare 2500 ml of 0.100M solution? *Ans.* 62.4 g.

15.50. How many molecules of sugar are there in 1.00 ml of a ten per cent sugar solution, density 1.20 g per ml ? *Ans.* 2.1×10^{20}.

15.51. What is the molarity of a solution prepared by dissolving 100 *l* of HCl (S.C.) in sufficient water to make 800 ml of solution?

Ans. 5.59M.

15.52. How many grams of calcium chloride would be required to prepare 750 ml of a solution containing 1.00 millimole of calcium chloride per milliliter? *Ans.* 83.3 g.

15.53. How many grams of a 5.0 per cent solution of calcium chloride would have to be evaporated in order to obtain 25.0 g of calcium chloride as a residue? *Ans.* 500 g.

15.54. A solution containing 10.0 mg of Cu^{+2} ion per milliliter is desired. How many grams of $CuSO_4 \cdot 5\ H_2O$ must be dissolved per liter of solution in order to obtain the desired solution? *Ans.* 39.2 g.

15.55. A 12-liter bottle is to be filled with 6.0M H_2SO_4. How much 18M H_2SO_4 must be added to the bottle before filling with water?

Ans. 4.0 *l*.

15.56. What is the molarity of Fe^{+2} ion in a solution containing 100 g of $FeCl_2$ per liter of solution? *Ans.* 0.789M.

15.57. What is the concentration of Cd^{+2} ion in milligrams per milliliter in a 1.00M $CdCl_2$ solution? *Ans.* 112.4.

15.58. How many grams of KOH would be required to prepare 2.5 *l* of 6.0M KOH solution? *Ans.* 842 g.

15.59. At 25° C., 0.200 g of $CaSO_4$ will dissolve in 100 ml of solution. What is the molarity of the solution? *Ans.* 0.0147M.

15.60. How many grams of NH_4NO_3 would be required to prepare 500 ml of 0.25M solution? *Ans.* 10.0 g.

15.61. Calculate (a) the molarity, and (b) the molality, of a solution of K_2CO_3 which contains 22 per cent of the salt by weight and has a density of 1.21 g per ml. *Ans.* (a) 1.93M; (b) 2.04m.

15.62. The acid solution in a fully charged lead storage cell contains 33 per cent H_2SO_4 by weight and has a density of 1.25 g per ml. Calculate (a) the mole fraction composition, (b) the molarity, and (c) the molality of the acid solution.

Ans. (a) $N_{H_2O} = 0.917$; (b) 4.21M; (c) 5.02m.

15.63. How many molecules of sugar would there be in 1.00 ml of 1.00M solution? *Ans.* 6.024×10^{20}.

15.64. How many grams of $CuSO_4 \cdot 5\ H_2O$ would be required to prepare one liter of 2.0M $CuSO_4$ solution? *Ans.* 499.4 g.

15.65. What is the molarity of a solution containing 25.0 g of K_2CrO_4 dissolved in water to make 300 ml of solution? *Ans.* 0.429M.

15.66. Make a table showing the number of grams of each of the following compounds required to make one liter of 1.0M, 1.0N, 2.0N, and 3.0N solutions: NaOH, $Ca(OH)_2$, $Al(OH)_3$, HCl, H_2SO_4, and H_3PO_4. *Ans.* NaOH 40.0, 40.0, 80.0, 120.0; $Ca(OH)_2$ 74.1, 37.1, 74.1, 111.2; $Al(OH)_3$ 78.0, 26.0, 52.0, 78.0; HCl 36.5, 36.5, 72.9, 109.4; H_2SO_4 98.1, 49.0, 98.1, 147.1; H_3PO_4 98.0, 32.7, 65.3, 98.0.

15.67. Concentrated NH_4OH has a density of 0.90 g per ml and is 28 per cent NH_3 by weight. What is its normality? *Ans.* 15N.

15.68. Water was added to 25.0 ml of 98 per cent H_2SO_4, density 1.84 g per ml, to make 100 ml of solution. Calculate the normality and molarity of the solution. *Ans.* 9.20N, 4.60M.

15.69. How many grams of zinc would be dissolved by the action of one liter of 1.00N HCl? *Ans.* 32.69 g.

15.70. How many liters of CO_2, measured at standard conditions, would be liberated by the action of 1500 ml. of 2N H_2SO_4 on $CaCO_3$? *Ans.* 33.6 *l*.

15.71. It was found that 0.3031 g of Mg required 27.40 ml of 0.910N HCl to react to form $MgCl_2$. Calculate the atomic weight of magnesium, knowing that it is approximately 25. *Ans.* 24.32.

15.72. How many grams are there in 2.50 gram-equivalent weights of $BaCl_2 \cdot 2\ H_2O$? *Ans.* 306 g.

15.73. A five per cent solution of phosphoric acid, H_3PO_4, has a density of 1.03 g per ml. What is the strength of the solution in terms of normality? *Ans.* 1.57N.

15.74. What is the per cent by weight of NH_3 in 12.0N NH_4OH, the density being 0.90 g per ml? *Ans.* 22.7%.

15.75. How many gram-equivalent weights of Fe^{+3} ion are there in 100 g of $FeCl_3$? *Ans.* 1.85.

15.76. What is the normality of the Ba^{+2} ion in a solution containing 20.0 mg of Ba^{+2} ion per milliliter? *Ans.* 0.291N.

15.77. How many milliequivalents of HCl are there in 1.00 ml of 6.0N HCl? *Ans.* 6.0.

15.78. What is the normality of a solution prepared by dissolving 20.0 g of Na_2CO_3 in sufficient water to make 600 ml of solution? *Ans.* 0.629N.

15.79. What is the normality of a solution of $AgNO_3$ containing 10.0 mg of Ag^+ ion per milliliter? *Ans.* 0.0926N.

15.80. How many grams of $KMnO_4$ per liter of solution would be required to prepare 0.01N $KMnO_4$ based upon each of the reactions:

a. $MnO_4^- + Fe^{+2} \rightarrow Mn^{+2} + Fe^{+3}$. *Ans.* 0.316 g.

b. $KMnO_4 + KOH \rightarrow K_2MnO_4 + H_2O$. *Ans.* 1.58 g.

15.81. How many milligrams of Al^{+3} ion are there in 1.00 ml of 0.25N $AlCl_3 \cdot 6\,H_2O$? *Ans.* 2.25 mg.

15.82. How many Al^{+3} ions and how many Cl^- ions are there in 1.00 ml of 0.001N $AlCl_3 \cdot 6\,H_2O$? *Ans.* 2.0×10^{17} Al^{+3}, 6.0×10^{17} Cl^-.

15.83. A $K_2Cr_2O_7$ solution was prepared by dissolving 10.0 g of $K_2Cr_2O_7$ in sufficient water to make one liter of solution. What was the normality of the solution based on the reaction:

$$Cr_2O_7^{-2} + Fe^{+2} \rightarrow Cr^{+3} + Fe^{+3}?$$

Ans. 0.204N.

15.84. How many gram-equivalent weights of HNO_3 are there in 1500 ml of 0.80N HNO_3? *Ans.* 1.2.

15.85. How many gram-equivalent weights of $KClO_3$ are there in 2500 ml of a 0.100N solution? *Ans.* 0.25.

15.86. How many milliequivalents of solute are there in 0.25 ml of 0.100N solution? *Ans.* 0.025.

15.87. With how many grams of zinc will 250 ml of 6.00N HCl react? *Ans.* 49.0 g.

15.88. With how many grams of CuS will 50.0 ml of 3.0M HNO_3 react according to the reaction:

$$CuS + HNO_3 \rightarrow Cu(NO_3)_2 + S + H_2O + NO?$$

Ans. 5.38 g.

16

Volumetric Analysis

Methods for expressing the strengths of solutions in terms of the solute and solvent were discussed in Chapter 15. Use will now be made of these methods, particularly normality. *Titration* is the process of analysis involving standard solutions. In each titration some method must be available for determining the end point of the chemical reaction taking place in the solution. Chemical indicators such as litmus and phenolphthalein are used to indicate the end point in an acid-base titration. In titrating oxidants and reductants the color of the solution is sometimes an indication of the end point. The method used for determining the end point of a titration does not affect the calculations.

In neutralization reactions such as:

$$\begin{array}{cccl} NaOH & + & HCl & \rightarrow NaCl + H_2O \\ 40.0 \text{ g} & & 36.5 \text{ g} & \\ 1.00 \text{ g-eq. wt.} & & 1.00 \text{ g-eq. wt.} & \end{array}$$

$$\begin{array}{cccl} 2\ NaOH & + & H_2SO_4 & \rightarrow Na_2SO_4 + 2\ H_2O \\ 80.0 \text{ g} & & 98.1 \text{ g} & \\ 2.00 \text{ g-eq. wt.} & & 2.00 \text{ g-eq. wt.} & \end{array}$$

and in oxidation-reduction equations such as:

$$\begin{array}{cccl} Mg & + & H_2SO_4 & \rightarrow MgSO_4 + H_2 \\ 24.3 \text{ g} & & 98.1 \text{ g} & \\ 2.00 \text{ g-eq. wt.} & & 2.00 \text{ g-eq. wt.} & \end{array}$$

$$\begin{array}{cccl} KMnO_4 & +8\ HCl+ & 5\ FeCl_2 & \rightarrow KCl + MnCl_2 + 5\ FeCl_2 + 4\ H_2O \\ 158.0 \text{ g} & & 5 \times 126.8 \text{ g} & \\ 5.00 \text{ g-eq. wt.} & & 5.00 \text{ g-eq. wt.} & \end{array}$$

we see that, at the end point of the titration, the same number of gram-equivalent weights of acid and base are involved.

That is: g-eq. wt. of acid = g-eq. wt. of base.

At the end point of an oxidation-reduction equation:

g-eq. wt. of oxidant = g-eq. wt. of reductant.

We will make use of these relationships repeatedly in the examples that follow.

In Sec. 15.6 we saw that Normality (N) $= \frac{\text{g-eq. wt. solute (E)}}{l \text{ solution (V)}}$,

or: $N = \frac{E}{V}$, $E = NV$, and $V = \frac{E}{N}$. All these expressions are valid for both neutralization and oxidation-reduction equations.

From the above it follows that, in a neutralization reaction, $N_A V_A = N_B V_B$, where the subscript A refers to acid, and subscript B to base. The volume V is in liters. In an oxidation-reduction reaction $N_O V_O = N_R V_R$, where the subscript O refers to the oxidant and subscript R to the reductant. In either case NV stands for gram-equivalent weights of solute. In titrations involving solutions of both titrants, the above formulas are quite useful.

16.1. Dimensions in Volumetric Analysis. In volumetric work the milliliter is more commonly used than is the liter as a unit of volume. For this reason the milliequivalent (Sec. 15.4) is a more practical unit to use than is the gram-equivalent weight. In terms of milliequivalents (meq.) and milliliters:

$$\text{Normality (N)} = \frac{\text{meq. solute}}{\text{ml solution}}.$$

That is, normality has the dimensions $\frac{\text{g-eq. wt. solute}}{l \text{ solution}}$ or $\frac{\text{meq. solute}}{\text{ml solution}}$.

It follows that: meq. = N × ml solution, and ml solution $= \frac{\text{meq.}}{\text{N}}$.

When using meq. and ml the formulas given earlier become $N_A v_A = N_B v_B$ and $N_O v_O = N_R v_R$, where v is the volume in ml. In this case the product Nv = meq.

In the examples that follow observe that the dimensional value of N is used rather than the symbol N.

Another dimensional quantity that is sometimes convenient to use is $\frac{\text{mg}}{\text{meq.}}$. That is:

$$\frac{\text{g}}{\text{g-eq. wt.}} = \frac{\text{mg}}{\text{meq.}}.$$

That is, the number of grams of a substance per gram-equivalent weight is the same numerically as the number of milligrams of the substance per milliequivalent. For example, there are $23.0 \frac{\text{g Na}}{\text{g-eq. wt. Na}}$ and $23.0 \frac{\text{mg Na}}{\text{meq. Na}}$.

Example 16.1. In a titration experiment, 23.05 ml of 0.100N NaOH was required to neutralize 10.00 ml of a solution of H_2SO_4 of unknown strength. What was the normality of the acid solution?

Solution. In a titration using NaOH and H_2SO_4

$$\text{meq. of NaOH} = \text{meq. of } H_2SO_4.$$

The meq. of NaOH used in the titration would be:

$$23.05 \cancel{\text{ml NaOH}} \times 0.100 \frac{\text{meq. NaOH}}{\cancel{\text{ml NaOH}}} = 2.31 \text{ meq. NaOH}$$

and the normality of the H_2SO_4 solution would be:

$$\frac{2.31 \text{ meq. } H_2SO_4}{10.00 \text{ ml } H_2SO_4} = 0.231 \frac{\text{meq. } H_2SO_4}{\text{ml } H_2SO_4} = 0.231\text{N } H_2SO_4.$$

A second method of solution for this particular problem would be to use the formula $N_A v_A = N_B v_B$. Then:

$$N_A = \frac{N_B v_B}{v_A} = \frac{0.100\text{N} \times 23.05 \cancel{\text{ml}}}{10.00 \cancel{\text{ml}}} = 0.231\text{N } H_2SO_4.$$

Example 16.2. A solution containing 0.275 g of NaOH required 35.4 ml of HCl for neutralization. What was the normality of the HCl?

Solution. Again, from the above equations we see that:

$$\text{meq. of NaOH} = \text{meq. of HCl}.$$

At the end point, the meq. of NaOH would be:

$$\frac{0.275 \cancel{\text{g NaOH}}}{40.0 \frac{\cancel{\text{g NaOH}}}{\cancel{\text{g-eq. wt. NaOH}}}} \times 1000 \frac{\text{meq. NaOH}}{\cancel{\text{g-eq. wt. NaOH}}} = 6.88 \text{ meq. NaOH}.$$

And the normality of the HCl would be:

$$\frac{6.88 \text{ meq. HCl}}{35.4 \text{ ml HCl}} = 0.194\text{N HCl}.$$

Example 16.3. What volume of 0.250N acid would be required to react with 0.500 g of $Ca(OH)_2$?

Solution. Since the name of the acid is not given, we cannot write the equation for the reaction. However, we know that:

$$\text{meq. of } Ca(OH)_2 = \text{meq. of acid.}$$

The meq. of $Ca(OH)_2$ would be:

$$\frac{0.500 \cancel{\text{ g } Ca(OH)_2}}{0.03705 \dfrac{\cancel{\text{g } Ca(OH)_2}}{\text{meq. } Ca(OH)_2}} = 13.5 \text{ meq. } Ca(OH)_2$$

and the volume of acid required would be:

$$\frac{13.5 \cancel{\text{ meq. acid}}}{0.250 \dfrac{\cancel{\text{meq. acid}}}{\text{ml acid}}} = 54.0 \text{ ml acid.}$$

Example 16.4. A 0.311 g sample of crude NaOH, when dissolved in water, required 46.1 ml of 0.122N H_2SO_4 to neutralize the NaOH in the sample. Calculate the per cent of NaOH in the sample.

Solution. From the equations given previously we know that:

$$\text{meq. of NaOH} = \text{meq. of } H_2SO_4.$$

The meq. of H_2SO_4 would be:

$$0.122 \frac{\text{meq. } H_2SO_4}{\cancel{\text{ml solution}}} \times 46.1 \cancel{\text{ ml solution}} = 5.62 \text{ meq. } H_2SO_4$$

and the 5.62 meq. of NaOH would be:

$$5.62 \cancel{\text{ meq. NaOH}} \times 0.040 \frac{\text{g NaOH}}{\cancel{\text{meq. NaOH}}} = 0.225 \text{ g NaOH.}$$

The per cent of NaOH in the sample would be:

$$\frac{0.225 \text{ g NaOH}}{0.311 \text{ g sample}} \times 100 = 72.3\% \text{ NaOH.}$$

Example 16.5. A volume of 22.5 ml of 2.50N NaOH was required to neutralize 10.5 ml of a solution of H_2SO_4 of unknown strength. The density of the H_2SO_4 solution was 1.16 g per ml. Calculate the per cent by weight of H_2SO_4 in the solution.

Solution. Again we will use the relationship that

$$\text{meq. of NaOH} = \text{meq. of } H_2SO_4.$$

The meq. of NaOH would be:

$$2.50 \frac{\text{meq. NaOH}}{\cancel{\text{ml solution}}} \times 22.5 \cancel{\text{ ml solution}} = 56.3 \text{ meq. NaOH.}$$

and the 56.3 meq. of H_2SO_4 would be:

$$56.3 \cancel{\text{ meq. } H_2SO_4} \times 0.0491 \frac{\text{g } H_2SO_4}{\cancel{\text{meq. } H_2SO_4}} = 2.76 \text{ g } H_2SO_4.$$

The weight of the 10.5 ml of H_2SO_4 used would be:

$$10.5 \cancel{ml} \times 1.16 \frac{g}{\cancel{ml}} = 12.2 \text{ g}$$

and the per cent H_2SO_4 would be:

$$\frac{2.76 \text{ g } H_2SO_4}{12.2 \text{ g solution}} \times 100 = 22.6\% \ H_2SO_4 \text{ by weight.}$$

16.2. Miscellaneous Types of Titration. Standard solutions other than of acid and base may react chemically. The end point of the titration may be determined providing there is a color change, precipitate, gas, or other visual effect to indicate the end point. In the problems given it must be assumed that the end point can be determined.

Example 16.6. How many liters of hydrogen, at standard conditions, would be liberated by the action of 250 ml of 6.0N H_2SO_4 on magnesium?

Solution.

$$\begin{array}{cccc} Mg + & H_2SO_4 & \rightarrow MgSO_4 + & H_2\ (g) \\ & 98.1 \text{ g} & & 2.02 \text{ g} \\ & 2.00 \text{ g-eq. wt.} & & 2.00 \text{ g-eq. wt.} \\ & & & 22.4\ l\ (S.C.) \end{array}$$

From the above equation we see that equal gram-equivalent weight quantities of H_2SO_4 and H_2 are involved. Therefore the g-eq. wt. H_2SO_4 would be:

$$6.0 \frac{\text{g-eq. wt. } H_2SO_4}{\cancel{l \text{ solution}}} \times 0.250 \cancel{l \text{ solution}} = 1.50 \text{ g-eq. wt. } H_2SO_4.$$

Since there are 11.2 l of hydrogen in 1.00 g-eq. wt., then:

$$1.50 \cancel{\text{g-eq. wt. } H_2} \times 11.2 \frac{l\ H_2}{\cancel{\text{g-eq. wt. } H_2}} = 16.8\ l\ H_2\ (S.C.).$$

Example 16.7. An excess of $BaCl_2$ solution was added to 25.0 ml of a solution of Na_2SO_4 of unknown strength. The precipitated $BaSO_4$ weighed 0.864 g. What was the molarity of the Na_2SO_4 solution?

Solution.

$$\begin{array}{cccc} BaCl_2 + & Na_2SO_4 & \rightarrow 2\ NaCl + & BaSO_4\ (s) \\ & 1.00 \text{ mole} & & 1.00 \text{ mole} \\ & 142.1 \text{ g} & & 233.4 \text{ g} \end{array}$$

From the above equation we see that equal mole quantities of Na_2SO_4 and $BaSO_4$ are involved. The moles of $BaSO_4$ precipitated would be:

$$\frac{0.864 \cancel{\text{g } BaSO_4}}{233.4 \frac{\cancel{\text{g } BaSO_4}}{\text{mole } BaSO_4}} = 0.00370 \text{ mole } BaSO_4.$$

And the molarity (moles solute per l solution) of the Na_2SO_4 would be:

$$\frac{0.00370 \text{ mole } Na_2SO_4}{0.0250\ l \text{ solution}} = 0.150M\ Na_2SO_4.$$

Example 16.8. What must be the molarity of a solution of NaCl in order that 1.00 ml will precipitate (or be equivalent to) 20.0 mg of Ag^+ ion?

Solution. The Cl^- ion will precipitate the Ag^+ ion as insoluble AgCl. That is:

$$\begin{array}{ccc} Ag^+ & + \quad Cl^- & \rightarrow AgCl\ (s). \\ 1.00 \text{ mole} & 1.00 \text{ mole} & \\ 107.9 \text{ g} & 35.45 \text{ g} & \end{array}$$

From the equation we see that equal mole quantities of Ag^+ and Cl^- ions are involved. Then 20.0 mg of Ag^+ ion would be:

$$\frac{0.020 \cancel{\text{g } Ag^+}}{107.9 \dfrac{\cancel{\text{g } Ag^+}}{\text{mole } Ag^+}} = 1.85 \times 10^{-4} \text{ mole } Ag^+.$$

Since 1.85×10^{-4} mole of Cl^- ion would originate from 1.85×10^{-4} mole of NaCl, the molarity of the NaCl solution would be:

$$\frac{1.85 \times 10^{-4} \text{ mole NaCl}}{1.00 \times 10^{-3}\ l \text{ solution}} = 0.185M \text{ NaCl}.$$

Example 16.9. An excess of silver nitrate solution was added to 25.0 ml of an HCl solution of unknown strength. The precipitated AgCl weighed 2.125 g. What was the normality of the HCl solution?

Solution. From the equation:

$$\begin{array}{cccc} AgNO_3 + & HCl & \rightarrow HNO_3 + & AgCl\ (s). \\ & 1.00 \text{ g-eq. wt.} & & 1.00 \text{ g-eq. wt.} \\ & 36.45 \text{ g} & & 143.4 \text{ g} \end{array}$$

The HCl and AgCl appear in the equation in equal g-eq. wt. quantities. The g-eq. wt. of AgCl precipitated would be:

$$\frac{2.125 \cancel{\text{g AgCl}}}{143.4 \dfrac{\cancel{\text{g AgCl}}}{\text{g-eq. wt. AgCl}}} = 0.0148 \text{ g-eq. wt. AgCl}$$

and the normality of the HCl would be:

$$\frac{0.0148 \text{ g-eq. wt. HCl}}{0.025\ l \text{ solution}} = 0.592N \text{ HCl}.$$

16.3. Titration of Oxidants and Reductants. As in neutralization reactions, equal gram-equivalent quantities of oxidants and reductants react, as shown in the equation for Example 15.12(c).

$$\underset{\substack{316.08\text{ g}\\ 10\text{ g-eq. wt.}}}{\underline{2\,KMnO_4}} + \underset{\substack{1519.1\text{ g}\\ 10\text{ g-eq. wt.}}}{\underline{10\,FeSO_4}}\ 8\,H_2SO_4 \rightarrow K_2SO_4 + 2\,MnSO_4 + 5\,Fe_2(SO_4)_3 + 8\,H_2O.$$

Example 16.10. It required 33.6 ml of a 0.100N $K_2Cr_2O_7$ solution to oxidize the Fe^{+2} to Fe^{+3} in 21.6 ml of a solution of $FeSO_4$ of unknown strength.

(a) We will write the equation for the reaction.

$$\underset{\substack{1.00\text{ mole}\\ 6.00\text{ g-eq. wt.}}}{Cr_2O_7^{-2}} + 14\,H^+ + \underset{\substack{6.00\text{ moles}\\ 6.00\text{ g-eq. wt.}}}{6\,Fe^{+2}} \rightarrow 2\,Cr^{+3} + 6\,Fe^{+3} + 7\,H_2O$$

From the above equation, and the fact that 1.00 g-eq. wt. of Fe^{+2} ion will originate from 1.00 g-eq. wt. of $FeSO_4$, we have:

$$\text{meq. of } Cr_2O_7^{-2} = \text{meq. of } Fe^{+2} = \text{meq. of } FeSO_4.$$

Then meq. of $Cr_2O_7^{-2}$ would be:

$$0.100\,\frac{\text{meq. } Cr_2O_7^{-2}}{\cancel{\text{ml solution}}} \times 33.6\,\cancel{\text{ml solution}} = 3.36\text{ meq. } Cr_2O_7^{-2}$$

and the normality of the $FeSO_4$ solution would be:

$$\frac{3.36\text{ meq. } FeSO_4}{21.6\text{ ml solution}} = 0.156N\ FeSO_4.$$

(b) Since the $FeSO_4$ solution is 0.156N, it also is 0.156N with respect to Fe^{+2} ion, and the number of mg Fe^{+2} per ml would be:

$$0.156\,\frac{\cancel{\text{meq. } Fe^{+2}}}{\text{ml solution}} \times 55.85\,\frac{\text{mg } Fe^{+2}}{\cancel{\text{meq. } Fe^{+2}}} = 8.71\,\frac{\text{mg } Fe^{+2}}{\text{ml solution}}.$$

Example 16.11. A solution was prepared by dissolving 0.865 g of $FeCl_2$ in water. What volume of 0.150N $KMnO_4$ solution would react with the solution of $FeCl_2$?

Solution. From the equation:

$$\underset{\substack{1.00\text{ mole}\\ 5.00\text{ g-eq. wt.}}}{KMnO_4} + 8\,HCl + \underset{\substack{5.00\text{ moles}\\ 5.00\text{ g-eq. wt.}}}{5\,FeCl_2} \rightarrow KCl + MnCl_2 + 5\,FeCl_3 + 4\,H_2O$$

we see that: meq. of $KMnO_4$ = meq. of $FeCl_2$.

The meq. of $FeCl_2$ would be:

$$\frac{0.865\,\cancel{\text{g } FeCl_2}}{0.1268\,\frac{\cancel{\text{g } FeCl_2}}{\text{meq. } FeCl_2}} = 6.82\text{ meq. } FeCl_2.$$

Then the volume of 0.150N $KMnO_4$ would be:

$$\frac{6.82 \text{ meq. KMnO}_4}{0.150 \dfrac{\text{meq. KMnO}_4}{\text{ml solution}}} = 45.5 \text{ ml solution } (KMnO_4).$$

16.4. Dilution of Standard Solutions. In the laboratory it frequently is necessary to prepare a solution of a given strength by diluting a standard solution. Or, in the course of a laboratory experiment it will be necessary to mix two standard solutions.

In dilution problems the relationship $N_1V_1 = N_2V_2$ may be used, where N_1V_1 represents equivalents of solute before dilution and N_2V_2 represents equivalents of solute after dilution. As in previous calculations, when V is given in liters the product NV represents gram-equivalent weights of solute; and when V is given in milliliters the product NV represents milliequivalents of solute.

Example 16.12. What would be the normality of a solution obtained by diluting 100 ml of 12.0N HCl to a volume of 500 ml?

Solution. The 100 ml of 12.0N HCl must contain the same number of milliequivalents of HCl as the 500 ml of solution.

The number of meq. in 100 ml of 12.0N HCl would be:

$$12.0 \frac{\text{g-eq. wt. HCl}}{l \text{ solution}} \times 0.100 \, l \text{ solution} = 1.2 \text{ g-eq. wt. HCl}.$$

Then the normality of the diluted solution would be:

$$\frac{1.2 \text{ g-eq. wt. HCl}}{0.500 \, l \text{ solution}} = 2.4\text{N HCl}.$$

Example 16.13. How could 50.0 ml of 4.0M H_2SO_4 solution be prepared from 18.0M H_2SO_4 solution?

Solution. The two solutions must contain the same number of moles of H_2SO_4. The number of moles of H_2SO_4 in 50.0 ml of 4.0M H_2SO_4 would be:

$$4.0 \frac{\text{mole } H_2SO_4}{l \text{ solution}} \times 0.050 \, l \text{ solution} = 0.20 \text{ mole } H_2SO_4.$$

The volume of 18.0M H_2SO_4 containing 0.20 mole of H_2SO_4 would be:

$$\frac{0.20 \text{ mole } H_2SO_4}{18.0 \dfrac{\text{moles } H_2SO_4}{l \text{ solution}}} = 0.0111 \, l \text{ solution}.$$

And: $0.0111 \, l \text{ solution} = 11.1 \text{ ml solution}.$

Problems

Part I

16.1. In a titration, 32.8 ml of 0.255N H_2SO_4 was required to neutralize 42.3 ml of a solution of NaOH of unknown strength. Calculate (a) the normality of the NaOH solution, and (b) the grams of Na_2SO_4 formed. *Ans.* (a) 0.198N; (b) 0.594 g.

16.2. A 2.34 g sample of impure H_2SO_4 required 42.3 ml of 0.100N NaOH to neutralize the acid in the sample. What was the per cent purity of the acid? *Ans.* 8.86%.

16.3. Commercial lye is principally NaOH. A 0.564 g sample of lye required 41.6 ml of 0.251N H_2SO_4 to neutralize the NaOH it contained. Calculate the per cent of NaOH in the lye. *Ans.* 74.1%.

16.4. A solution of sodium hydroxide was prepared by dissolving 12.54 g of NaOH in water to make 250 ml of solution. There was required 18.23 ml of the standard NaOH solution to neutralize 12.56 ml. of a solution of sulfuric acid of unknown strength. Calculate (a) the normality, and (b) the molarity of the acid solution.
Ans. (a) 1.83N; (b) 0.915M.

16.5. How many milliliters of 6.0N HCl would be required to neutralize 50.0 ml of 0.100M $Ba(OH)_2$? *Ans.* 1.67 ml.

16.6. How many grams of zinc would react with 1.00 *l* of 1.00N HCl?
Ans. 32.7 g.

16.7. The addition of excess barium chloride solution to 25.0 ml of dilute sulfuric acid resulted in the precipitation of 2.675 g of $BaSO_4$. What was the normality of the acid solution? *Ans.* 0.92N.

16.8. A sample of marble weighing 1.750 g required 19.5 ml of 1.50N HCl to react with the $CaCO_3$ in the marble. What per cent of the marble was $CaCO_3$? *Ans.* 83.4%.

16.9. A solution contains 20.0 mg of Ag^+ ion per milliliter. What volume of 0.10N HCl would be required to precipitate the silver in 25.0 ml of the solution as AgCl? *Ans.* 46.3 ml.

16.10. It required 26.4 ml of 0.175N H_2SO_4 to precipitate the barium ions as $BaSO_4$ from 20.0 ml of a $BaCl_2$ solution. What was the molarity of the $BaCl_2$ solution? *Ans.* 0.116M.

16.11. In a titration 61.33 ml of 0.2N $K_2Cr_2O_7$ was required to react with 47.65 ml of $FeSO_4$ solution. Calculate (a) the normality of the $FeSO_4$ solution, and (b) the number of grams of $K_2Cr_2O_7$ in one liter of the solution. *Ans.* (a) 0.257N; (b) 9.81 g.

16.12. Iodine reacts with $Na_2S_2O_3$ according to the equation:

$$I_2 + 2\ Na_2S_2O_3 \rightarrow 2\ NaI + Na_2S_4O_6.$$

A 0.376 g sample of crude iodine required 17.1 ml of 0.100N $Na_2S_2O_3$

to react with the iodine in the sample. Calculate the per cent purity of the crude iodine. *Ans.* 57.7%.

16.13. A solution was prepared by dissolving 1.765 g of $FeSO_4$ in water. What volume of 0.150N $KMnO_4$ solution would be required to react with the $FeSO_4$ solution? *Ans.* 77.5 ml.

16.14. How many gram-equivalent weights of $KMnO_4$ will react with 500 ml of 0.250M $FeSO_4$ solution? *Ans.* 0.125.

16.15. How many milliliters of a 0.100N solution of any oxidant will react with 250 ml of a 0.250N solution of any reductant? *Ans.* 625 ml.

16.16. How many milliliters of 0.100N $KMnO_4$ are equivalent to 1.00 ml of a solution of $FeSO_4$ containing 40.0 mg of Fe^{+2} ion per milliliter? *Ans.* 7.16 ml.

16.17. What volume of H_2SO_4, density 1.84 g per ml and containing 98 per cent acid by weight, would be required to prepare one liter of 6N acid? *Ans.* 164 ml.

16.18. Concentrated HCl has a density of 1.20 g per ml and is 39 per cent acid by weight. What is its normality? *Ans.* 12.8N.

16.19. What volume of 1.75N KOH solution, diluted to one liter, would give a 1.00N solution of KOH? *Ans.* 571 ml.

16.20. An experiment calls for 300 ml of 1.00N HCl. The only acid available is 6.0N HCl. How much water and 6.0N HCl must be used? *Ans.* 250 ml water, 50 ml acid.

16.21. An excess of silver nitrate solution was added to 10.0 g of a solution of sodium chloride of unknown strength. The AgCl which precipitated out was found to weigh 1.22 g. What was the per cent by weight of NaCl in the solution? *Ans.* 5.0%.

16.22. How many milliliters of 6.0M $CuSO_4$ would be required to prepare 100 ml of 0.25N $CuSO_4$? *Ans.* 2.08 ml.

16.23. What is the molarity in terms of KNO_3 and Na_2SO_4 of a solution prepared by mixing 100 ml of 3.0N KNO_3 and 250 ml of 5.0M Na_2SO_4? Assume the volumes to be additive. That is, the resulting solution will have a volume of 350 ml. *Ans.* 0.86M KNO_3, 3.57M Na_2SO_4.

Part II

16.24. If 22.1 ml of 0.25N H_2SO_4 is required to neutralize 32.1 ml of a solution of NaOH, calculate the normality of the NaOH solution. *Ans.* 0.172N.

16.25. A solution containing 6.00 per cent HCl by weight, and density 1.028 g per ml., was used to determine the strength of a solution of NaOH. It was found that 23.41 ml of the HCl solution neutralized 28.20 ml of the NaOH solution. Calculate:

a. The normality of the NaOH solution. *Ans.* 1.40N.

b. The number of milliequivalents of NaOH used. *Ans.* 39.6.

c. The number of grams of NaOH used. *Ans.* 1.58 g.

16.26. How many grams of NaCl would be formed by mixing 50.0 ml of 1.54N HCl with 50.0 ml of 1.12N NaOH? *Ans.* 3.27 g.

16.27. A 10.00 ml sample of vinegar, density 1.01 g per ml, was diluted to 100 ml volume. It was found that 25.0 ml of the diluted vinegar required 24.15 ml of 0.0976N NaOH to neutralize it. Calculate the strength of the vinegar in terms of (a) normality, (b) grams of $HC_2H_3O_2$ per liter, and (c) per cent $HC_2H_3O_2$ in the vinegar. *Ans.* (a) 0.943N; (b) 56.6 g ; (c) 5.60%.

16.28. How many milliliters of 12.0N HCl would be required to neutralize 125 ml of 0.050M $Ba(OH)_2$? *Ans.* 1.04 ml.

16.29. How many milliliters of 0.500M H_2SO_4 will react with 25.00 ml of 0.100M $Ca(OH)_2$? *Ans.* 5.00 ml.

16.30. It was found that 18.5 ml of a 1.00N base reacted with 27.2 ml of a solution of phosphoric acid. Calculate:
a. The normality of the acid. *Ans.* 0.680N.
b. The molarity of the acid. *Ans.* 0.227M.

16.31. A sample of vinegar had a density of 1.06 g per ml. A 3.21 ml aliquot of the vinegar required 35.4 ml of 0.100N NaOH to neutralize the acetic acid, $HC_2H_3O_2$, in the vinegar. Calculate the per cent by weight of acetic acid in the vinegar. *Ans.* 6.23%.

16.32. What volume of 6.0N HCl would be required to neutralize 1.250 g of NaOH? *Ans.* 5.22 ml.

16.33. A 42.2 ml sample of KOH solution required 21.1 ml of a solution of 2.50N H_2SO_4 for neutralization.
a. What was the normality of the KOH? *Ans.* 1.25N.
b. How many mg per ml of KOH were there in the solution? *Ans.* 70.1 mg.

16.34. With how many grams of $Ca(OH)_2$ will 250 ml of 6.0N HCl react? *Ans.* 55.6 g.

16.35. How many ml of 1.00M H_2SO_4 would be required to react with 100 ml of 1.00N KOH? *Ans.* 50 ml.

16.36. A sample of 25.0 ml of concentrated H_2SO_4 containing 95.0 per cent acid by weight, and having a density of 1.84 g per ml, is diluted with water to 500 ml volume. What volume of the acid solution would be required to neutralize 50.0 ml of 0.50N NaOH solution? *Ans.* 14.0 ml.

16.37. A sample of impure NaOH weighing 0.764 g required 116 ml of 0.15N H_2SO_4 for neutralization. What was the per cent by weight of NaOH in the sample? *Ans.* 91.1%.

16.38. A solution of H_2SO_4 has a density of 1.60 g per ml. A 2.50 ml sample of the acid neutralized 27.8 ml of 2.00N NaOH. What is the per cent by weight of acid in the solution? *Ans.* 68%.

16.39. How many liters of CO_2, at S.C., would be liberated by the action of 1500 ml of 2.0N H_2SO_4 acting on Na_2CO_3? *Ans.* 33.6 *l*.

16.40. The addition of a solution of silver nitrate in excess to 25.0 ml of a solution of HCl resulted in the precipitation of 1.986 g of AgCl. What is the normality of the HCl solution? *Ans.* 0.552N.

16.41. A volume of 33.6 ml of a solution of nitric acid of unknown strength was used to react with 0.636 g of Na_2CO_3. What was the normality of the acid solution? *Ans.* 0.357N.

16.42. A sample of limestone weighing 1.680 g required 53.8 ml of 0.50N HCl to react with the $CaCO_3$ in the limestone. What was the per cent by weight of $CaCO_3$ in the limestone? *Ans.* 80%.

16.43. How many milliliters of 0.75N H_2SO_4 would be required to react with 2.50 g of soda ash containing 72.5 per cent of Na_2CO_3? Assume that no other constituent reacted with the acid. *Ans.* 45.6 ml.

16.44. A 1.000 g sample of limestone is dissolved in 50.00 ml of 1.00N HCl. The excess acid required 59.33 ml of 0.60N NaOH for neutralization. Calculate the per cent of $CaCO_3$ in the sample. *Ans.* 72%.

16.45. One gram of a silver alloy was dissolved in HNO_3. It required 72.5 ml of 0.075N HCl to precipitate the silver as AgCl. Calculate:

a. The number of grams of AgCl precipitated. *Ans.* 0.78 g.

b. The per cent of silver in the alloy. *Ans.* 58.8%.

16.46. What volume of 0.35N HCl would be required to precipitate as AgCl the silver in 100 ml of 0.50M $AgNO_3$? *Ans.* 143 ml.

16.47. How many milliliters of 6N HCl would be required to react with 1.45 g of Mg? *Ans.* 20 ml.

16.48. An experiment calls for 300 ml of 1.00N HCl. The only acid available is 6.0N HCl. How much water and 6.0N HCl must be used? *Ans.* 250 ml water, 50 ml acid.

16.49. How many milliequivalents of zinc will react with 320 ml of 0.100M H_2SO_4? *Ans.* 64 milliequivalents.

16.50. It required 26.2 ml of a solution of $AgNO_3$ to react with 25.0 ml of 0.253M NaCl. What was the concentration of the $AgNO_3$ solution in milligrams of Ag^+ ion per milliliter? *Ans.* 26 mg.

16.51. There was required 16.5 ml of 0.325N H_2SO_4 to precipitate the Ba^{+2} ion as $BaSO_4$ from 25.0 ml of a $BaCl_2$ solution. What was the molarity of the $BaCl_2$ solution? *Ans.* 0.107M.

16.52. It is desired to prepare one liter of 0.100N HCl from a solution which is known to be approximately 2N. It required 47.68 ml of 1.123N NaOH to neutralize 25.00 ml of the 2N HCl. How many milliliters of the approximately 2N HCl would be required to prepare the desired solution? *Ans.* 46.7 ml.

16.53. How many milliliters of 6.00N HCl would be required to react with FeS in order to prepare 3.00 *l* (S.C.) of H_2S? *Ans.* 44.6 ml.

16.54. What must be the normality of an acid in order that 1.00 ml of the acid is equivalent to 20.0 mg of NaOH? *Ans.* 0.500N.

16.55. A solution was prepared containing 50.0 g of $BaCl_2$ per liter of solution. How many milliliters of 0.65M Na_2SO_4 would be required to react with 25.0 ml of the $BaCl_2$ solution? *Ans.* 9.23 ml.

16.56. How many grams of $KClO_3$ will react with 500 ml of 12.0N HCl according to the equation:

$$KClO_3 + 6\ HCl \rightarrow 3\ H_2O + KCl + 3\ Cl_2?$$ *Ans.* 122.6 g.

16.57. A solution was prepared by dissolving 2.168 g of $FeSO_4$ in water. What volume of 0.500N $K_2Cr_2O_7$ would react with the $FeSO_4$ solution? *Ans.* 28.6 ml.

16.58. What volume of H_2S, at standard conditions, would react with 100 ml of 0.100N HNO_3? *Ans.* 112 ml.

16.59. How many grams of $FeCl_2$ would be oxidized to $FeCl_3$ by 500 ml of 0.100N $KMnO_4$? *Ans.* 6.34 g.

16.60. How many gram-equivalent weights of $K_2Cr_2O_7$ will react with 50.0 g of $FeSO_4$, according to the equation:

$$K_2Cr_2O_7 + 7\ H_2SO_4 + 6\ FeSO_4 \rightarrow K_2SO_4 + Cr_2(SO_4)_3 + 3\ Fe_2(SO_4)_3 + 7\ H_2O?$$ *Ans.* 0.329.

16.61. How many grams of $KMnO_4$ would be required to oxidize 100 g of $FeCl_2$? *Ans.* 24.9 g.

16.62. How many gram-equivalent weights of HNO_3 will react with 10.00 g of iodine? *Ans.* 0.0788.

16.63. A solution was prepared by dissolving 500 ml of H_2S at S.C. in 3000 ml of water. How many milliliters of chlorine at S.C. would react with 100 ml of the H_2S solution? *Ans.* 16.7 ml.

16.64. How many milliliters of 0.200N $KMnO_4$ solution are equivalent to 1.00 ml of $FeSO_4$ solution containing 20 mg of Fe^{+2} ion per milliliter? *Ans.* 1.79 ml.

16.65. It is desired to prepare a 0.100N solution of $Na_2S_2O_3$. It was found that 25.0 ml of a given solution of $Na_2S_2O_3$ was equivalent to 0.470 g of iodine. To what volume would 1000 ml of the $Na_2S_2O_3$ solution have to be diluted in order to obtain a 0.100N solution? *Ans.* 1480 ml.

16.66. An iodine solution is prepared by dissolving 0.562 g of crude iodine in a strong solution of KI in water. The solution of iodine required 6.96 ml of 0.150N $Na_2S_2O_3$ for complete reaction. What was the per cent of iodine in the crude sample? *Ans.* 23.6%.

16.67. How many milliliters of a solution containing 25.00 g of $KMnO_4$ per liter will react with 1.50 g of $FeSO_4$? *Ans.* 12.5 ml.

16.68. A solution is known to contain 150.0 mg of $ZnSO_4$ per milliliter of solution. How many milliliters of the $ZnSO_4$ solution would be required to prepare 500 ml of 0.100N $ZnSO_4$? *Ans.* 26.9 ml.

17

Determination of the Molecular Weights of Compounds

Physical methods will be described by means of which molecular weights may be determined for substances in water solution. The methods are limited to nonvolatile, nonionizable solutes which do not react chemically with water. Physical methods for the determination of molecular weights give approximate values. Such approximate values are used in conjunction with the more accurate values obtained by chemical methods to determine the molecular weights of compounds (Chapter 13). An understanding of the experimental methods for the determination of molecular weights gives one a greater appreciation of the meaning and usage of the concept of molecular weight.

17.1. Introduction. Solutions of the same molality contain the same ratio of solute to solvent molecules. For example, in a one-molal water solution there is one solute molecule for every 55.5 molecules of water, since:

$$1.00 \text{ mole of solute} = \text{N molecules}$$

$$\text{and } 1000 \text{ g } H_2O = \frac{1000 \cancel{\text{g } H_2O}}{18.02 \frac{\cancel{\text{g } H_2O}}{\text{mole } H_2O}} = 55.5 \text{ moles} = 55.5\text{N molecules.}$$

The addition of an impurity to a liquid such as water will affect certain physical properties of the liquid in direct proportion to the number of particles present. The vapor pressure is a physical property of water so affected. The vapor pressure of water in turn determines the freezing point and boiling point of the solution. The discussion which follows will show how the molecular weight of the solute may be determined from the *freezing-point lowering* and the *boiling-point elevation* of water solutions.

17.2. Determination of the Molecular Weight of the Solute from the Freezing-Point Lowering. The addition of one mole of solute to 1000 grams of water produces a solution freezing at −1.86° C. This quantity, 1.86° C., is called the *molal freezing-point lowering constant* for water. The constant is different for different liquids. The addition of one mole of methyl alcohol, CH_3OH = 32 g., or one mole of sugar, $C_{12}H_{22}O_{11}$ = 342 g., to 1000 g. of water produces a solution which freezes at −1.86° C. The freezing point of a 0.100m solution would be one-tenth this amount, or −0.186° C. This would be expected since the number of solute molecules in the latter solution is only one-tenth that in a one-molal solution.

The above relationships hold only for very dilute solutions. Solutions actually containing one mole of solute in 1000 grams of water would give experimental values quite different from the calculated values. Solutions of such strength are used only as reference concentrations.

The freezing point depression, ΔT_f, is directly proportional to the *molal* concentration, m, of the solute. That is, $\Delta T_f = K_f m$, where K_f is the molal freezing-point lowering constant for the particular solvent used (Table 17.1). The constant K_f has the dimensions $\frac{°C \times \text{kg solvent}}{\text{mole}}$.

TABLE 17.1

SOME CONSTANTS FOR SEVERAL SOLVENTS

Solvent	*f.p.* (°C)	K_f	*b.p.* (°C)	K_b
Water	0	−1.86	100	0.52
Benzene	5.51	−4.90	80.10	2.53
Acetic acid	16.6	−3.90	118.1	3.07

Freezing point depression and boiling point elevation of a solvent are most useful in determining the molecular weight of the solute. We will use the formula $\Delta T_f = K_f m$ from which to derive a working formula for molecular weight determination.

Since $m = \frac{\text{mole solute}}{\text{kg solvent}}$ and $\text{mole solute} = \frac{\text{g solute}}{\text{mol. wt. solute}}$, then:

$$m = \frac{\text{g solute}}{\text{kg solvent} \times \text{mol. wt. solute}}.$$

Substituting the above value of m in the equation $\Delta T_f = K_f m$ gives:

$$\Delta T_f = \frac{K_f \times \text{g solute}}{\text{mol. wt. solute} \times \text{kg solvent}}.$$

Solving the above equation for mol. wt. solute gives:

$$\text{mol. wt. solute} = \frac{K_f \times \text{g solute}}{\Delta T_f \times \text{kg solvent}}.$$

Example 17.1. Calculate the molecular weight of urea, given that 4.00 g dissolved in 1000 g of water produces a solution freezing at −0.124° C.

Solution. Two principles are involved: (1) the lowering of the freezing point is directly proportional to the molal concentration, and (2) one mole of urea dissolved in 1000 g of water would give a solution freezing at −1.86° C.

We will now substitute in the above formula to determine the molecular weight of urea using the data given in the problem. Since: 1000 g H_2O = 1.00 kg H_2O, then:

$$\text{mol. wt. urea} = \frac{-1.86 \dfrac{\cancel{{}^\circ\text{C}} \times \cancel{\text{kg H}_2\text{O}}}{\text{mole}} \times 4.00 \text{ g urea}}{-0.124 \cancel{{}^\circ\text{C}} \times 1.00 \cancel{\text{kg H}_2\text{O}}} = 60 \frac{\text{g urea}}{\text{mole}}.$$

Example 17.2. Calculate the molecular weight of a substance A, 2.50 g of which dissolved in 300 g of benzene produced a solution freezing at 5.17° C.

Solution. We will use the same equation as in Example 17.1. Since 300 g benzene = 0.300 kg benzene, and $\Delta T_f = 5.17 - 5.51 = -0.34°$ C, then:

$$\text{mol. wt. A} = \frac{-4.90 \dfrac{\cancel{{}^\circ\text{C}} \times \cancel{\text{kg benzene}}}{\text{mole}} \times 2.50 \text{ g A}}{-0.34 \cancel{{}^\circ\text{C}} \times 0.300 \cancel{\text{kg benzene}}} = 120 \frac{\text{g A}}{\text{mole}}.$$

17.3. Determination of the Molecular Weight of the Solute from the Boiling-Point Elevation. The vapor pressure of a solution is always less than that of the pure solvent, provided the solute is nonvolatile. The boiling point of a solution is, therefore, always higher than the boiling point of the solvent. One mole of solute dissolved in 1000 grams of water will result in a solution boiling at 100.52° C, at one atmosphere pressure. This increase in boiling point, 0.52° C, is called the *molal boiling-point elevation constant* for water. The same relationships exist between the freezing-point lowering and concentration as exist between the boiling-point elevation and concentration.

The boiling point elevation, ΔT_b, is directly proportional to the molal concentration, m, of the solute. That is, $\Delta T_b = K_b m$, where K_b is the molal boiling-point elevation constant (Table 17.1). The dimension of ΔT_b is °C. The constant K_b has the dimensions $\frac{°C \times \text{kg solvent}}{\text{mole}}$. Since the formula $\Delta T_b = K_b m$ is symmetrical with the formula $\Delta T_f = K_f m$, we may assume that the working formulas derived from them would be symmetrical. That is:

$$\Delta T_b = \frac{K_b \times \text{g solute}}{\text{mol. wt. solute} \times \text{kg solvent}}$$

and:

$$\text{mol. wt. solute} = \frac{K_b \times \text{g solute}}{\Delta T_b \times \text{kg solvent}}.$$

Example 17.3. Calculate the molecular weight of a substance A, 4.80 g of which dissolved in 240 g of water gave a solution boiling at 100.065° C at 1.00 atmosphere pressure.

Solution. Since the boiling point of water is 100.000° C at 1.00 atm, then $\Delta T_b = 100.065 - 100.000 = 0.065°$ C, also: 240 g H_2O = 0.240 kg H_2O. Then:

$$\text{mol. wt. A} = \frac{0.52 \frac{°\cancel{C} \times \cancel{\text{kg } H_2O}}{\text{mole}} \times 4.80 \text{ g A}}{0.065° \cancel{C} \times 0.240 \cancel{\text{kg } H_2O}} = 160 \frac{\text{g A}}{\text{mole}}.$$

Example 17.4. A nonvolatile, nonionizable compound A has the empirical formula CH_2O. A solution consisting of 2.80 g of A in 250 g of benzene boiled at 80.33° C at 1.00 atm. Determine the molecular formula for the compound.

Solution. The boiling point of benzene at 1.00 atm is 80.10° C. Therefore $\Delta T_b = 0.23°$ C, also: 250 g benzene = 0.250 kg benzene.

Then:

$$\text{mol. wt. A} = \frac{2.53 \frac{°\cancel{C} \times \cancel{\text{kg benzene}}}{\text{mole}} \times 2.80 \text{ g A}}{0.23° \cancel{C} \times 0.250 \cancel{\text{kg benzene}}}$$

$$= 123 \frac{\text{g A}}{\text{mole}}.$$

The empirical formula CH_2O corresponds to a formula weight of 30.

The true formula would be $\frac{123}{30} = 4$ times 30 or 120, which would correspond to the formula $C_4H_8O_4$. Remember, the above methods are not as accurate as are the methods of chemical analysis.

Problems

Part I

17.1. A solution containing 2 g of sugar in 100 g of water freezes at $-0.11°$ C. What would be the freezing point of a solution containing 4 g of sugar in 100 g of water? *Ans.* $-0.22°$ C.

17.2. Calculate the molecular weight of a substance, 0.41 g of which dissolved in 1000 g of water lowered the freezing point 0.016° C. *Ans.* 48.

17.3. Calculate the molecular weight of a substance, 1.34 g of which dissolved in 125 g of water elevated the boiling point 0.071° C. *Ans.* 78.

17.4. Calculate the freezing point of the solution given in problem 17.3. *Ans.* $-0.26°$ C.

17.5. Calculate the boiling point of a solution containing 4.0 g of sugar, $C_{12}H_{22}O_{11}$, in 250 ml of water at one atmosphere pressure. *Ans.* 100.024° C.

17.6. Calculate the freezing point of a solution containing 6.0 g of urea, CON_2H_4, dissolved in 500 g of water. *Ans.* $-0.37°$ C.

17.7. What would be the ratio of solute to water molecules in a 0.100m solution? *Ans.* 1 : 555.

17.8. How would the ratio of solute to water molecules compare in 0.5m and 1.0m solutions? *Ans.* 1 : 2.

Part II

17.9. Determine the freezing point of a solution prepared by dissolving 2.50 g of urea, CON_2H_4, in 400 g of water. *Ans.* $-0.194°$ C.

17.10. What would be the elevation of the boiling point in problem 17.9? *Ans.* 0.054° C.

17.11. What would be the freezing point of a solution containing 0.0500 mole fraction of sugar in water? *Ans.* $-5.44°$ C.

17.12. What would be the boiling point of a water solution containing 0.050 mole fraction of a nonvolatile, nonionizable solute in water at 760 mm of Hg pressure? *Ans.* 101.52° C.

17.13. Calculate the molecular weight of a substance, 2.12 g of which dissolved in 180 g of water freezes at $-0.118°$ C. *Ans.* 186.

17.14. Calculate the boiling point of the solution given in problem 17.13, at one atmosphere pressure. *Ans.* 100.033° C.

17.15. Find the molecular weight of a compound, 1.21 g of which dissolved in 60 g of water gave a solution boiling at 212.38° F at a pressure of one atmosphere. *Ans.* 50.

17.16. Calculate the freezing point of the solution given in problem 17.15. *Ans.* 30.67° F.

17.17. What weight of urea, CON_2H_4, must be dissolved in 500 g of water to produce the same lowering of the freezing point as 1.50 g of sugar, $C_{12}H_{22}O_{11}$, dissolved in 250 g of water? *Ans.* 0.53 g.

17.18. How much glycerine, $C_3H_8O_3$, must be dissolved in 800 g of water to produce a freezing-point lowering of 0.186° C? *Ans.* 7.36 g.

17.19. A 1 per cent solution of each of two compounds A and B is prepared. The freezing-point lowering of A is twice that of B. How do their molecular weights compare? *Ans.* B twice that of A.

17.20. Calculate the boiling point of a water solution containing one-tenth mole of a substance dissolved in 400 ml of water, pressure one atmosphere. *Ans.* 100.13° C.

17.21. Calculate the freezing point of the solution in problem 17.20. *Ans.* −0.465° C.

17.22. Calculate the freezing points of the following solutions:

a. Two quarts of alcohol, C_2H_5OH, density 0.79 g per ml, mixed with 8 quarts of water. *Ans.* 17.6° F.

b. Two quarts of glycol, $C_2H_6O_2$, density 1.10 g per ml, mixed with 8 quarts of water. *Ans.* 17.1° F.

17.23. Calculate the boiling point of a 5 per cent sugar solution, $C_{12}H_{22}O_{11}$. *Ans.* 100.080° C.

17.24. How many grams of methyl alcohol, CH_3OH, would have to be dissolved in 5000 g of water to produce a solution with a freezing-point lowering of 10° C? *Ans.* 860 g.

17.25. What would be the freezing point of a solution containing 0.40 mole of CH_3OH dissolved in 800 g of water? *Ans.* −0.93° C.

17.26. What would be the boiling point at 760 mm of Hg of a solution containing one mole of solute dissolved in 55.5 moles of water? *Ans.* 100.52° C.

17.27. When one mole of a substance was dissolved in 1000 g of water, the resultant volume was 1030 ml. How many molecules of solute would there be in one ml. of the solution? *Ans.* 5.8×10^{20}.

17.28. How many molecules of solvent are there per milliliter in the solution in problem 17.27? *Ans.* 3.2×10^{22}.

17.29. What is the freezing point of a 5.00 per cent by weight sugar solution? *Ans.* −0.286° C.

17.30. Assuming complete ionization, how many particles in the form of ions would there be in a solution consisting of 1.00 formula weight of NaCl and 1000 g of water? *Ans.* 1.2048×10^{24}.

17.31. What would be the freezing point of the solution in problem 17.30? *Ans.* −3.72° C.

17.32. Assuming complete ionization, how many particles in the form of ions would there be in a solution consisting of 1.00 formula weight of $CaCl_2$ and 1000 g of water? *Ans.* 1.8072×10^{24}.

17.33. What would be the boiling-point elevation of the solution in problem 17.32? *Ans.* 1.56° C.

17.34. A nonelectrolyte has the empirical formula CH_2O. A solution was prepared consisting of 8.00 g of the nonelectrolyte dissolved in 250 g of water. The freezing point of the solution was found to be −0.331° C. What is the molecular formula of the compound? *Ans.* $C_6H_{12}O_6$.

17.35 What per cent solution of glucose, $C_6H_{12}O_6$, would have the same freezing point as a 1.00 per cent solution of urea, $CO(NH_2)_2$? *Ans.* 3.00%.

18

Chemical Equilibria

Many chemical processes are reversible. That is, the products of a reaction may themselves interact, thus setting up a reversible process. Under constant conditions, such as temperature and concentration, such interaction of products will result in an equilibrium being established. Such an equilibrium is a dynamic one. That is, the equilibrants are constantly interacting even though at any instant there exists a given mass ratio among the substances involved. A knowledge of the concentrations of the reacting substances at equilibrium is essential in industries which depend upon a chemical change to obtain a marketable product. Chemical equilibria may involve substances in the solid, liquid, or gaseous states. All plant and animal tissues depend upon specific chemical equilibria to function properly.

Reactions involving ions occur rapidly. The reaction between HCl and NaOH in solution is essentially instantaneous. On the other hand, the reaction between covalent compounds is slow. The reaction, $CCl_4 + 2\ H_2O \rightarrow CO_2 + 4\ HCl$, may require months for measurable quantities to react. The slowness in reaction of covalent compounds is partially explained by bonds that must be broken before reaction occurs. For the above reason molecular equilibria and ionic equilibria will be discussed separately.

Dimensional units will be omitted in most of the solved problems involving chemical equilibria. Since the unit of concentration is *moles per liter*, the dimensional units may be supplied if desired.

The discussion of molecular equilibria will be limited to substances in the gaseous state. The concentration of a solid, and usually a liquid, has a constant value (Example 18.5).

Molecular Equilibria in Gases

18.1. The Equilibrium Constant. Chemical equilibria involve two opposing processes occurring simultaneously and at the same rate.

By *rate of reaction* is meant the amount of reacting material converted to products in a given period of time. Amounts are usually given in moles, and time in seconds.

A chemical equilibrium may be represented by the general equation:

$$A + B \underset{S_2}{\overset{S_1}{\rightleftharpoons}} C + D.$$

In the above reaction S_1 represents the rate of the forward reaction, and S_2 the rate of the reverse reaction.

At equilibrium $S_1 = S_2$. The equilibrium constant for the above reaction would be:

$$K = \frac{(C)(D)}{(A)(B)},$$

where K is called the *equilibrium constant.* The notation () represents concentration in *moles per liter.*

A more general equilibrium equation would be:

$$aA + bB \underset{S_2}{\overset{S_1}{\rightleftharpoons}} cC + dD,$$

for which

$$K = \frac{(C)^c(D)^d}{(A)^a(B)^b}.$$

The value of K for any given reaction is essentially a constant at a given temperature. Note that the value of K is independent of the concentrations of the reacting substances. Any change in the concentrations of the reacting substances by the addition or removal of reacting substances will so shift the equilibrium as to maintain K constant for that temperature. It is evident that the addition of C or D will shift the equilibrium to the left, and the removal of C or D will shift the equilibrium to the right, in either case the value of K remaining constant. Evidently when K is large, C and D predominate; when K is small, A and B predominate.

Example 18.1. A closed reaction chamber containing PCl_5 was heated to 230° C at one atmosphere pressure until equilibrium had been established. Analysis showed the following concentrations in the reaction chamber: $(PCl_5) = 0.45$ mole per liter, $(PCl_3) = (Cl_2) = 0.096$ mole per liter. Calculate K for the reaction $PCl_5 \rightleftharpoons PCl_3 + Cl_2$.

Solution. The equilibrium constant K for the reaction is given by the expression:

$$K = \frac{(PCl_3)(Cl_2)}{(PCl_5)}.$$

Substituting the given concentrations in the above, we have:

$$K = \frac{(0.096)(0.096)}{(0.45)} = 0.0205 \text{ at } 230° \text{ C.}$$

Example 18.2. Quantities of PCl_3 and Cl_2 were placed in a reaction chamber and heated to 230° C at one atmosphere pressure. At equilibrium (PCl_5) = 0.235 mole per liter, and (PCl_3) = 0.174 mole per liter. Calculate (Cl_2).

Solution. For the given reaction at 230° C, K = 0.0205. Solving the equilibrium constant expression for (Cl_2), and substituting concentration values, we have:

$$(Cl_2) = \frac{K \times (PCl_5)}{(PCl_3)} = \frac{(0.0205)(0.235)}{(0.174)} = 0.028 \text{ mole per liter.}$$

Example 18.3. One liter of HI was heated at 500° C and constant pressure until equilibrium had been established according to the equation $2\ HI \rightleftharpoons H_2 + I_2$. Analysis showed the following concentrations in the reaction chamber: (H_2) = 0.42 mole per liter, (I_2) = 0.42 mole per liter, and (HI) = 3.52 mole per liter. Calculate the value of K for the above equation at 500° C.

Solution. For the given equation:

$$K = \frac{(H_2)(I_2)}{(HI)^2} = \frac{(0.42)(0.42)}{(3.52)^2} = 0.014 \text{ at } 500° \text{ C.}$$

Example 18.4. One mole of HI is introduced into the system in equilibrium in Example 18.3. Calculate the concentrations of H_2, I_2, and HI after the system has again reached equilibrium at 500° C.

Solution. The addition of one mole of HI would tend to increase the denominator in the expression for K, thereby decreasing the value of K. In order to retain a constant value for K, the system reacts by further dissociation of HI.

Let x = moles of HI dissociating to maintain K a constant. Then, for each mole of HI dissociating, there will be formed 0.50 mole each of H_2 and I_2. Therefore:

$0.50x$ = moles each of H_2 and I_2 formed to maintain K constant.

At the new equilibrium:

$$(H_2) = (I_2) = 0.42 + 0.50x$$

and

$$(HI) = 3.52 + 1.00 - x.$$

Substituting the above values in the equilibrium equation gives:

$$0.014 = \frac{(0.42 + 0.50x)(0.42 + 0.50x)}{(3.52 + 1.00 - x)^2}.$$

Simplifying and collecting like terms gives:

$$0.24x^2 + 0.55x - 0.11 = 0.$$

Solving the quadratic equation for x gives:

$$x = \frac{-0.55 \pm \sqrt{(0.55)^2 - (4)(0.24)(-0.11)}}{(2)(0.24)}$$
$$= 0.17 \text{ or } -2.3.$$

A negative value has no significance. Therefore, 0.17 mole of HI would dissociate to maintain K constant. Then, under the new equilibrium conditions:

$$(H_2) = 0.42 + 0.50x \quad = 0.42 + 0.09 \quad = 0.51 \text{ mole per liter,}$$
$$(I_2) = 0.42 + 0.50x \quad = 0.42 + 0.09 \quad = 0.51 \text{ mole per liter,}$$
$$(HI) = 3.52 + 1.00 - x = 3.52 + 1.00 - 0.17 = 4.35 \text{ moles per liter.}$$

The validity of these values may be checked by substituting in the original equation. Then:

$$\frac{(0.51)(0.51)}{(4.35)^2} = 0.014 = K.$$

That is, the value of K has been maintained constant by the system in equilibrium.

Example 18.5. Iron filings and water were placed in a 5.0 liter tank and sealed. The tank was heated to 1000° C. Upon analysis the tank was found to contain 1.10 g of hydrogen and 42.50 g of water vapor. The following reaction occurred in the tank:

$$3\ Fe + 4\ H_2O \rightleftharpoons Fe_3O_4 + 4\ H_2.$$

Calculate the equilibrium constant for the reaction.

Solution. First, set up the formula for the equilibrium constant.

$$K = \frac{(Fe_3O_4)(H_2)^4}{(Fe)^3(H_2O)^4}.$$

Since concentration $= \frac{\text{moles}}{\text{volume}}$, the concentration of a pure solid in a chemical equilibrium has a constant value. Since Fe_3O_4 and Fe are solids, their concentrations will be represented by the constants k and k^1. Then:

$$K = \frac{k(H_2)^4}{k^1(H_2O)^4} \quad \text{or} \quad K \times \frac{k^1}{k} = K^1 = \frac{(H_2)^4}{(H_2O)^4}.$$

Also $\qquad (H_2) = \frac{1.10/2.02}{5.0}$ moles per liter = 0.11 mole per *l.*

and $\qquad (H_2O) = \frac{42.50/18.02}{5.0}$ moles per liter = 0.47 mole per *l.*

Then $$K^1 = \frac{(0.11)^4}{(0.47)^4} = 0.0030.$$

Example 18.6. Given that K = 0.395 at 350° C for the equilibrium:

$$2\ NH_3 \rightleftharpoons N_2 + 3\ H_2.$$

Find (NH_3), (N_2), and (H_2) in a 5.00 liter container in which had been placed 15.0 g of NH_3 and heated to 350° C.

Solution. The expression for $K = \frac{(N_2)(H_2)^3}{(NH_3)^2}$.

The initial concentration of NH_3 would be:

$$\frac{15.0 \cancel{\text{g NH}_3}/5.0\ l}{17.0 \frac{\cancel{\text{g NH}_3}}{\text{mole NH}_3}} = 0.177 \frac{\text{mole NH}_3}{l}.$$

All references to concentration will now be in terms of *moles per liter*.

Let x equal the number of moles of ammonia that decompose.

Then: $$(NH_3) = 0.177 - x$$
$$(N_2) = 0.5x$$
and: $$(H_2) = 1.5x.$$

Substituting in the equilibrium equation gives:

$$\frac{(0.5x)(1.5x)^3}{(0.177 - x)^2} = 0.395.$$

With such a relatively large equilibrium constant we are not justified in making the approximation that $0.177 - x = 0.177$. Therefore we must solve the quadratic equation. Then, by partial simplification we have:

$$\frac{1.69x^4}{(0.177 - x)^2} = 0.395.$$

As a second simplification step we will extract the square root of the expression. Then:

$$\sqrt{\frac{1.69x^4}{(0.177 - x)^2}} = \sqrt{0.395} \text{ or } \frac{1.30x^2}{0.177 - x} = 0.628,$$

and $$1.30x^2 + 0.628x - 0.111 = 0$$
$$x = 0.140.$$

Then $$(NH_3) = 0.177 - 0.140 = 0.037 \text{ mole per liter}$$
$$(N_2) = (0.5)(0.140) = 0.070 \text{ mole per liter}$$
and $$(H_2) = (1.5)(0.140) = 0.210 \text{ mole per liter}$$

Problems

Part I

18.1. Given the equation $A \rightleftharpoons B + C$, calculate K if, at equilibrium, A = 4.6 moles per liter, and B = C = 2.3 moles per liter. *Ans.* 1.15.

18.2. Two moles of B were introduced into the system in equilibrium in problem 18.1. Calculate (A), (B), (C) after equilibrium has again been attained. *Ans.* 5.2, 3.6, 1.6.

18.3. Two moles of A were removed from the system in equilibrium in problem 18.1. Calculate (A), (B), (C) at the new equilibrium. *Ans.* 3.0, 1.9, 1.9.

18.4. Analysis showed the following concentrations at 440° C for the equilibrium $2\ HI \rightleftharpoons H_2 + I_2$. Compare the two values of K at 440° C.

	(H_2)	(I_2)	(HI)
Trial 1	0.0317	0.0806	0.347
Trial 2	0.0097	0.333	0.390

Ans. 0.0213, 0.0212.

18.5. The following concentrations were obtained at 5000° C for the reaction in problem 18.4: $(H_2) = 0.22$, $(I_2) = 0.22$, and $(HI) = 1.38$. Compare the value of K with those found in problem 18.4. *Ans.* 0.0255.

18.6. K is equal to 69 for the reaction $N_2 + 3\ H_2 \rightleftharpoons 2\ NH_3$ at 500° C. Analysis of a 7-liter container showed that, at 500° C, there were present 3.71 moles of hydrogen and 4.55 moles of ammonia. How many moles of nitrogen were there in the container? *Ans.* 0.288 mole.

18.7. The equilibrium constant for the reaction $N_2 + O_2 \rightarrow 2\ NO$ is 1.2×10^{-4} at 2000° C. What would be the concentrations of each of the three equilibrants, after attaining equilibrium, if a charge of 100 g each of nitrogen and oxygen were placed in a 25 *l* sealed flask and heated to 2000° C? *Ans.* $N_2 = 0.142$; $O_2 = 0.124$; $NO = 0.00146$.

18.8. Solid carbon and 1.00 g of hydrogen were placed in a 5-liter tank, sealed, and heated to 1000° C. At equilibrium the tank was found to contain 0.22 g of CH_4. Calculate the equilibrium constant for the reaction C (solid) $+ 2\ H_2 \rightleftharpoons CH_4$ at 1000° C. *Ans.* 0.31.

Part II

18.9. PCl_5 is 40 per cent dissociated at a given temperature according to the equation $PCl_5 \rightleftharpoons PCl_3 + Cl_2$. What is the percentage increase in the number of particles in the container due to the dissociation of the PCl_5? *Ans.* 40%.

18.10. At 1500° C water vapor is 5 per cent dissociated according to the equation $2\ H_2O \rightleftharpoons 2\ H_2 + O_2$. Calculate K for water vapor at

1500° C Note: assume a concentration of one mole of water per liter before dissociation. *Ans.* 0.000069.

18.11. If there had been one mole of water vapor in the reaction chamber in problem 18.10, calculate the number of molecules of H_2O, H_2, and O_2 present at equilibrium at 1500° C.
Ans. 0.95N, 0.05N, 0.025N.

18.12. Nitrogen and hydrogen were added to a 5-liter flask under pressure, sealed, and heated. The equilibrium mixture contained 19.0 g of ammonia, 0.160 g of hydrogen, and 3.40 g of nitrogen. Calculate the equilibrium constant for the reaction. *Ans.* 8.1×10^3.

18.13. A 2-liter flask contained an equilibrium mixture consisting of 0.050 mole of SO_3, 0.100 mole of SO_2, and 0.200 mole of O_2 at 300° C. Calculate the equilibrium constant for the reaction $2\ SO_2 + O_2 \rightleftharpoons 2\ SO_3$.
Ans. 2.5.

18.14. The equilibrium constant for the reaction $N_2 + O_2 \rightleftharpoons 2\ NO$ is 2.6×10^{-4} at 4000° C. What would be the concentration in moles of nitrogen and oxygen at equilibrium if 0.250 mole of NO were placed in a 1-liter closed container and heated to 4000° C ?
Ans. 0.124 mole each of N_2 and O_2.

18.15. The equilibrium constant for the reaction $N_2O_4 \rightleftharpoons 2\ NO_2$ is 0.18 at 25° C. What would be the concentration of N_2O_4 and NO_2 in a 500 ml flask in which had been placed 0.025 mole of N_2O_4 at 25° C ?
Ans. $N_2O_4 = 0.020$; $NO_2 = 0.060$.

18.16. At 1500° K the gaseous mixture represented by the equation C (solid) + $CO_2 \rightleftharpoons 2\ CO$ contains 2.0 moles of CO_2 and 8.0 moles of CO per liter. Calculate the equilibrium constant for the reaction at 1500° K. *Ans.* 32.

18.17. What is the equilibrium constant for the reaction $PCl_3 + Cl_2 \rightleftharpoons PCl_5$ at 230° C , given that K = 0.0205 for the reaction $PCl_5 \rightleftharpoons PCl_3 + Cl_2$?
Ans. 48.8.

18.18. One mole of PCl_5 was placed in a 5-liter flask, sealed, and heated to 230° C. Calculate (PCl_5), (PCl_3), and (Cl_2) at equilibrium.
Ans. (PCl_5) = 0.145; (PCl_3) = (Cl_2) = 0.055.

Ionic Equilibria in Solution

18.2. The Ionization Constants of Acids and Bases. Certain covalent compounds such as $HC_2H_3O_2$ and NH_4OH undergo partial ionization when in water solution. An equilibrium is thus established between ionized and nonionized molecules. The equilibrium constant principle may, therefore, be applied to such ionization equations. For example, the ionization equations for acetic acid, $HC_2H_3O_2$, and ammonium hydroxide, NH_4OH, and the equilibrium expressions would be:

$$HC_2H_3O_2 \rightleftharpoons H^+ + C_2H_3O_2^-,$$

and $$K_a = \frac{(H^+)(C_2H_3O_2^-)}{(HC_2H_3O_2)} = 1.84 \times 10^{-5} \text{ at } 25° \text{ C};$$

$$NH_4OH \rightleftharpoons NH_4^+ + OH^-,$$

and $$K_b = \frac{(NH_4^+)(OH^-)}{(NH_4OH)} = 1.8 \times 10^{-5} \text{ at } 25° \text{ C}.$$

K_a is called the *acid constant*, and K_b the *basic constant*.

Example 18.7. Analysis of a solution of acetic acid, $HC_2H_3O_2$, at 25° C showed the following concentrations:

$$(H^+) = (C_2H_3O_2^-) = 0.00150 \text{ mole per liter},$$

and $$(HC_2H_3O_2) = 0.122 \text{ mole per liter}.$$

Calculate K_a for acetic acid at 25° C.

Solution.

$$K_a = \frac{(H^+)(C_2H_3O_2^-)}{(HC_2H_3O_2)} = \frac{(0.00150)(0.00150)}{(0.122)} = 1.84 \times 10^{-5} \text{ at } 25° \text{ C}.$$

TABLE 18.1

IONIC EQUILIBRIUM CONSTANTS AT 25° C

Substance	*Equilibrium Equation*	*Constant*
Acetic acid	$HC_2H_3O_2 \rightleftharpoons H^+ + C_2H_3O_2^-$	1.8×10^{-5}
Carbonic acid	$H_2CO_3 \rightleftharpoons H^+ + HCO_3^-$	4.5×10^{-7}
	$HCO_3^- \rightleftharpoons H^+ + CO_3^{-2}$	6.0×10^{-11}
Hydrocyanic acid	$HCN \rightleftharpoons H^+ + CN^-$	7.0×10^{-10}
Hydrosulfuric acid	$H_2S \rightleftharpoons H^+ + HS^-$	1.1×10^{-7}
	$HS^- \rightleftharpoons H^+ + S^{-2}$	1.0×10^{-15}
Nitrous acid	$HNO_2 \rightleftharpoons H^+ + NO_2^-$	4.5×10^{-4}
Phosphoric acid	$H_3PO_4 \rightleftharpoons H^+ + H_2PO_4^-$	7.5×10^{-3}
	$H_2PO_4^- \rightleftharpoons H^+ + HPO_4^{-2}$	2.0×10^{-7}
	$HPO_4^{-2} \rightleftharpoons H^+ + PO_4^{-3}$	1.0×10^{-12}
Phosphorous acid	$H_3PO_3 \rightleftharpoons H^+ + H_2PO_3^-$	1.7×10^{-2}
Sulfurous acid	$H_2SO_3 \rightleftharpoons H^+ + HSO_3^-$	1.2×10^{-2}
	$HSO_3^- \rightleftharpoons H^+ + SO_3^{-2}$	1.0×10^{-7}
Water	$H_2O \rightleftharpoons H^+ + OH^-$	1.0×10^{-14}
Ammonium hydroxide	$NH_4OH \rightleftharpoons NH_4^+ + OH^-$	1.8×10^{-5}

Example 18.8. At 25° C acetic acid, $HC_2H_3O_2$, is 1.34 per cent ionized in 0.100M solution. Calculate K_a.

Solution. In one liter of 1.00M $HC_2H_3O_2$ there would be one mole. Therefore, in 0.100M $HC_2H_3O_2$ which is 1.34 per cent ionized:

$$(H^+) = (C_2H_3O_2^-) = 0.100 \times 0.0134 = 0.00134 \text{ mole per liter,}$$

and $$(HC_2H_3O_2) = 0.10000 - 0.00134 = 0.09866 \text{ mole per liter.}$$

Then $$K_a = \frac{(0.00134)(0.00134)}{(0.09866)} = 1.82 \times 10^{-5} \text{ at } 25° \text{ C.}$$

Example 18.9. Calculate (H^+) in 1.000M $HC_2H_3O_2$ at 25°C.

Solution. (See Example 18.7.) $(H^+) = (C_2H_3O_2^-) = x$, and $(HC_2H_3O_2) = 1.000 - x$. *Ans.* 4.29×10^{-2} mole per liter.

Example 18.10. Calculate (H^+), $(H_2PO_4^-)$, (HPO_4^{-2}), and (PO_4^{-3}) in 0.100M H_3PO_4. Given, at 25° C:

$$H_3PO_4 \rightarrow H^+ + H_2PO_4^- \qquad K_1 = 7.5 \times 10^{-3}$$
$$H_2PO_4^- \rightarrow H^+ + HPO_4^{-2} \qquad K_2 = 2.0 \times 10^{-7}$$
$$HPO_4^{-2} \rightarrow H^+ + PO_4^{-3} \qquad K_3 = 1.0 \times 10^{-12}.$$

Solution. Since K_2 and K_3 are extremely small compared to K_1, we will assume that essentially all of the H^+ ion originates from the primary ionization step, and:

$$K_1 = \frac{(H^+)(H_2PO_4^-)}{(H_3PO_4)} = 7.5 \times 10^{-3}.$$

Let $x = (H^+) = (H_2PO_4^-)$. Then $(H_3PO_4) = 0.100 - x$. Since K_1 is relatively large, we are not justified in assuming that $0.100 - x = 0.100$. Therefore:

$$\frac{x^2}{0.100 - x} = 7.5 \times 10^{-3}$$

or $$x^2 + 0.0075x - 0.00075 = 0$$

and $$x = 2.4 \times 10^{-2} \frac{\text{mole}}{l}$$

and $$(H^+) = (H_2PO_4^-) = 2.4 \times 10^{-2} \frac{\text{mole}}{l}.$$

Assuming the second ionization step to supply all the $H_3PO_4^{-2}$ ions we have:

$$K_2 = \frac{(H^+)(HPO_4^{-2})}{(H_2PO_4^-)}$$

or: $$(HPO_4^{-2}) = \frac{K_2\,(H_2PO_4^-)}{(H^+)}$$

$$= \frac{(2.0 \times 10^{-7})(2.4 \times 10^{-2})}{(2.4 \times 10^{-2})}$$

$$= 2.0 \times 10^{-7} \frac{\text{mole}}{l}.$$

All PO_4^{-3} ions originate from the step three ionization.

$$K_3 = \frac{(H^+)(PO_4^{-3})}{(HPO_4^{-2})}$$

$$(PO_4^{-3}) = \frac{K_3\,(HPO_4^{-2})}{(H^+)}$$

$$= \frac{(1.0 \times 10^{-12})(2.0 \times 10^{-7})}{(2.4 \times 10^{-2})}$$

$$= 8.3 \times 10^{-18} \frac{\text{mole}}{l}.$$

18.3. The *p*H and *p*OH Values of Water Solutions. The ionic dissociation of water may be represented by the equilibrium:

$$H_2O \rightleftharpoons H^+ + OH^-.$$

Applying the equilibrium constant principle gives:

$$K_1 = \frac{(H^+)(OH^-)}{(H_2O)}.$$

K_1 is known as the *ionization constant* for water.

It has been shown experimentally that, at room temperature, 10,000,000 liters of water contain one mole of H^+, 1.008 g, and one mole of OH^-, 17.007 g.
Therefore:

$$(H^+) = (OH^-) = \frac{1}{10{,}000{,}000} = 10^{-7} \text{ mole per liter in water.}$$

A quantity called the *ion product*, K_w, for water is given as:

$$K_w = (H^+)(OH^-) = 10^{-7} \times 10^{-7} = 10^{-14} \text{ at } 25° \text{ C.}$$

Since K_w is constant at any given temperature, in any water solution the product of the concentrations of the H^+ and OH^- ions must be 10^{-14} at 25° C when concentration is expressed as moles per liter. The addition of an acid to water will increase the hydrogen ion concentration; the addition of a base will increase the hydroxyl ion concentration. The increase in either case will be accompanied by a corresponding decrease in the other ion such that K_w is maintained constant. It is evident that in an acid solution the H^+ ion concentration is greater than 10^{-7} mole per liter, and in a solution of a base the H^+ ion concentration is less than 10^{-7} mole per liter.

Concentrations of H^+ and OH^- ions as expressed above are somewhat cumbersome to handle. Because of this, a system has been devised in which the acidity or alkalinity of solutions may be expressed as *p*H or *p*OH values. Table 18.2 shows the relationship among the quantities *p*H, *p*OH, (H^+), and (OH^-). In Table 18.2 observe that $pH + pOH = 14$ at 25° C.

*p*H is defined as the logarithm of the reciprocal of the hydrogen ion concentration. That is:

$$pH = \log \frac{1}{(H^+)}.$$

TABLE 18.2

HYDROGEN ION AND HYDROXYL ION CONCENTRATIONS AT 25° C

(H^+)	*p*H	(OH^-)	*p*OH
10^{0}	0	10^{-14}	14
10^{-1}	1	10^{-13}	13
10^{-2}	2	10^{-12}	12
10^{-3}	3	10^{-11}	11
10^{-7}	7	10^{-7}	7
10^{-12}	12	10^{-2}	2
10^{-13}	13	10^{-1}	1
10^{-14}	14	10^{0}	0

Example 18.11. Calculate the *p*H and *p*OH of a water solution containing 1.0×10^{-6} mole of H^+ ion per liter at 25° C.

Solution. Substituting the concentration of the hydrogen ion in the above formula gives:

$$pH = \log \frac{1}{1.0 \times 10^{-6}} = \log \frac{1}{0.000001} = \log 1{,}000{,}000 = 6.$$

Since $pH + pOH = 14$ at 25° C, then:

$$pOH = 14 - 6 = 8.$$

*p*H may be defined as the logarithm of the number of liters of solution containing one mole of H^+ ion.

Example 18.12. Calculate the pH of a water solution containing 1.0×10^{-6} mole of H^+ ion per liter at 25° C.

Solution. This problem is identical with Example 18.9.

Given that $1.0 \times 10^{-6} = 0.000001$ mole of H^+ ion is contained in 1.0 liter of water, then 1.0 mole of H^+ ion would be contained in $\frac{1}{0.000001} = 1{,}000{,}000$ liters of water Therefore:

$$pH = \log 1{,}000{,}000 = 6.$$

Observe that the above solution is essentially the same as that given for Example 18.9.

Example 18.13. A 0.050M solution of hydrocyanic acid, HCN, has a pH of 5.4 at 25° C. Calculate K_a for HCN at 25° C.

Solution.

$$HCN \rightleftharpoons H^+ + CN^-. \text{ Therefore:}$$

$$K_a = \frac{(H^+)(CN^-)}{(HCN)}$$

and
$$pH = \log \frac{1}{(H^+)}$$

or
$$5.4 = \log \frac{1}{(H^+)}$$

and
$$\log (H^+) = -5.4$$

or
$$(H^+) = 4.0 \times 10^{-6} = (CN^-).$$

Since the degree of ionization is so small the approximation will be made that $(HCN) = 0.050 = 5.0 \times 10^{-2}$. Then:

$$K_a = \frac{(4.0 \times 10^{-6})(4.0 \times 10^{-6})}{5.0 \times 10^{-2}} = 3.2 \times 10^{-10} \text{ at } 25° \text{ C.}$$

Example 18.14. Calculate the pH of 0.10M H_3PO_4 at 25° C.

Solution. From Example 18.10 we see that (H^+) in 0.100M H_3PO_4 at 25° C is equal to $2.4 \times 10^{-2} \frac{\text{mole}}{l}$.

Then:
$$pH = \log \frac{1}{2.4 \times 10^{-2}}$$
$$= 1.62.$$

Example 18.15. Calculate the per cent of ionization of 0.050M acetic acid.

Solution.

Let $(H^+) = (C_2H_3O_2^-) = x$ mole per liter.

Then $(HC_2H_3O_2) = 0.05 - x$. Since x is very small, $(HC_2H_3O_2)$ may be assumed to be equal to 0.05.

$$K_a = \frac{(H^+)(C_2H_3O_2^-)}{(HC_2H_3O_2)}$$

and $1.8 \times 10^{-5} = \dfrac{x^2}{0.05}$

or $x = 9.5 \times 10^{-4}$ mole per liter of H^+ and $C_2H_3O_2^-$,

and per cent ionization $= \dfrac{9.5 \times 10^{-4} \text{ mole per liter}}{0.050 \text{ mole per liter}} \times 100 = 1.9$ per cent

Problems

Part I

18.19. An acid dissociates according to the equation $HA \rightleftharpoons H^+ + A^-$. A one-molar solution of the acid is 1 per cent ionized. What is the value of K_a? *Ans.* 0.0001.

18.20. An acid dissociates according to the equation $H_2A \rightleftharpoons H^+ + HA^-$. A 0.100M solution of the acid is 1 per cent ionized. What is the value of K_a? *Ans.* 1.0×10^{-5}.

18.21. The ionization of a one-molar solution of HCN is 0.010 per cent at 18° C. Calculate K_a. *Ans.* 1.0×10^{-8}.

18.22. Calculate the value of K_a for a 0.0010M solution of acetic acid which is 12.6 per cent ionized at 25° C. *Ans.* 1.8×10^{-5}.

18.23. Solve Example 18.11 using $(HCN) = 0.050 - (2.51 \times 10^{-5})$. *Ans.* 1.3×10^{-8}.

18.24. Solve Example 18.12 assuming the approximation valid that $(H_3PO_4) = 0.10$. *Ans.* 2.7×10^{-2}.

18.25. What is the per cent ionization in 0.10M HNO_2 at 25° C? *Ans.* 6.7%.

18.26. What is the hydrogen ion concentration in 0.10M NaOH, assuming 100 per cent ionization? *Ans.* 1.0×10^{-13}.

18.27. The concentration of hydrogen ion in a solution is 0.001 mole per liter. Calculate: (a) the grams of hydrogen ion per liter, (b) the liters of solution containing one mole of hydrogen ion, and (c) the *p*H and *p*OH values. *Ans.* (a) 0.001008 g; (b) 1000 *l*; (c) 3, 11.

18.28. Calculate the *p*H value of 0.100M HCl, assuming complete ionization. *Ans.* 1.

18.29. Calculate the *p*OH value of 0.100M NaOH, assuming complete ionization. *Ans.* 1.

18.30. Calculate the *p*H and *p*OH of 0.30M HCl which is 88 per cent ionized. *Ans.* 0.6, 13.4.

18.31. Following three successive additions of HCl to water, the respective hydrogen ion concentrations were 10^{-6}, $10^{-4.8}$, and $10^{-1.65}$. Calculate: (a) the concentration of the hydroxyl ion, (b) the pH, and (c) the pOH of each of the solutions. *Ans.* (a) 10^{-8}, $10^{-9.2}$, $10^{-12.35}$; (b) 6, 4.8, 1.65; (c) 8, 9.2, 12.35.

18.32. Following three successive additions of NaOH to water, the respective hydroxyl ion concentrations were 10^{-4}, $10^{-2.15}$, and $10^{-0.364}$. Calculate (a) the concentration of the hydrogen ion, (b) the pH, and (c) the pOH of each of the solutions.
Ans. (a) 10^{-10}, $10^{-13.85}$, $10^{-13.636}$; (b) 10, 13.85, 13.636; (c) 4, 2.15, 0.364.

Part II

18.33. Calculate K_a for hydrofluoric acid, HF, given that it is 9.0 per cent ionized in 0.10M solution. *Ans.* 8.9×10^{-4}.

18.34. Calculate the percentage of ionization of 0.050M H_2S, the secondary ionization being negligible. *Ans.* 0.15%.

18.35. What is the molarity of an HCN solution having a CN^- ion concentration of 2.0×10^{-5} mole per liter? *Ans.* 0.57M.

18.36. A 0.50M H_3PO_4 solution is 12 per cent ionized according to the equation, $H_3PO_4 \rightleftharpoons H^+ + H_2PO_4^-$, at 25° C. Calculate K_a for the above reaction. *Ans.* 8.2×10^{-3}.

18.37. In a 0.10M solution of H_2S at 25° C the primary ionization is 0.080 per cent. Calculate the concentration of H^+ due to the primary ionization. *Ans.* 8.0×10^{-5}.

18.38. At 25° C ammonium hydroxide is 1.33 per cent ionized in 0.20M solution. Calculate K_b for ammonium hydroxide at 25° C.
Ans. 3.5×10^{-5}.

18.39. Calculate the hydrogen ion concentration in moles per liter of a 0.010M solution of acetic acid at 25° C. *Ans.* 4.2×10^{-4}.

18.40. Calculate the value of K_a for a 0.010M solution of acetic acid, 4.2 per cent ionized, at 25° C. *Ans.* 1.8×10^{-5}.

18.41. A 0.10M solution of ammonium hydroxide is 1.3 per cent ionized at 25° C. What is the value of K_b? *Ans.* 1.8×10^{-5}.

18.42. A one-molar solution of HNO_2 is about 2 per cent ionized at room temperature. Calculate K_a for HNO_2. *Ans.* 4.0×10^{-4}.

18.43. Calculate (OH^-) of 0.10M NH_4OH.
Ans. 1.3×10^{-3} mole per liter.

18.44. What is the per cent ionization of 0.0010M HNO_2 solution at 25° C?
Ans. 48%.

18.45. What is the pH of the solution in problem 18.44? *Ans.* 3.32.

18.46. A 0.010M solution of acetic acid has a pH of 3.4. What is the per cent ionization? *Ans.* 3.98%.

18.47. What is the per cent ionization in a solution which contains 100 g of acetic acid dissolved in water to make 1000 ml of solution? *Ans.* 0.33%.

18.48. What is the per cent ionization of 6.0M acetic acid at 25° C ? *Ans.* 0.173%.

18.49. What is the *p*H of the solution in problem 18.48? *Ans.* 1.98.

18.50. A solution contains 25.0 g of ammonium hydroxide in 500 ml of solution. What is the *p*H of the solution at 25° C ? *Ans.* 11.70.

18.51. In 0.010M solution HNO_2 has a *p*H of 2.7 at 25° C. Calculate K_a for HNO_2 at 25° C. *Ans.* 4.0×10^{-4}.

18.52. Calculate the hydrogen ion concentration in 1.00M hydrocyanic acid solution. *Ans.* 2.6×10^{-5}.

18.53. Calculate the per cent ionization of the HCN solution in problem 18.52. *Ans.* 2.6×10^{-3}.

18.54. Calculate the concentration of H^+ and OH^- in 1.00M NH_4OH solution at 25° C. *Ans.* $(H^+) = 2.4 \times 10^{-12}$; $(OH^-) = 4.2 \times 10^{-3}$.

18.55. Calculate the *p*H and the *p*OH of the solution in problem 18.54. *Ans.* *p*H = 11.6; *p*OH = 2.4.

18.56. The OH^- concentration in 0.10M NH_4OH solution is 1.34×10^{-3} mole per liter at 25° C. Calculate K_b for NH_4OH at 25° C. *Ans.* 1.8×10^{-5}.

18.57. What concentration of ammonium hydroxide is one per cent ionized at 25° C ? *Ans.* 0.17M.

18.58. What concentration of acetic acid is one per cent ionized at 25° C ? *Ans.* 0.17M.

18.59. Calculate the concentration of hydrogen ion in moles per liter of a water solution having a *p*H of 4.5. *Ans.* 3.2×10^{-5}.

18.60. A solution has a hydrogen ion concentration of 2.5×10^{-4} mole per liter. Calculate the *p*H of the solution. *Ans.* 3.6.

18.61. Calculate the *p*H, (OH^-), and (H^+) of 0.001M HCl, assuming complete ionization. *Ans.* 3, 10^{-11}, 10^{-3}.

18.62. Calculate the *p*H, (OH^-), and (H^+) of 0.001M KOH, assuming complete ionization. *Ans.* 11, 10^{-3}, 10^{-11}.

18.63. Calculate the hydroxyl ion concentration in moles per liter of a solution (a) the *p*H of which is 4.0, and (b) the *p*OH of which is 4.0. *Ans.* (a) 10^{-10}; (b) 10^{-4}.

18.64. What is the concentration of ammonia in grams per liter in 0.10M NH_4OH solution at 25° C ? *Ans.* 1.68 g.

18.65. What is the *p*H value of a 0.00010N solution of acid, assuming 100 per cent ionization? *Ans.* 4.0.

18.66. What is the *p*H value of a 0.00010N solution of base, assuming 100 per cent ionization? *Ans* 10.0.

18.67. What is the HS^- ion concentration of 0.0010N H_2S solution? *Ans.* 1.1×10^{-5}.

The Solubility Product

18.4. The Solubility Product Principle. The *solubility product principle* applies to saturated solutions of very slightly soluble electrolytes. Such electrolytes may be assumed to be 100 per cent ionized in solution, the ions being in equilibrium with the undissolved solid according to the equation:

$$\underset{\text{solid}}{BaSO_4} \rightleftharpoons \underbrace{Ba^{+2} + SO_4^{-2}}_{\text{ions in solution}}.$$

The equilibrium constant expression for the above equation is:

$$K = \frac{(Ba^{+2})(SO_4^{-2})}{(BaSO_4)}.$$

Since the concentration of an undissolved solid such as $BaSO_4$ is a constant, the above expression becomes:

$$K(BaSO_4) = K_{S.P.} = (Ba^{+2})(SO_4^{-2}),$$

where $K_{S.P.}$ is called the *solubility product constant.* The solubility product is therefore the product of the concentrations of the ions originating from a salt. Evidently, if the product of the concentrations of the ions is less than $K_{S.P.}$, the solution is unsaturated. If two solutions are mixed, one of which contains Ba^{+2} ions and the other SO_4^{-2} ions, precipitation of $BaSO_4$ will occur only if $(Ba^{+2})(SO_4^{-2}) > K_{S.P.}$ for $BaSO_4$.

$$K_{S.P.} = (Ba^{+2})(SO_4^{-2}) = 1.1 \times 10^{-10} \text{ at } 25° \text{ C}.$$

$K_{S.P.}$ varies with temperature, most values being given at 25° C.

TABLE 18.3

SOLUBILITY PRODUCTS AT 25° C

Substance	*Formula*	$K_{S.P.}$
Aluminum hydroxide	$Al(OH)_3$	3.7×10^{-15}
Calcium carbonate	$CaCO_3$	8.7×10^{-9}
Copper sulfide (18° C)	CuS	8.5×10^{-45}
Silver bromide	$AgBr$	7.7×10^{-13}
Silver chloride	$AgCl$	1.6×10^{-10}
Silver iodide	AgI	1.5×10^{-16}
Silver sulfide (18° C)	Ag_2S	1.6×10^{-49}
Zinc sulfide (18° C)	ZnS	1.2×10^{-23}

Example 18.16. A solution of AgCl in equilibrium with the solid contained 1.3×10^{-5} mole of Ag^+ ion per liter and 1.3×10^{-5} mole of Cl^- ion per liter. Calculate $K_{S.P.}$ for AgCl.

Solution.

$$K_{S.P.} = (Ag^+)(Cl^-) = (1.3 \times 10^{-5})(1.3 \times 10^{-5}) = 1.7 \times 10^{-10}.$$

Example 18.17. The solubility of $BaSO_4$ in water at 18° C is 0.00233 g per liter. Calculate $K_{S.P.}$ for $BaSO_4$.

Solution. The concentrations of the Ba^{+2} and SO_4^{-2} ions must be found. Since 1 mole of barium sulfate yields 1 mole each of the two ions, then:

$$(Ba^{+2}) = (SO_4^{-2}) = (BaSO_4) = \frac{0.00233 \cancel{\text{ g } BaSO_4}}{233 \frac{\cancel{\text{g } BaSO_4}}{\text{mole}}} = 0.00001 \text{ mole.}$$

Therefore:

$K_{S.P.} = (0.00001)(0.00001) = 1.0 \times 10^{-10}$ = solubility product constant for $BaSO_4$ at 18° C.

Example 18.18. The solubility of $Mg(OH)_2$ in water at 25° C is 0.00912 g per liter. Calculate $K_{S.P.}$ for $Mg(OH)_2$, assuming complete ionization.

Solution. $Mg(OH)_2$ ionizes according to the equation:

$$\underset{\text{1 mole}}{Mg(OH)_2} \rightleftharpoons \underset{\text{1 mole}}{Mg^{+2}} + \underbrace{OH^- + OH^-}_{\text{2 moles}}.$$

Therefore:

$$K_{S.P.} = (Mg^{+2})(OH^-)(OH^-) = (Mg^{+2})(OH^-)^2$$

and concentration of $Mg(OH)_2 = \frac{0.00912 \cancel{g}}{58.3 \frac{\cancel{g}}{\text{mole}}} = 0.000156$ mole per liter.

Since one mole of $Mg(OH)_2$ yields one mole of Mg^{+2} ion and two moles of OH^- ion, then:

$(Mg^{+2}) = 0.000156$ mole per liter,

and $(OH^-) = 2 \times 0.000156 = 0.000312$ mole per liter.

Substituting the above values in the expression for $K_{S.P.}$ gives:

$K_{S.P.} = (0.000156)(0.000312)^2 = 1.52 \times 10^{-11}$ for $Mg(OH)_2$ at 25° C.

Example 18.19. One liter of solution was prepared containing 0.00408 mole of $Pb(NO_3)_2$. To this solution was added 0.0105 mole of NH_4Cl. Given that $K_{S.P.}$ for $PbCl_2$ is equal to 2.4×10^{-4}, determine whether or not $PbCl_2$ will precipitate from the solution.

Solution. If the ionic product as given by the expression:

$$K_{S.P.} = (Pb^{+2})(Cl^-)^2$$

is greater than 2.4×10^{-4}, precipitation of $PbCl_2$ will occur.

Since one mole of $Pb(NO_3)_2$ yields one mole of Pb^{+2} ion and one mole of NH_4Cl yields one mole of Cl^- ion, then:

$$(Pb^{+2}) = 0.00408 \text{ mole per liter,}$$

and

$$(Cl^-) = 0.0105 \text{ mole per liter.}$$

Therefore, $(Pb^{+2})(Cl^-)^2 = (0.00408)(0.0105)^2 = 4.5 \times 10^{-7}$.

Since 4.5×10^{-7} is less than 2.4×10^{-4}, the value of $K_{S.P.}$ for $PbCl_2$, precipitation of $PbCl_2$ will not occur.

Example 18.20. Given that $K_{S.P.}$ for $Al(OH)_3$ is equal to 3.7×10^{-15}, calculate the solubility of $Al(OH)_3$ in grams per liter.

Solution. From the equation:

$$\underset{1\ \text{mole}}{Al(OH)_3} \rightleftharpoons \underset{1\ \text{mole}}{Al^{+3}} + \underset{3\ \text{moles}}{3\ OH^-}$$

we have that:

$$K_{S.P.} = (Al^{+3})(OH^-)^3 = 3.7 \times 10^{-15}.$$

Let x = moles of $Al(OH)_3$ dissolved per liter.

Then x = moles of Al^{+3} ion per liter,

and $3x$ = moles of OH^- ion per liter.

That is, $(x)(3x)^3 = 3.7 \times 10^{-15}$

or $x = 1.1 \times 10^{-4}$ moles of $Al(OH)_3$ per liter.

Since one mole of $Al(OH)_3$ is equal to 78 g., then:

$$1.1 \times 10^{-4} \text{ mole of } Al(OH)_3 = (1.1 \times 10^{-4})(78) = 8.6 \times 10^{-3}\ \text{g.}$$

That is, the solubility of $Al(OH)_3$ is 8.6×10^{-3} gram per liter.

Example 18.21. What must be the concentration of Ag^+ ion in a solution containing 2.0×10^{-6} mole of Cl^- ion per liter to just start precipitation of AgCl, given that $K_{S.P.}$ for AgCl is equal to 1.7×10^{-10}?

Solution. In a saturated solution of AgCl:

$$(Ag^+)(Cl^-) = 1.7 \times 10^{-10}$$

or

$$(Ag^+) = \frac{1.7 \times 10^{-10}}{(Cl^-)}$$

and

$$(Ag^+) = \frac{1.7 \times 10^{-10}}{2.0 \times 10^{-6}} = 8.5 \times 10^{-5} \text{ mole per liter.}$$

The solubility product involves the use of very small quantities. However, such quantities are quite significant and find wide application in qualitative and quantitative analysis.

18.5. Hydrolysis Constants. Hydrolysis is the process in which the ions of a salt and water interact to form a slightly ionized acid or base. The hydrolysis of sodium acetate, $NaC_2H_3O_2$, results in the formation of acetic acid, $HC_2H_3O_2$, which is a weak acid. Sodium acetate is completely ionized in water solution.

$$NaC_2H_3O_2 \rightarrow Na^+ + C_2H_3O_2^-.$$

The acetate ions, $C_2H_3O_2^-$, combine with H^+ ions from water to form acetic acid. The equation of hydrolysis for sodium acetate is obtained by adding the following equilibrium equations.

$$\begin{array}{c} H_2O \rightleftharpoons \cancel{H^+} + OH^- \\ \cancel{H^+} + C_2H_3O_2^- \rightleftharpoons HC_2H_3O_2 \\ \hline C_2H_3O_2^- + H_2O \rightleftharpoons HC_2H_3O_2 + OH^- \end{array}$$

A water solution of sodium acetate is basic because of the excess of OH^- ions present.

The hydrolysis constant, K_h, for a salt which hydrolyzes to yield a weak acid may be obtained from the equilibrium equations for water and the weak acid HX. First solve each equation for $[H^+]$, and equate. Then:

$$K_w = [H^+][OH^-] \quad \text{or} \quad [H^+] = \frac{K_w}{[OH^-]}$$

and $$K_a = \frac{[H^+][X^-]}{[HX]} \quad \text{or} \quad [H^+] = \frac{K_a[HX]}{[X^-]}.$$

Equating gives:

$$\frac{K_w}{[OH^-]} = \frac{K_a[HX]}{[X^-]}$$

or $$\frac{[HX][OH^-]}{[X^-]} = \frac{K_w}{K_a} = K_h.$$

Example 18.22. What is the K_h value for sodium acetate at 25° C?

Solution. From the previous discussion we see that $K_h = \frac{K_w}{K_a}$. In Sec. 18.3 K_w was given as equal to 1.0×10^{-14} at 25° C, and in Table 18.1 K_a for acetic acid was given as equal to 1.8×10^{-5} at 25° C. Therefore, for sodium acetate at 25° C:

$$K_h = \frac{1.0 \times 10^{-14}}{1.8 \times 10^{-5}} = 5.6 \times 10^{-10}.$$

The hydrolysis of ammonium chloride, NH_4Cl, results in the formation of the weak base ammonium hydroxide, NH_4OH. Ammonium chloride is completely ionized in water solution.

$$NH_4Cl \rightarrow NH_4^+ + OH^-.$$

The ammonium ions, NH_4^+, combine with OH^- ions from water to form ammonium hydroxide. The equation of hydrolysis for ammonium chloride is obtained by adding the following equilibrium equations.

$$\begin{array}{r} H_2O \rightleftharpoons H^+ + \cancel{OH^-} \\ NH_4^+ + \cancel{OH^-} \rightleftharpoons NH_4OH \\ \hline NH_4^+ + H_2O \rightleftharpoons NH_4OH + H^+ \end{array}$$

A water solution of ammonium chloride is acidic due to the excess of H^+ ions present.

The hydrolysis constant, K_h, for a salt which hydrolyzes to yield a weak base may be obtained from the equilibrium equations for water and the weak base MOH. Then:

$$K_w = [H^+][OH^-] \quad \text{or} \quad [OH^-] = \frac{K_w}{[H^+]}.$$

and

$$K_b = \frac{[M^+][OH^-]}{[MOH]} \quad \text{or} \quad [OH^-] = \frac{K_b[MOH]}{[M^+]}.$$

Equating gives:

$$\frac{K_w}{[H^+]} = \frac{K_b[MOH]}{[M^+]}$$

or

$$\frac{[MOH][H^+]}{[M^+]} = \frac{K_w}{K_b} = K_h.$$

Example 18.23. What is the K_h value for ammonium chloride at 25° C ?

Solution. From the previous discussion we see that $K_h = \frac{K_w}{K_b}$. In Table 18.1 we see that K_b for ammonium hydroxide is 1.8×10^{-5} at 25° C. Therefore,

$$K_h = \frac{1.0 \times 10^{-14}}{1.8 \times 10^{-15}} = 5.6 \times 10^{-10} \text{ at } 25° \text{ C}.$$

It is purely coincidence that K_a for acetic acid and K_b for ammonium hydroxide are numerically the same at 25° C.

Example 18.24. Calculate the *p*H of 0.100M $NaHCO_3$ in water solution at 25° C. For H_2CO_3 at 25° C, $K_1 = 4.5 \times 10^{-7}$ and $K_2 = 6.0 \times 10^{-11}$.

Solution. This problem requires the manipulation of a number of constants. To find *p*H we must know (H^+). Also, since we are dealing with a water solution, the equation $K_w = (H^+)(OH^-) = 1.0 \times 10^{-14}$ must be

satisfied. The HCO_3^- ions undergo two reactions in water solution.
Ionization: $HCO_3^- \rightleftharpoons H^+ + CO_3^{-2}$

$$K_2 = \frac{(H^+)(CO_3^{-2})}{(HCO_3^-)} = 6.0 \times 10^{-11}.$$

Hydrolysis: $HCO_3^- + H_2O \rightleftharpoons OH^- + H_2CO_3$

$$K_h = \frac{K_w}{K_1} = \frac{(OH^-)(H_2CO_3)}{(HCO_3^-)} = \frac{1.0 \times 10^{-14}}{4.5 \times 10^{-7}} = 2.2 \times 10^{-8}.$$

In order to maintain electrical neutrality, the same number of HCO_3^- ions must be involved in each of the above reactions.

Let $x = (CO_3^{-2}) = (H_2CO_3)$. Since x will be extremely small, then:

$$(HCO_3^-) = 0.100 - x = 0.100.$$

We will now substitute the value of x in the equations for K_2 and K_h:

$$K_2 = \frac{(H^+)\ x}{0.100} = 6.0 \times 10^{-11}$$

and

$$K_h = \frac{(OH^-)\ x}{0.100} = 2.2 \times 10^{-8}.$$

Next, we will multiply $K_2 \times K_h$ and substitute 1.0×10^{-14} for the product $(H^+)(OH^-)$.

$$\frac{(1.0 \times 10^{-14})\ x^2}{(0.100)^2} = 1.32 \times 10^{-18}$$

$$x = 1.16 \times 10^{-3}.$$

We can now find the value of the pH by solving K_2 for (H^+), substituting the value of x found above.

$$(H^+) = \frac{(6.0 \times 10^{-11})(0.100)}{1.16 \times 10^{-3}} = 5.2 \times 10^{-9}\ \frac{\text{mole}}{l}$$

$$pH = \log \frac{1}{5.2 \times 10^{-9}} = 8.28$$

Problems

Part I

18.68. How many grams are there in 4.0×10^{-7} mole of Ag^+?

Ans. 4.3×10^{-5} g.

18.69. How many moles are there in 6.4×10^{-6} g. of S^{-2}?

Ans. 2.0×10^{-7} mole.

18.70. At 25° C the solubility of AgCl is 1.8×10^{-3} gram per liter. Calculate $K_{S.P.}$ for AgCl, assuming complete ionization.

Ans. 1.6×10^{-10}.

18.71. A saturated solution of Ag_2CrO_4 was prepared by shaking the pure compound with water. How much of the salt would dissolve in 500 ml of water, given that $K_{S.P.}$ for Ag_2CrO_4 is 9.0×10^{-12}?
Ans. 0.022 g.

18.72. $K_{S.P.}$ for ZnS is 1.2×10^{-23}. Calculate the S^{-2} ion concentration necessary to just start precipitation of ZnS from a 0.005M solution of $ZnSO_4$.
Ans. 2.4×10^{-21}.

18.73. $K_{S.P.}$ for $CaSO_4$ is equal to 6.1×10^{-5}. How many grams of $CaCl_2$ must be added to 500 ml of 0.01M H_2SO_4 to just start precipitation of $CaSO_4$?
Ans. 0.34 g.

Part II

18.74. The solubility of ZnS at 25° C is 3.5×10^{-12} mole per liter. Calculate $K_{S.P.}$ for ZnS.
Ans. 1.2×10^{-23}.

18.75. $K_{S.P.}$ for CaC_2O_4 is equal to 2.6×10^{-9}. Will CaC_2O_4 precipitate from 100 ml of a solution containing 100 mg of Ca^{+2} ion, to which has been added 0.02 mole of $(NH_4)_2C_2O_4$?
Ans. Yes.

18.76. A solution in equilibrium with solid Ag_2S was found to contain 1.8×10^{-16} mole of S^{-2} ion per liter and 1.5×10^{-18} mole of Ag^+ ion per liter. Calculate $K_{S.P.}$ for Ag_2S.
Ans. 4.1×10^{-52}.

18.77. A solution in equilibrium with solid Bi_2S_3 contained 9.6×10^{-20} mole of Bi^{+3} ion per liter and 2.5×10^{-12} mole of S^{-2} ion per liter. Calculate $K_{S.P.}$ for Bi_2S_3.
Ans. 1.4×10^{-73}.

18.78. How many milligrams of $C_2O_4^{-2}$ ion must be present in 100 ml of solution containing 0.050 mg of Ba^{+2} ion in order to just start precipitation of BaC_2O_4, given that $K_{S.P.}$ for BaC_2O_4 is 1.5×10^{-7}?
Ans. 363 mg.

18.79. $K_{S.P.}$ for AgCl is equal to 1.7×10^{-10}. How many grams of AgCl will dissolve in one liter of 0.001M KCl, assuming the KCl to be completely ionized?
Ans. 2.4×10^{-5} g.

18.80. $K_{S.P.}$ for lithium carbonate, Li_2CO_3, is 1.7×10^{-3} at 25° C. How many grams of lithium carbonate will dissolve in 2500 ml of water at 25° C?
Ans. 13.9 g.

18.81. Will precipitation of calcium carbonate occur if 0.100 ml of 0.100N $CaCl_2$ is added to 1.00 *l* of 0.001M Na_2CO_3?
Ans. Yes

18.82. What is the concentration of sulfide ion, in moles per liter, above which precipitation of Ag_2S will take place in 0.000100M $AgNO_3$ solution?
Ans. 1.6×10^{-41}.

18.83. The concentration of Ag^+ ion in a saturated solution of silver chromate, Ag_2CrO_4, at 25° C is 1.6×10^{-4} mole per liter. Calculate $K_{S.P.}$ for Ag_2CrO_4 at 25° C.
Ans. 2.0×10^{-12}.

18.84. The concentration of Mg^{+2} ion in a solution is 0.010M. Above what concentration of OH^- ion in the given solution will $Mg(OH)_2$ precipitate, given that $K_{S.P.}$ for $Mg(OH)_2$ is equal to 1.2×10^{-11} at 18° C?
Ans. 3.4×10^{-5}.

18.85. What is the value of K_h for dilute water solutions of:

a. Potassium acetate at 25° C ? *Ans.* 5.6×10^{-10}.

b. Ammonium chlorate at 25° C ? *Ans.* 5.6×10^{-10}.

c. Sodium cyanide at 25° C ? *Ans.* 1.4×10^{-5}.

d. Potassium nitrite at 25° C ? *Ans.* 2.2×10^{-11}.

18.86. Calculate $[C_2H_3O_2^-]$, $[OH^-]$, and $[HC_2H_3O_2]$ in a 0.100M water solution of sodium acetate.
Ans. $[C_2H_3O_2^-] = [OH^-] = [HC_2H_3O_2] = 7.5 \times 10^{-6}$ mole per liter.

18.87. What percentage of the sodium acetate is hydrolyzed in 0.100M $NaC_2H_3O_2$? *Ans.* $(7.5 \times 10^{-3})\%$.

18.88. What is the *p*H of 0.100M $NaC_2H_3O_2$ solution? *Ans.* 8.90.

18.89. Calculate the *p*H of 0.100M NH_4Cl in water solution at 25° C. $K = 1.8 \times 10^{-5}$. *Ans.* 5.13.

18.90. Calculate (H^+), (OH^-), (HCO_3^-), and the *p*H of 0.250M $NaHCO_3$ in water solution at 25° C. *Ans.* *p*H = 8.34.

18.6. Theory of Acid-Base Titration. Acid-base titration was discussed in Chapter 16. When small increments of a base are added to a water solution of an acid, there are corresponding increases in the *p*H of the solution. Fig. 18.1 shows the curve obtained in such a titration using 0.10N acid.

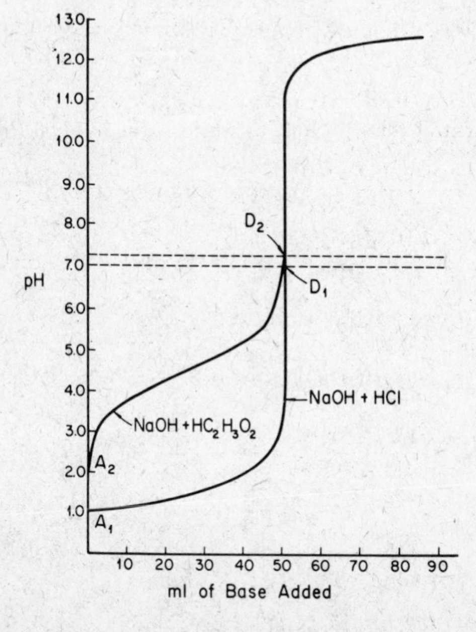

Fig. 18.1.

The point of deflection of the curve is the end point of the titration, and occurs at the *p*H of the salt formed. When a strong acid and strong base are titrated, such as HCl and NaOH, the point of deflection, D_1, corresponds to a *p*H of approximately 7, since NaCl undergoes essentially no hydrolysis. For a 0.100N strong acid such as HCl, the initial *p*H value at A_1 is approximately one. When a 0.100N solution of a weak acid such as acetic acid is used, the initial *p*H value at A_2 will be much higher, due to the slight degree of ionization of the acid. For a weak acid the end point, D_2, has a *p*H value greater than 7 due to the hydrolysis of the salt formed, in this case $NaC_2H_3O_2$. Beyond the end point A_2, the two curves coincide since the change in *p*H is due solely to the addition of NaOH solution.

Example 18.25. Would the *p*H at the end point of the titration be greater than 7 or less than 7 when HCl and NH_4OH are titrated?

Solution. The salt formed would be NH_4Cl. Therefore, due to the hydrolysis of the NH_4Cl the *p*H would be less than 7 at the end point.

It is possible to calculate the *p*H of a solution containing known amounts of standard solutions of acid and base.

Example 18.26. A 0.100N NaOH solution was added to 50.0 ml of 0.100N $HC_2H_3O_2$ solution. What would be the *p*H of the solution after the addition of (a) 50.0 ml of NaOH solution, and (b) 60.0 ml of NaOH solution?

Solution. (a) The addition of 50.00 ml of 0.100N NaOH solution represents the end point of the titration. Therefore, the solution would be:

$$\frac{0.00500 \text{ mole } NaC_2H_3O_2}{0.100\ l \text{ solution}} = 0.0500\text{M with respect to } NaC_2H_3O_2.$$

$$\text{Since } K_h = \frac{K_w}{K_b} = \frac{(HC_2H_3O_2)(OH^-)}{(C_2H_3O_2^-)}, \text{ let } x = (HC_2H_3O_2) = (OH^-).$$

Since a salt is essentially 100% ionized, then:

$$(C_2H_3O_2^-) = 0.05\ \frac{\text{mole}}{l}$$

and:

$$K_h = \frac{1.0 \times 10^{-14}}{1.8 \times 10^{-5}} = \frac{x^2}{5.0 \times 10^{-2}}.$$

Then:

$$x = 5.3 \times 10^{-6}\ \frac{\text{mole } OH^-}{l}.$$

Since: $(H^+)(OH^-) = 1.0 \times 10^{-14}$ at 25° C, then:

$$(H^+) = \frac{1.0 \times 10^{-14}}{5.3 \times 10^{-6}} = 1.9 \times 10^{-9}$$

and: $$pH = \log \frac{1}{1.9 \times 10^{-9}} = 8.7.$$

(b) All OH^- ions will originate from the excess of 10.0 ml of 0.100N NaOH. Therefore:

$$(OH^-) = \frac{(0.010)(0.100) \text{ mole}}{0.110\ l} = 9.1 \times 10^{-3} \frac{\text{mole}}{l}$$

and: $$pOH = \log \frac{1}{9.1 \times 10^{-3}} = 2.04$$

and: $$pH = 14.00 - 2.04 = 11.96.$$

Chemical indicators and pH meters are used to determine the end point of a titration. The ideal indicator for a specific titration should change color at the pH of the solution at the end point. That is, the choice of an indicator is determined by the pH of the solution resulting from the hydrolysis of the salt formed. Chemical indicators are usually weak acids or bases. However, the quantity of indicator used is so small that their contribution of H^+ or OH^- ions is negligible.

18.7. Complex Ion Equilibria. Most cations, particularly heavy metal cations, react with many reagents and anions to form *complex ions*. All complex ions dissociate in water. Some complex ions dissociate so slightly that stable salts of the ions may be prepared. The equilibrium constant for the dissociation of complex ions is called the *instability constant*, K_I. The charge on a complex ion is the algebraic sum of the charges on the cation and the *ligand* forming the complex ion. In the following examples, NH_3, CN^-, and F^- are the ligands.

$$Ag(NH_3)_2^+ \rightleftharpoons Ag^+ + 2\ NH_3 \qquad K_I = 6.0 \times 10^{-8} \frac{\text{mole}^2}{l^2}$$

$$Cu(CN)_2^- \rightleftharpoons Cu^+ + 2\ CN^- \qquad K_I = 1.0 \times 10^{-16} \frac{\text{mole}^2}{l^2}$$

$$Cu(CN)_4^{-2} \rightleftharpoons Cu^{+2} + 4\ CN^- \qquad K_I = 5.0 \times 10^{-31} \frac{\text{mole}^4}{l^4}$$

$$AlF_6^{-3} \rightleftharpoons Al^{+3} + 6\ F^- \qquad K_I = 1.4 \times 10^{-20} \frac{\text{mole}^6}{l^6}$$

Observe the dimensional values given for K_I in the above examples. A comparison of the stability of two or more complex ions is valid only when the K_I values of the ions have the same dimensions.

Example 18.27. What is the (Ag^+) in a solution consisting of 0.001M $AgNO_3$ and 2.000M NH_3 at 25° C?

Solution. From the above equation we see that:

$$K_I = \frac{(Ag^+)(NH_3)^2}{(Ag(NH_3)_2{}^+)}.$$

Let $x = (Ag^+)$. Then, since most of the Ag^+ would be in the form of the complex ion $(Ag(NH_3)_2{}^+) = 0.001 \frac{\text{mole}}{l}$. By approximation, since the complex ion contains 2 NH_3, we have:

$$(NH_3) = 2.000 - 2(0.001 - x) = 2.000 \frac{\text{mole}}{l}.$$

Substituting the values in the above formula for K_I we have:

$$6.0 \times 10^{-8} \frac{\text{mole}^2}{l^2} = \frac{(Ag^+)\left(2.000 \frac{\text{mole}}{l}\right)^2}{\left(0.001 \frac{\text{mole}}{l}\right)}$$

or

$$(Ag^+) = 1.5 \times 10^{-11} \frac{\text{mole}}{l}.$$

Problems

18.91. What would be the initial pH value on the y-axis if 0.050N NaOH were titrated with 0.100N formic acid, HCOOH? K_a for formic acid $= 1.78 \times 10^{-4}$. *Ans.* 2.38.

18.92. Calculate the pH of a solution containing 25.0 ml of 0.100N $HC_2H_3O_2$ and 10.0 ml of 0.100N NaOH. *Ans.* 4.6.

18.93. Calculate the pH of a solution containing 100.0 ml 0.010N NaOH and 25.0 ml of 0.040N HCN. $K_a = 4.9 \times 10^{-10}$. *Ans.* 11.1.

18.94. Calculate the pH of a solution containing 5.00 ml of 0.050N HCl and 50.0 ml of 0.010N NaOH. *Ans.* 10.5.

18.95. What is the (Ag^+) in 500 ml of 1.0M NH_3 containing 0.050 mole of $AgNO_3$? *Ans.* $9.4 \times 10^{-9} \frac{\text{mole}}{l}$.

18.96. The $Cu(NH_3)_4{}^{+2}$ ion dissociates according to the equation $Cu(NH_3)_4{}^{+2} \rightleftharpoons Cu^{+2} + 4\ NH_3$ $(K_I = 5.0 \times 10^{-14})$. What is (Cu^{+2}) if a solution is 0.50M with NH_3 and 0.10M with $CuSO_4$? *Ans.* $5.0 \times 10^{-11} \frac{\text{mole}}{l}$.

18.97. A solution is 0.01M with $Cd(NO_3)_2$ and 2.0M with NH_3. Find (Cd^{+2}) given that K_I for the $Cd(NH_3)_4{}^{+2}$ ion is equal to $5.5 \times 10^{-8} \frac{\text{mole}^4}{l^4}$. *Ans.* $3.7 \times 10^{-11} \frac{\text{mole}}{l}$.

18.98. How many grams of AgI will dissolve in 500 ml of 1.0M NH_3? $K_{S.P.}$ for AgI = 1.0×10^{-16} and for $Ag(NH_3)_2^+$ $K_I = 6.0 \times 10^{-8}$ $\frac{\text{mole}^2}{l^2}$. *Ans.* **4.77 g.**

19

Electrochemistry

The passage of an electric current through a solution brings about chemical changes which obey the laws and principles of chemical changes occurring in test tube reactions. First, electric terms are defined; then follows a discussion of the changes brought about in solution due to the passage of an electric current, and, conversely, the production of an electric current by chemical change as in the lead storage battery and dry cell. As in previous discussions the quantitative aspects are stressed.

Units Associated with the Measurement of Electricity

19.1. Introduction. In many chemical reactions the energy is liberated in the form of electricity rather than heat. Such is the case in a dry cell or storage battery. Also, electric energy may be used to bring about chemical changes.

Electrochemistry involves the study of electric energy, either in its effect in bringing about a chemical change, or as a product of a chemical change.

19.2. Practical Units of Electricity. Certain arbitrary units have been established for the measurement of electric energy. Only the practical units of electricity will be discussed. One *faraday* of electricity is defined as the amount of electric energy required to liberate one gram-equivalent weight of an element from solution. A smaller unit of quantity, the *coulomb*, is defined as the quantity of electricity required to deposit 0.001118 g of silver from a solution containing Ag^+ ions. Since:

1.00 gram-equivalent weight of silver = 107.87 g, then:

$$1.00 \text{ faraday} = \frac{107.87 \cancel{g}}{0.001118 \frac{\cancel{g}}{\text{coulomb}}} = 96{,}490 \text{ coulombs.}$$

An *ampere* is a rate of flow of one coulomb per second. An *ohm* is the resistance of a column of mercury one square millimeter in cross section and 106.300 centimeters in length at 0° C. A *volt* is the potential necessary to drive a current of one ampere through a resistance of one ohm. Ohm's law expresses the relationship among the ampere, volt, and ohm:

$$\text{amperes (I)} = \frac{\text{volts (V)}}{\text{ohms (R)}}.$$

Laws Relating to Conduction in Solution

19.3. Faraday's Laws of Electrolysis. *Electrolysis* is the process resulting from the passage of an electric current through a solution of an electrolyte.

Faraday stated two laws relating to electrolysis. The first law states that the amount of electrochemical change at an electrode is directly proportional to the quantity of electricity flowing through the solution. The number of grams of an element liberated by one coulomb is called the *electrochemical equivalent* of the element; the number of grams liberated by one faraday is called the ***gram-equivalent weight*** of the element. Below is the mathematical statement of the first law:

$$m = eQ.$$

- Q: quantity of electricity in coulombs
- e: electrochemical equivalent in grams
- m: weight in grams of substance liberated

When Q is expressed in faradays, then e becomes the gram-equivalent weight.

Example 19.1. A current of 0.050 ampere was allowed to pass through a solution of silver nitrate for 30 minutes. How much silver was deposited?

Solution. One coulomb will deposit 0.001118 g of silver. A coulomb is one ampere per second. Since the current of 0.050 ampere flowed for $30 \times 60 = 1800$ seconds:

$$1800 \times 0.050 = 90 \text{ coulombs} = \text{electric energy used.}$$

And $0.001118 \frac{\text{g}}{\cancel{\text{coulombs}}} \times 90 \cancel{\text{coulombs}} = 0.10 \text{ g}$ of silver deposited.

Faraday's second law states that, for a given quantity of electricity, the weights of the elements liberated from solution are directly proportional to their electrochemical equivalents or their gram-equivalent weights. That is:

$$\frac{m_1}{m_2} = \frac{e_1}{e_2},$$

where m is the mass of an element deposited and e its electrochemical equivalent or gram-equivalent weight.

Example 19.2. Two electrolytic cells were placed in series. One contained a solution of $AgNO_3$ and the other a solution of $CuSO_4$. Electricity was passed through the cells until 1.273 g of Ag had been deposited. How much copper was deposited at the same time?

Solution. The amounts of the two elements deposited would be in direct proportion of their gram-equivalent weights.

One faraday will deposit 107.87 g of Ag and 31.77 g of Cu (because copper is present as Cu^{+2} ions). Therefore 1.273 g of Ag would require:

$$\frac{1.273 \cancel{\text{ g Ag}}}{107.87 \dfrac{\cancel{\text{g Ag}}}{\text{faraday}}} = \frac{1.273}{107.87} \text{ faraday}$$

and the copper deposited would be:

$$\frac{1.273}{107.87} \cancel{\text{faraday}} \times 31.77 \frac{\text{g Cu}}{\cancel{\text{faraday}}} = 0.3749 \text{ g Cu.}$$

Example 19.3. How many grams each of Ag^+ and Cu^{+2} ions would be deposited by 0.040 faraday?

Solution. Since one faraday deposits one gram-equivalent weight of an element, then 0.040 faraday would deposit 0.040 gram-equivalent weight of an element.

$$107.87 \frac{\text{g Ag}}{\cancel{\text{faraday}}} \times 0.040 \cancel{\text{ faraday}} = 4.3 \text{ g Ag}$$

$$31.77 \frac{\text{g Cu}}{\cancel{\text{faraday}}} \times 0.040 \cancel{\text{ faraday}} = 1.3 \text{ g Cu}$$

Example 19.4. Calculate the atomic weight of calcium, given that 0.0324 faraday liberated 0.651 g of the element. The approximate atomic weight of calcium is 40.

Solution. The weight of an element deposited by one faraday is the gram-equivalent weight of the element.

Then the gram-equivalent weight of calcium is:

$$\frac{0.651 \text{ g Ca}}{0.0324 \text{ faraday}} = 20.1 \frac{\text{g Ca}}{\text{faraday}} = \text{g-eq. wt. of Ca.}$$

$$\text{Oxidation number of Ca} = \frac{40}{20.1} = 2. \text{ Therefore:}$$

$$\text{atomic weight of Ca} = 20.1 \times 2 = 40.2.$$

Example 19.5. It was found that 0.172 g of chromium was deposited by 0.0761 ampere in 3 hours and 30 minutes. Calculate (a) the electrochemical equivalent of chromium, and (b) the gram-equivalent weight of chromium.

Solution. (a) Coulombs used = 12,600 sec $\times$ 0.0761 amp = 959.

By definition, one coulomb deposits one electrochemical equivalent of an element. Therefore:

$$0.172 \text{ g} \div 959 \text{ coulombs} = 0.000179 \text{ g per coulomb}$$
$$= \text{electrochemical equivalent of chromium.}$$

(b) By definition, one gram-equivalent weight or 96,490 electrochemical equivalents of chromium would be deposited from solution by one faraday. Therefore:

$$0.000179 \frac{\text{g}}{\cancel{\text{coulomb}}} \times 96{,}490 \cancel{\text{coulomb}} = 17.3 \text{ g}$$
$$= \text{gram-equivalent weight of chromium.}$$

19.4. The Significance of the Avogadro Number in Electrolysis. In Sec. 6.3 it was shown that one gram-atom of an element contains N atoms, where N is the Avogadro number equal to 6.023×10^{23}. In Sec. 9.3 it was shown that one gram-equivalent weight of an element involves N electrons of that element when it enters into chemical combination with another element. Therefore, in the process of electrolysis N electrons pass through the solution for each faraday used. That is, one faraday contains 6.023×10^{23} electrons.

Example 19.6. What is the charge in coulombs on the N^{-3} ion?

Solution.

$$6.023 \times 10^{23} \text{ electrons} = 96{,}490 \text{ coulombs. Therefore:}$$
$$1.00 \text{ electron} = 1.60 \times 10^{-19} \text{ coulombs,}$$

and

$$3.00 \text{ electrons} = 4.80 \times 10^{-19} \text{ coulombs}$$
$$= \text{charge in coulombs on the } N^{-3} \text{ ion.}$$

Example 19.7. How many atoms of calcium will be deposited from a solution of $CaCl_2$ by a current of 25 milliamperes flowing for 60 seconds?

Solution. Coulombs = $0.025 \times 60 = 1.50$. Since:

$$1.00 \text{ coulomb} = 6.023 \times 10^{23} \div 96{,}490 = 6.24 \times 10^{18} \text{ electrons,}$$

then

$$1.50 \text{ coulombs} = (1.50)(6.24 \times 10^{18}) = 9.36 \times 10^{18} \text{ electrons.}$$

Two electrons are required for each Ca^{+2} ion deposited. Therefore:

$$\frac{9.36 \times 10^{18}}{2} = 4.68 \times 10^{18} \text{ atoms of calcium deposited.}$$

Problems

Part I

19.1. An electric motor uses 7.80 amperes of current. How many coulombs of electricity would be used by the motor per hour?
Ans. 2.81×10^4.

19.2. What is the resistance of the filament in a light bulb which uses 1.00 ampere at 32 volts? *Ans.* 32.

19.3. How much time would be required to use 100,000 coulombs of electricity in an electric iron drawing 10.0 amperes?
Ans. 10,000 sec.

19.4. How much would a copper plate increase in weight if silver-plated by a current of 650 milliamperes for 24 hours? *Ans.* 62.8 g.

19.5. It was found that 0.287 g of nickel was deposited by a current of 175 milliamperes in 90 minutes. Calculate (a) the electrochemical equivalent of nickel, and (b) the gram-equivalent weight of nickel.
Ans. (a) 0.000304 g ; (b) 29.33 g.

19.6. How much hydrogen at standard conditions would be deposited by a current of one ampere flowing for one minute? *Ans.* 6.96 ml.

19.7. A given quantity of electricity was passed through each of two cells containing Cu^{+2} ions and Ag^{+} ions, respectively. It was found that 0.637 g of copper had been deposited in the one cell. How much silver was deposited in the other cell? *Ans.* 2.16 g.

19.8. How much mercurous and mercuric mercury would be deposited by 0.100 ampere in 30 minutes? *Ans.* 0.374 g , 0.187 g.

19.9. How many electrons are there in one coulomb? *Ans.* 6.2×10^{18}.

19.10. How many molecules of chlorine would be deposited from a solution containing Cl^{-} ions in one minute by a current of 300 milliamperes?
Ans. 5.6×10^{19}.

Part II

19.11. How many grams each of Ag^{+} ion, Cu^{+2} ion, and Fe^{+3} ion would be deposited by 50,000 coulombs?
Ans. 56.0 g Ag, 16.5 g Cu, 9.65 g Fe.

19.12. A current of 500 milliamperes flowing for exactly one hour deposited 0.6095 g of zinc. Determine the gram-equivalent weight of zinc.
Ans. 32.67.

19.13. The current in a silver-plating bath was only 80 per cent efficient in depositing silver. How many grams of silver could be deposited in 30 minutes by a current of 0.250 ampere? *Ans.* 0.403 g.

19.14. How much current would flow through a heating coil of 60 ohms resistance at 110 volts? *Ans.* 1.8 amp.

19.15. A 100-watt electric bulb draws about 0.90 ampere at 110 volts. What is the resistance of the filament in the bulb? *Ans.* 123 ohms.

19.16. Calculate the quantities of chlorine, calcium, and aluminum that would be deposited by 1500 coulombs.
Ans. 0.553 g , 0.312 g , 0.140 g.

19.17. Calculate the quantities of ferrous and ferric ions that would be deposited by 1.000 faraday. *Ans.* 27.9 g , 18.6 g.

19.18. Chlorine is prepared commercially by the electrolysis of a brine solution. A current of 2500 amperes was passed through a brine solution for 24 hours. Calculate (a) the volume of hydrogen liberated at S.C., (b) the volume of chlorine liberated at S.C., and (c) the amount of NaOH formed at the cathode.
Ans. (a) 2.5×10^4 *l* ; (b) 2.5×10^4 *l* ; (c) 89.6 kg.

19.19. If a given quantity of electricity deposits 1.952 g of platinic ion, how much auric ion would be deposited by the same amount of electricity? *Ans.* 2.629 g.

19.20. What current strength would be required to deposit 1.50 g of silver per hour? *Ans.* 0.373 amp.

19.21. A metal A forms the oxide AO. A given quantity of electricity deposited 0.862 g of silver and 0.321 g of the metal A. Calculate the atomic weight of the element A. *Ans.* 80.4.

19.22. What weight of water would be decomposed by a current of 100 amperes in 12 hours? *Ans.* 403 g.

19.23. A current of 1.46 amperes was found to liberate 203 ml of chlorine at S.C. in 20 minutes. What is the gram-equivalent weight of chlorine? *Ans.* 35.5 g.

19.24. How long must a current of one ampere flow through acidulated water in order to liberate one gram of hydrogen? *Ans.* 26.8 hr.

19.25. How many ampere-hours would be required to deposit one gram-equivalent weight of an element? *Ans.* 26.8.

19.26. What weight of sodium would be deposited in one hour with a potential of 100 volts and a resistance of 50 ohms? *Ans.* 1.72 g.

19.27. How much each of Cu^+ and Cu^{+2} would be deposited as copper by a current of 0.25 ampere flowing for 60 minutes?
Ans. 0.593 g , 0.297 g.

19.28. What volumes each of hydrogen and oxygen would be obtained at 27° C and 740 mm of Hg by passing a current of 25 amperes through acidulated water for 24 hours? *Ans.* 284 *l* , 142 *l*.

19.29. How many electrons are lost by one gram of Cl^- ions as the result of electrolysis? *Ans.* 1.70×10^{22}.

19.30. How many electrons are gained by one gram of Cu^{+2} ion as the result of electrolysis? *Ans.* 1.90×10^{22}.

19.31. What is the charge in coulombs on a S^{-2} ion? *Ans.* 3.2×10^{-19}.

19.32. How much antimony would be deposited from a solution of $SbCl_3$ by a current of 100 milliamperes in 10.0 minutes? *Ans.* 0.025 g.

19.33. How many minutes would a current of 50 milliamperes have to flow in order to deposit 1.00 gram-equivalent weight of oxygen?

Ans. 32,160 min.

19.34. A bar measuring 10.0 cm by 2.00 cm by 5.0 cm was silver plated by a current of 75 milliamperes for three hours. What was the thickness of the silver deposit on the bar, given that the density of silver is 10.5 g per cm^3? *Ans.* 0.0054 mm.

19.35. How many atoms of copper, as Cu^{+2} ions, would be deposited by a current of 1.00 milliampere in 1.00 second? *Ans.* 3.12×10^{15}.

19.36. How many grams of Fe^{+2} iron could be oxidized to Fe^{+3} iron by a current of 0.100 ampere in 1.00 hour? *Ans.* 0.208 g.

19.37. A current of 5.00 amperes was allowed to flow through acidified water for 250 minutes. What weight of water was decomposed?

Ans. 7.00 g.

19.38. How many electrons will pass through a copper wire if a current of 0.00100 milliampere is allowed to flow for 0.00100 second?

Ans. 6.24×10^9.

19.39. What volume each of oxygen and chlorine, at standard conditions, will be deposited by the passage of 0.0010 faraday of electricity?

Ans. 5.60 ml O_2, 11.2 ml Cl_2.

19.40. How many minutes would be required to deposit the copper in 500 ml of 0.25N $CuSO_4$ by a current of 75 milliamperes?

Ans. 2680 min.

Reactions Occurring during Electrolysis

19.5. Electrolysis. When a source of direct current, such as a battery, is connected to electrodes inserted into a water solution of an electrolyte, current will flow through the solution. The electrical energy is used to bring about chemical changes. During the process of electrolysis cations (+ ions) migrate to the cathode where they acquire electrons, and thus undergo reduction; anions (− ions) migrate to the anode where they lose electrons, and thus undergo oxidation. That is, during electrolysis, *reduction* occurs at the cathode and *oxidation* at the anode. During the electrolysis of water solutions, water molecules may undergo either oxidation or reduction, as shown by the following equations.

Anodic oxidation: $2\,H_2O \rightarrow 4\,H^+ + O_2\uparrow + 4\,e^-$.

Cathodic reduction: $2\,H_2O + 2\,e^- \rightarrow H_2\uparrow + 2\,OH^-$.

That is, *during electrolysis of a water solution oxygen may be released at the anode or hydrogen at the cathode.* A discussion of the electrolysis of a solution resolves itself into a description of the chemical reactions occurring at the anode and at the cathode.

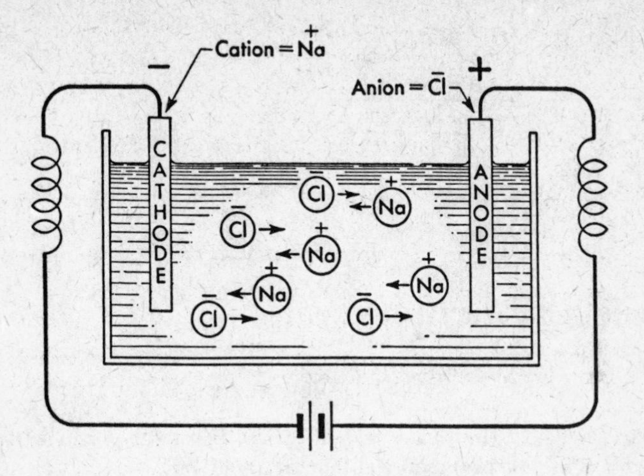

Fig. 19.1. Electric conduction in solutions of electrolytes.

19.6. Electrolysis of Fused Sodium Chloride. When sodium chloride is melted, the liquid contains Na^+ and Cl^- ions. When electrolyzed, sodium is obtained at the cathode and chlorine at the anode. The equations are:

$$
\begin{array}{ll}
\text{Anode:} & 2\,Cl^- \rightarrow Cl_2\uparrow + \cancel{2e^-}. \\
\text{Cathode:} & 2\,Na^+ + \cancel{2e^-} \rightarrow 2\,Na. \\
\hline
\text{Over-all reaction:} & 2\,Na^+ + 2\,Cl^- \rightarrow 2\,Na + Cl_2\uparrow.
\end{array}
$$

19.7. Electrolysis of a Water Solution of Sodium Chloride. In the electrolysis of a water solution of sodium chloride, chlorine is obtained at the anode and hydrogen at the cathode. The equations are:

$$
\begin{array}{ll}
\text{Anode:} & 2\,Na^+ + 2\,Cl^- \rightarrow Cl_2\uparrow + \cancel{2e^-} + 2\,Na^+. \\
\text{Cathode:} & 2\,H_2O + \cancel{2e^-} \rightarrow H_2\uparrow + 2\,OH^-. \\
\hline
\text{Over-all reaction:} & 2\,Na^+ + 2\,Cl^- + 2\,H_2O \rightarrow Cl_2\uparrow + H_2\uparrow + \\
 & \qquad 2\,Na^+ + 2\,OH^-.
\end{array}
$$

Example 19.8. How many pounds of sodium could be obtained by the electrolysis of 1000 pounds of fused sodium chloride?

Solution. The solution is based on the equation for the electrolysis of fused NaCl.

$$
\underbrace{2\,Na^+ + 2\,Cl^-}_{\substack{2.00\text{ lb-moles}\\116.9\text{ lb}}} \rightarrow \underbrace{2\,Na + Cl_2}_{\substack{2.00\text{ lb-moles}\\46.0\text{ lb}}}
$$

Then 1000 lb of NaCl would be:

$$\frac{1000 \cancel{\text{lb NaCl}}}{58.4 \dfrac{\cancel{\text{lb NaCl}}}{\text{lb-mole NaCl}}} = \frac{1000}{58.4} \text{ lb-mole NaCl.}$$

From the equation we see that 2.00 lb-moles of NaCl yield 2.00 lb-moles of Na. Therefore, $\frac{1000}{58.4}$ lb-moles of Na would be produced. Since there are 23.0 lb of Na in 1.00 lb-mole, then:

$$\frac{1000}{58.4} \cancel{\text{lb-mole Na}} \times 23.0 \frac{\text{lb Na}}{\cancel{\text{lb-mole Na}}} = 393 \text{ lb Na.}$$

Example 19.9. How many pounds of NaOH could be obtained by the electrolysis of 1000 pounds of NaCl in water solution?

Solution. Again, the solution is based on the equation of electrolysis.

$$\underbrace{2\,Na^+ + 2\,Cl^-}_{\substack{2.00 \text{ lb-moles} \\ 116.9 \text{ lb}}} + 2\,H_2O \rightarrow Cl_2 + H_2 + \underbrace{2\,Na^+ + 2\,OH^-}_{\substack{2.00 \text{ lb-moles} \\ 80.0 \text{ lb}}}$$

From Example 19.8 we see that 1000 lb of NaCl $= \frac{1000}{58.4}$ lb-mole NaCl.

The equation shows that 2.00 lb-moles of NaCl yield 2.00 lb-moles of NaOH. Since there are 40.0 lb of NaOH in 1.00 lb-mole, then:

$$\frac{1000}{58.4} \cancel{\text{lb-mole NaOH}} \times 40.0 \frac{\text{lb NaOH}}{\cancel{\text{lb-mole NaOH}}} = 684 \text{ lb NaOH.}$$

19.8. Electrolysis of a Water Solution of Sulfuric Acid. A water solution of H_2SO_4 is very conductive because of the ionization of the acid.

$$H_2SO_4 \rightleftharpoons H^+ + HSO_4^-.$$

When an electric current is passed through a water solution of H_2SO_4, hydrogen ions are reduced at the cathode. At the anode, water is oxidized.

$$\begin{array}{ll} \text{Anode:} & 2\,H_2O \rightarrow \cancel{4\,H^+} + O_2\uparrow + \cancel{4\,e^-}. \\ \text{Cathode:} & \cancel{4\,H^+} + \cancel{4\,e^-} \rightarrow 2\,H_2\uparrow. \\ \hline \text{Over-all reaction:} & 2\,H_2O \rightarrow 2\,H_2\uparrow + O_2\uparrow. \end{array}$$

That is, the net result of the electrolysis is the decomposition of water into hydrogen and oxygen.

19.9. Electrolysis of a Water Solution of Sodium Sulfate. The electrolysis of a water solution of sodium sulfate results in the oxidation of water at the anode, and the reduction of water at the cathode.

$$\begin{array}{ll} \text{Anode:} & 2\,H_2O \rightarrow 4\,H^+ + O_2\uparrow + \cancel{4\,e^-}. \\ \text{Cathode:} & 4\,H_2O + \cancel{4\,e^-} \rightarrow 2\,H_2\uparrow + 4\,OH^-. \\ \hline & 6\,H_2O \rightarrow 2\,H_2\uparrow + O_2\uparrow + 4\,H^+ + 4\,OH^-. \end{array}$$

However, the H^+ and OH^- ions combine to form water. Therefore the overall reaction would be:

$$\begin{array}{ll} & 6\,H_2O \rightarrow 2\,H_2\uparrow + O_2\uparrow + \cancel{4\,H^+} + \cancel{4\,OH^-}. \\ & \cancel{4\,H^+} + \cancel{4\,OH^-} \rightarrow 4\,H_2O. \\ \hline \text{Over-all reaction:} & 2\,H_2O \rightarrow 2\,H_2\uparrow + O_2\uparrow. \end{array}$$

Again, the net result of the electrolysis is the decomposition of water.

Example 19.10. How many grams of water would be decomposed by a current of 0.500 ampere flowing through a water solution of H_2SO_4 for 6 hours?

Solution. One gram-equivalent weight of hydrogen, or 1.008 g of hydrogen, and 1.00 gram-equivalent weight of oxygen, or 8.000 g of oxygen, will be released for each faraday of electricity that passes through the solution. The amount of electricity passing through the solution would be:

$$\frac{0.500\ \cancel{\text{amp}} \times 6\ \cancel{\text{hr}} \times 60\ \dfrac{\cancel{\text{min}}}{\cancel{\text{hr}}} \times 60\ \dfrac{\cancel{\text{sec}}}{\cancel{\text{min}}}}{96{,}490\ \dfrac{\cancel{\text{coulombs}}}{\text{faraday}}} = 0.112\ \text{faraday}.$$

Since amp × sec = coulombs, then amp, sec, and coulombs cancel in the above expression.

Since 1.00 faraday would decompose:

$$1.008\ \text{g}\ H_2 + 8.000\ \text{g}\ O_2 = 9.008\ \text{g}\ H_2O,$$

then 0.112 faraday would decompose:

$$9.008\ \frac{\text{g}\ H_2O}{\cancel{\text{faraday}}} \times 0.112\ \cancel{\text{faraday}} = 1.009\ \text{g}\ H_2O.$$

19.10. Electrolysis of Water Solutions. If the cation is a fairly active metal, such as in Group IA of the periodic table, the cation will not be deposited during electrolysis and hydrogen will be liberated at the cathode (Sec. 19.7). On the other hand, if the cation is an inactive metal such as copper, silver, gold, etc., the cation will accept electrons at the cathode and be deposited as the free metal (Sec. 19.3).

If the anion has only a slight attraction for electrons, such as the elements in Group VIIA, the ions will lose their electrons at the anode and be deposited as the free element (Sec. 19.7). Anions that have a strong attraction for electrons, such as SO_4^{-2} and NO_3^-, will not release their electrons at the anode, and oxygen will be liberated as a result of the oxidation of water (Secs. 19.8 and 19.9).

Problems

Part I

19.41. By means of equations account for the origin of the following substances during the electrolysis of a water solution of a compound.
  a. The formation of hydrogen at the cathode.
  b. The formation of oxygen at the anode.

19.42. Write the equations for the anode and cathode reactions that occur when an electric current is passed through water solutions of the following compounds. Write the over-all reaction for each.
  a. KBr.
  b. HNO_3.
  c. K_2SO_4.
  d. $CuSO_4$.
  e. $AgNO_3$.

19.43. Write the equations for the reactions that occur at the anode and cathode during the electrolysis of fused KBr. Write the over-all reaction.

19.44. How many grams of sodium could be obtained by the electrolysis of 100 g of fused NaCl? *Ans.* 39.3 g.

19.45. How many liters of chlorine at S.C. could be obtained by the electrolysis of 2.50 g of NaCl in water solution? *Ans.* 0.48 *l.*

19.46. An electric current was passed through a water solution of LiCl until 500 ml of chlorine had been collected at S.C.
  a. How many grams of LiCl were decomposed? *Ans.* 1.89 g.
  b. How many grams of LiOH were formed? *Ans.* 1.07 g.

19.47. A direct current of 25.0 amperes was passed through a water solution of $CrCl_3$ for 10 hours. Write the over-all reaction. How many liters of chlorine at S.C. were obtained? *Ans.* 105 *l.*

19.48. A solution was prepared by dissolving 5.00 g of NaCl in water. The solution was electrolyzed for a short time. Titration with a standard acid showed that the solution contained 23.0 milliequivalents of NaOH. What per cent of the NaCl had been decomposed by the electric current? *Ans.* 26.9%.

19.49. How many moles each of hydrogen and oxygen could be obtained by passing 0.400 ampere through a water solution of HCl for 30.0 minutes? *Ans.* 0.00373 mole of each.

Part II

19.50. How many grams each of potassium and bromine could be obtained by the electrolysis of 25.0 g of fused KBr?

Ans. 8.21 g K, 16.79 g Br_2.

19.51. Write the over-all reaction for the electrolysis of a water solution of $KClO_3$.

19.52. How much KOH could be obtained by the electrolysis of 1500 lb of KCl in water solution? *Ans.* 1129 lb.

19.53. How many liters each of hydrogen and oxygen could be obtained by the decomposition of 500 g of water by electrolysis?

Ans. 623 *l* H_2; 311 *l* O_2.

19.54. How much chromium could be obtained by the electrolysis of 5.00 g of $CrCl_3$ in water solution? *Ans.* 1.64 g.

19.55. A current was passed through fused LiCl until 1500 ml of chlorine had been collected at S.C. How much lithium was deposited simultaneously at the cathode? *Ans.* 0.929 g.

19.56. A current of 1.25 amperes was passed through a water solution of $CuSO_4$ for one hour. What substances were deposited at the electrodes and how much? *Ans.* 1.48 g Cu; 0.373 g O_2.

19.57. The over-all reaction for an electrolysis process is:

$$2\ RbI + 2\ H_2O \rightarrow I_2 + H_2\uparrow + 2\ RbOH.$$

Write the equations for the reactions at the anode and at the cathode.

Electrochemical Cells

19.11. Galvanic Cells. In the previous sections, the effect of an electric current was described when passing through a water solution of an electrolyte. In a galvanic cell, spontaneous chemical reactions occur at each of the electrodes of the cell to produce an electric current. As in electrolysis, oxidation occurs at the anode and reduction at the cathode, when the cell is in use.

The Daniell cell (Fig. 19.2) is a typical galvanic cell. The cell consists of a copper electrode in a water solution containing Cu^{+2} ions, and a zinc electrode in a water solution containing Zn^{+2} ions. The two solutions are separated by a porous partition such as unglazed porcelain. When the two electrodes are connected to an external circuit (Fig. 19.2), electrons will flow *from* the Zn electrode *to* the Cu electrode externally, and from the Cu to the Zn internally to complete the circuit. That is, when the cell is in operation, Zn undergoes oxidation and Cu undergoes reduction as shown by the following half-reactions.

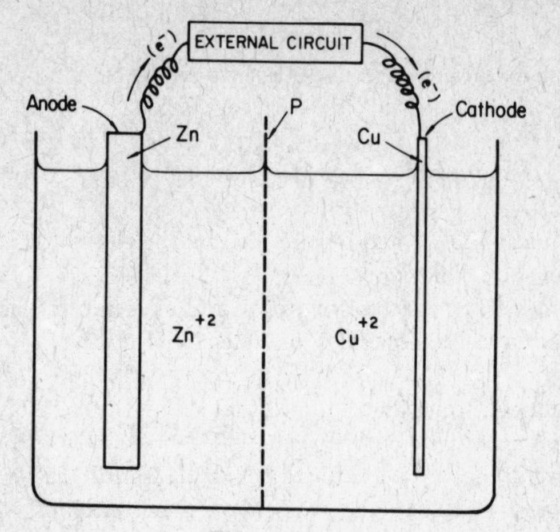

Fig. 19.2.

Anode (oxidation):	$Zn\ (s) \rightarrow Zn^{+2} + \cancel{2\ e^-}$.
Cathode (reduction):	$Cu^{+2} + \cancel{2\ e^-} \rightarrow Cu\ (s)$.
Cell reaction:	$Cu^{+2} + Zn\ (s) \rightarrow Cu\ (s) + Zn^{+2}$.

For the above cell reaction, ΔH is negative. The reaction proceeds spontaneously.

Many different galvanic cells are possible. The requirements for a galvanic cell are two half-cells which undergo spontaneous oxidation and reduction when constructed as in Fig. 19.2. A *half-cell* consists of a metal, nonmetal, or even a compound as an electrode in a solution of ions. Theoretically, any oxidation-reduction reaction may be used as the basis for a cell. The mechanical setup of a cell, particularly with gas electrodes, may require considerable ingenuity.

Example 19.11. Construct a cell based on the reaction:

$$3\ Fe\ (s) + 2\ Au^{+3} \rightarrow 3\ Fe^{+2} + 2\ Au\ (s) \quad \Delta H = \text{negative}.$$

Solution. First write the half-reactions for the given equation.

Anode (oxidation):	$Fe\ (s) \rightarrow Fe^{+2} + 2\ e^-$.
Cathode (reduction):	$Au^{+3} + 3\ e^- \rightarrow Au\ (s)$.

The anode half-cell would consist of an iron electrode in a solution containing Fe^{+2} ions, while the cathode half-cell would consist of a gold electrode in a solution containing Au^{+3} ions. A compound such as $FeSO_4$ would serve as a source of Fe^{+2} ions, and $AuCl_3$ as a source of Au^{+3} ions.

Example 19.12. Construct a cell based on the equation:

$$Zn\ (s) + Cl_2\ (g) \rightarrow Zn^{+2} + 2\ Cl^-.$$

Solution. Write the half-reactions.

Anode: $Zn\ (s) \rightarrow Zn^{+2} + 2\ e^-$

Cathode: $Cl_2\ (g) + 2\ e^- \rightarrow 2\ Cl^-$

The anode would consist of zinc in a solution of Zn^{+2} ions, and the cathode would consist of a chlorine electrode in a solution containing Cl^- ions. The construction of a gas electrode such as chlorine would involve some problems.

19.12. Standard Electrode Potentials. Each half-cell in a galvanic cell has the ability to release electrons (to undergo oxidation). However, each half-cell attains a maximum potential which varies with different half-cells. Therefore, in a cell consisting of two half-cells, the half-cell with the greater maximum potential will act as the anode (release electrons). This ability of a half-cell to release electrons can be measured in volts, and is called the *oxidation potential* of the half-cell. The voltage of a cell is the *algebraic difference* between the oxidation potentials of the two half-cells making up the cell.

Example 19.13. Write the half-cell reactions for each of the following oxidation-reduction reactions:

(a) $Zn\ (s) + Sn^{+2} \rightarrow Zn^{+2} + Sn\ (s)$.

(b) $Sn\ (s) + Cu^{+2} \rightarrow Sn^{+2} + Cu\ (s)$.

Solution.

(a) $Zn\ (s) \rightarrow Zn^{+2} + 2\ e^-$ Anode.

$Sn^{+2} + 2\ e^- \rightarrow Sn\ (s)$ Cathode.

(b) $Sn\ (s) \rightarrow Sn^{+2} + 2\ e^-$ Anode.

$Cu^{+2} + 2\ e^- \rightarrow Cu\ (s)$ Cathode.

In the first equation tin acts as the cathode (reduction), while in the second equation tin acts as the anode (oxidation). That is, the oxidation potentials are in the order: $Zn > Sn > Cu$ (see Table 19.1).

The magnitude of the oxidation potential varies with temperature, the concentration of the ions, and the material constituting the electrode. For this reason, a reference *standard state* has been adopted for oxidation potentials. The standard state has been chosen as 25° C, *unit activity* of the ions, and 1.00 atm pressure. For our purpose, unit activity of an ion may be defined as a *one molal* water solution of that ion. Oxidation potentials for the standard state are designated as E° (volts), and are called *standard oxidation potentials* (Table 19.1).

Electrode potentials may be expressed as reduction potentials. Some reduction potentials are given at the bottom of Table 19.1. In such tables the reduction reaction is written with the sign of E° changed. In the discussion that follows the term electrode will be used to refer to what has previously been described as a half-cell.

TABLE 19.1

Standard Oxidation Potentials

Reaction	*E° (volts)*
$Li \rightarrow Li^+ + e^-$	+3.045
$K \rightarrow K^+ + e^-$	+2.925
$Mg \rightarrow Mg^{+2} + 2e^-$	+2.363
$Al \rightarrow Al^{+3} + 3e^-$	+1.662
$Zn \rightarrow Zn^{+2} + 2e^-$	+0.763
$Fe \rightarrow Fe^{+2} + 2e^-$	+0.440
$Cd \rightarrow Cd^{+2} + 2e^-$	+0.403
$Ni \rightarrow Ni^{+2} + 2e^-$	+0.250
$Sn \rightarrow Sn^{+2} + 2e^-$	+0.136
$H_2 \rightarrow 2H^+ + 2e^-$	0.000
$Cu \rightarrow Cu^{+2} + 2e^-$	−0.337
$2\,I^- \rightarrow I_2 + 2e^-$	−0.536
$Fe^{+2} \rightarrow Fe^{+3} + e^-$	−0.771
$Hg \rightarrow Hg^{+2} + 2e^-$	−0.789
$Ag \rightarrow Ag^+ + e^-$	−0.799
$2\,Cl^- \rightarrow Cl_2 + 2e^-$	−1.360
$Au \rightarrow Au^{+3} + 3e^-$	−1.50
$2\,F^- \rightarrow F_2 + 2e^-$	−2.87

Standard Reduction Potentials

Reaction	*E° (volts)*
$Li^+ + e^- \rightarrow Li$	−3.045
$Zn^{+2} + 2e^- \rightarrow Zn$	−0.763
$2H^+ + 2e^- \rightarrow H_2$	0.000
$Cu^{+2} + 2e^- \rightarrow Cu$	+0.337
$F_2 + 2e^- \rightarrow 2\,F^-$	+2.87

Voltage is a relative term, and must be referred to a zero standard to have significance. Reference to Table 19.1 will show that the voltage of the standard hydrogen electrode is taken as the zero of voltage. Therefore, E° for an electrode is the voltage it will give when connected to a standard hydrogen electrode to form a cell.

Example 19.14. (a) Calculate E° for the following cells and (b) designate which electrode acts as the anode and which as the cathode.
(1) Cadmium and hydrogen.
(2) Silver and hydrogen.
(3) Cadmium and silver.

Solution. The anode would be the electrode having the higher oxidation potential (Table 19.1).

	Anode	*Cathode*	*E° (calculated)*
(1)	Cd	H_2	(+0.403) − (0.000) = +0.403 v
(2)	H_2	Ag	(0.000) − (−0.799) = +0.799 v
(3)	Cd	Ag	(+0.403) − (−0.799) = +1.202 v

19.13. The Nernst Equation. The effect of the concentration of the ions on the voltage of an electrode may be determined by means of the Nernst equation:

$$E = E^\circ - \frac{RT}{nF} \ln \frac{a_{\text{species on right}}}{a_{\text{species on left}}}$$

where E is the voltage of the electrode at the absolute temperature T; R is the gas law constant, in this case equal to 8.314 $\frac{\text{volt coulomb}}{\text{mole °K}}$; n is the number of electrons gained or lost as shown in the equation for the electrode; and F is the Faraday equal to 96,490 $\frac{\text{volt coulomb}}{\text{mole of } e^-}$. The letter a stands for the *activity*. For our purposes, the activities of the ions in solution will be considered as concentrations expressed as molality, m. Solids, such as zinc, and liquids, such as mercury, have an activity of one, and thus may be omitted from the expression for activity in the Nernst equation. Gases have an activity of one when they exist at a pressure of one atmosphere. At any other pressure gases must be included as part of the activity factor.

Converting the Nernst equation to $\log_{10}$ at 25° C gives:

$$E = E^\circ - \frac{(8.314)(298.1)}{n\,(96{,}490)} (2.303) \log \frac{a_{\text{species on right}}}{a_{\text{species on left}}}$$

$$= E^\circ - \frac{0.0592}{n} \log \frac{a_{\text{species on right}}}{a_{\text{species on left}}}.$$

Example 19.15. Calculate E for a Zn electrode in which Zn^{+2} = 0.025 m.

Solution. From Table 19.1, $E° = 0.763$ v, $n = 2$, and Zn^{+2} represents the oxidized species. For Zn, $a = 1$.

$$Zn \rightarrow Zn^{+2} + 2e^-$$

Then:
$$E = 0.763 - \left(\frac{0.0592}{n} \log 0.025\right)$$
$$= 0.763 - [(0.0296)(-1.602)] = 0.810 \text{ v.}$$

A conventional system has been devised to represent cells. For example, a Daniell cell may be written:

$$\underset{\text{(Anode)}}{Zn, Zn^{+2}\ (0.50m)}\ ||\ \underset{\text{(Cathode)}}{Cu^{+2}\ (0.20m), Cu}$$

The concentrations of the ions are given in parentheses.

The Nernst equation may be used to determine the voltage of a cell, E_{cell}, at other than the standard state.

Example 19.16. Determine E_{cell} for the Daniell cell given above at 25° C.

Solution. We will apply the Nernst equation to the cell reaction, where $E°_{anode} - E°_{cathode} = E°_{cell}$.

$$Zn + Cu^{+2} \rightarrow Zn^{+2} + Cu.$$

Then:
$$E_{cell} = (E°_{anode} - E°_{cathode}) - \left(0.0296 \log \frac{a_{Zn^{+2}}}{a_{Cu^{+2}}}\right)$$
$$= [(0.763) - (-0.337)] - \left(0.0296 \log \frac{0.50}{0.20}\right)$$
$$= 1.088 \text{ v.}$$

Example 19.17. Given the reaction at 25° C:

$$Zn\ (s) + I_2\ (s) \rightarrow Zn^{+2} + 2\ I^-$$

(a) Set up the cell for the reaction.

(b) Calculate E_{cell} given that $Zn^{+2} = 0.15m$ and $I^- = 0.01m$.

Solution. (a) From Table 19.1 we would assume that zinc would act as the anode and iodine as the cathode. Therefore the cell would be:

$$Zn, Zn^{+2}\ (0.15m)\ ||\ I_2, I^-\ (0.01m).$$

(b)
$$E_{cell} = [(0.763) - (-0.536)] - [0.0296 \log (0.15)(0.01)^2]$$
$$= 1.442 \text{ v.}$$

Observe that the activity of I^- is squared since there are 2 I^- in the equation. If the assumption concerning the anode and cathode is in error, the voltage will come out negative.

19.14. Quantitative Treatment of Electrodes. In a Daniell cell, zinc atoms go into solution as zinc ions, and copper ions precipitate on the copper electrode as copper atoms. This means that one faraday (N electrons) is associated with one equivalent weight of the cathode material.

Example 19.18. A Daniell cell has a zinc electrode weighing 500 g. Assuming sufficient copper ions at the copper electrode, for how many hours can the cell be used at a current of 50 milliamperes (ma)?

Solution. The cell would supply:

$$\frac{500 \text{ g Zn}}{32.69 \frac{\text{g Zn}}{\text{faraday}}} = 15.3 \text{ faradays of electricity.}$$

Then:

$$\frac{\dfrac{15.3 \text{ faraday} \times 96{,}490 \frac{\text{amp sec}}{\text{faraday}}}{0.050 \text{ amp}}}{3600 \frac{\text{sec}}{\text{hr}}} = 8.20 \times 10^3 \text{ hr.}$$

Problems

Part II

19.58. In Example 19.16 derive the equation for E_{cell} by applying the Nernst equation to the anode and cathode reactions for the cell.

19.59. A Daniell cell was used for 1,000 hr at 100 ma. How much copper was deposited at the cathode, and how many grams of zinc was lost by the anode? *Ans.* Zn = 122 g, Cu = 119 g.

19.60. For a magnesium-iodine cell:
(a) Calculate $E°_{cell}$, (b) write the equation for the cell reaction, and (c) what is the direction of flow of electrons?
Ans. (a) $E°_{cell} = 2.87$ v, (b) $Mg + I_2 \rightarrow Mg^{+2} + 2\,I^-$, (c) from Mg to I_2

19.61. Calculate E for a potassium electrode at 25° C for which the K ion (a) is 0.25m, and (b) 0.75m. *Ans.* (a) 2.96 v, (b) 2.93 v.

19.62. Calculate E at 25° C for a chlorine electrode for which the Cl^- ion is (a) 0.20m, and (b) 0.30m. The chlorine is at 1.00 atm pressure. *Ans.* (a) −1.278 v, (b) −1.298 v.

19.63. Calculate E_{cell} at 25° C for the cell:

$$Mg, Mg^{+2}\ (1.0m)\ ||\ F^-\ (1.0m), F_2\ (1.0 \text{ atm}).$$

Ans. 5.23 v.

19.64. Given the cell at 25° C:

$$I^-\ (0.01m), I_2\ ||\ F^-\ (0.001m), F_2\ (1.0 \text{ atm})$$

(a) Write the electrode reactions.
(b) Write the cell reaction.
(c) Calculate E_{cell}. *Ans.* 2.30 v.

19.65. Given a cell consisting of a cadmium electrode in 0.001m Cd^{+2} ions, and an iodine electrode in 0.01m I^- ions, calculate E_{cell}. *Ans.* 1.17 v.

19.66. A Daniell cell was used at 100 ma until 5.00 g of copper had deposited on the cathode. How many hours was the cell used? *Ans.* 42.2 hr.

19.67. Calculate E at 25° C for the electrode

$$H_2 \text{ (1.00 atm)}, H^+ \text{ (}1.0 \times 10^{-7}\text{m)} \,||.$$

Ans. 0.41 v.

19.68. a. Write the equation for the cell:

$$Fe^{+2} \text{ (1.00m)}, Fe^{+3} \text{ (1.00m)} \,||\, Cl^- \text{ (1.00m)}, Cl_2 \text{ (1.00 atm)}.$$

b. Calculate E for the cell given in part a. *Ans.* 0.589 v.

19.69. Calculate E_{cell} at 25° C for the cell:

$$I^- \text{ (0.02m)}, I_2 \,||\, Ag^+ \text{ (0.01m)}, Ag.$$

Ans. 0.311 v.

19.70. How many electrons could be obtained from an aluminum electrode for each gram of aluminum consumed. *Ans.* 6.7×10^{22} electrons.

19.71. Given the cell reaction:

$$Ni \text{ (s)} + 2\,Fe^{+3} \rightarrow Ni^{+2} + 2\,Fe^{+2}.$$

(a) Calculate $E°_{cell}$.

(b) Calculate E_{cell} given that $Fe^{+3} = 0.05$m, $Ni^{+2} = 0.01$m, and $Fe^{+2} = 0.03$m. *Ans.* (a) 1.021 v, (b) 1.093 v.

19.72. A solution was prepared containing Fe^{+2} (1.0m), Fe^{+3} (1.0m), I^- (1.0m), and solid I_2. Write the equation for the spontaneous reaction that would take place. *Ans.* $2\,Fe^{+3} + 2\,I^- \rightarrow 2\,Fe^{+2} + I_2$.

20

Nuclear Chemistry

In chemical reactions the electronic structure of the atom is involved, the nucleus remaining unchanged. Within recent years the scope of chemistry has been expanded to include changes within the nucleus of the atom, and the energy associated with such changes. Nuclear chemistry is the study of such changes. The discussion in this chapter will be limited to changes within the nuclei of atoms which have been taking place spontaneously during past ages, and changes within the atom brought about by the inventive genius of man.

In nuclear chemistry we deal with specific isotopes of elements. Isotopes are the atoms of a given element which differ from each other in the number of neutrons in the nucleus, their electronic structure being the same. Two or more isotopes are known for every element.

Natural Radioactivity

20.1. Introduction. Radioactivity involves the nuclei of atoms. The nuclei of certain isotopes are unstable, undergoing disintegration over which man has no control, whereas with stable isotopes no such process occurs. When nuclei undergo disintegration, two types of particles may be lost: (1) *alpha* (α) particles of atomic mass 4 and a +2 charge, and (2) *beta* (β) particles of negligible mass and a charge of −1. That is, when the nucleus of an atom loses an α-particle, its weight decreases by 4 atomic mass units, and the nuclear charge (atomic number) decreases by 2. When the nucleus of an atom loses a β-particle, there is no change in weight, and the nuclear charge increases by one. For example:

$$\underset{\text{Uranium}}{{}_{92}U^{238}} \rightarrow \underset{\text{Thorium}}{{}_{90}Th^{234}} + \underset{\alpha\text{-particle}}{{}_{2}He^{4}}.$$

$$\underset{\text{Thorium}}{{}_{90}Th^{234}} \rightarrow \underset{\text{Protactinium}}{{}_{91}Pa^{234}} + \underset{\beta\text{-particle}}{{}_{-1}e^{0}}$$

TABLE 20.1

PARTICLES ASSOCIATED WITH RADIOACTIVITY

Particle		*Description*
Hydrogen	($_1H^1$)	Hydrogen atom with mass number of 1.
Deuteron	($_1H^2$)	Hydrogen atom with mass number of 2.
Proton	($_1H^1$)	Hydrogen ion, H^+, with mass 1.00758.
Electron	($_{-1}e^0$)	Charge -1 and mass 0.0005486.
Positron	($_1e^0$)	Charge $+1$ and mass 0.0005486.
Neutron	($_0n^1$)	Charge 0 and mass 1.00897.
Alpha	($_2He^4$)	Helium ion, He^{+2}, with mass number 4.
Beta	($_{-1}e^0$)	An electron.

Gamma (γ) radiation is associated with natural radioactivity. Gamma rays are similar to *X*-rays and possess no mass.

20.2. Rate of Radioactive Disintegration. The products of disintegration of a naturally occurring radioactive substance appear always to be the same. Two important facts have been established concerning radioactive disintegration: (1) the number of nuclei disintegrating per unit of time for a given substance is directly proportional to the mass undergoing disintegration, and (2) each radioactive substance has a characteristic rate of disintegration.

The *half-life period* of a radioactive substance is the time required for one-half of a given mass of the substance to undergo disintegration. Half-life periods vary enormously with different substances, one product of uranium disintegration having a half-life period of 1.5×10^{-4} seconds and another a half-life period of 270,000 years.

Example 20.1. The half-life period of radon, $_{86}Rn^{222}$, is approximately 4 days. A tube containing 1.00 microgram (0.000001 g.) of radon was stored in a hospital clinic for 12 days. How much radon remained in the tube?

Solution. Every 4 days one-half the remaining radon would disintegrate. Then, according to the following scheme:

days: 0 – 4 – 8 – 12
radon left: 1.00 – 0.50 – 0.25 – 0.125 micrograms left.

Free electrons are not present as such in the nuclei of atoms. However, electrons from radioactive disintegration originate from the neutrons in the nuclei.

$$_0n^1 \rightarrow {}_1H^1 + {}_{-1}e^0 + \text{one neutrino.}$$

The rate of disintegration is given by the formula:

$$2.3 \log \frac{N_1}{N_2} \propto t \quad \text{or} \quad 2.3 \log \frac{N_1}{N_2} = kt$$

where N_1 represents the initial number of atoms subject to disintegration, and N_2 the number of such atoms remaining after time t. The half-life period, $t_{\frac{1}{2}}$, occurs when $N_2 = \frac{N_1}{2}$, in which case the above formula becomes:

$$kt_{\frac{1}{2}} = 2.3 \log 2 = 0.693.$$

The numerical value of the constant k is a characteristic of each radioactive substance.

Example 20.2. A tube contains 12.0 micrograms of $_{84}Po^{210}$, the half-life period of which is 138 days. How much $_{84}Po^{210}$ would remain after 90 days? The nuclear equation is:

$$_{84}Po^{210} \rightarrow {}_{82}Pb^{206} + {}_{2}He^{4}.$$

Solution. From the half-life period we can evaluate the constant k. Then:

$$k = \frac{0.693}{t_{\frac{1}{2}}} = \frac{0.693}{138 \text{ days}} = 5.02 \times 10^{-3} \frac{1}{\text{days}}.$$

Having determined the value of k we can now solve the general formula for N_2 as the unknown. Then:

$$2.3 \log \frac{N_1}{N_2} = kt$$

or

$$2.3 (\log N_1 - \log N_2) = kt$$

and

$$\log N_2 = \frac{2.3 \log N_1 - kt}{2.3}.$$

Substituting the known values in the above formula gives:

$$\log N_2 = \frac{(2.3 \log 12.0) - (5.02 \times 10^{-3} \times 90)}{2.3} = 0.88$$

and $N_2 = 7.6$ micrograms.

The Transmutation of Elements

20.3. Introduction. The development of artificial radioactivity has made possible the conversion of one element into another by means of nuclear changes within the atom. The bombardment of nuclei with certain high energy particles such as protons ($_1H^1$), deuterons ($_1H^2$), alpha particles ($_2H^4$), beta particles (e^-), and neutrons ($_0n^1$) may result in the capture of the particle by a nucleus, followed by elimination of a particle from the nucleus different from that captured. The result is the formation of unstable radioactive isotopes of the element, or transmutation into a new element.

20.4. Nuclear Reactions. Transmutation of elements and the artificial preparation of unstable radioactive isotopes involve nuclear changes within the atom. For example, when aluminum is bombarded with alpha particles, the products are radioactive phosphorus and neutrons.

$$_{13}Al^{27} + {_2He^4} \rightarrow {_{15}P^{30}} + {_0n^1}.$$

When lithium is bombarded with deuterons the product is helium.

$$_3Li^6 + {_1H^2} \rightarrow 2\,{_2He^4}.$$

The bombardment of lithium 7 with deuterons produces a different type of reaction.

$$_3Li^7 + {_1H^2} \rightarrow {_4Be^6} + {_0n^1}.$$

Any process in which a nucleus reacts with another nucleus, or with the elementary particles previously mentioned, to produce new nuclei is called a *nuclear reaction.*

Following are a number of nuclear reactions known to scientists which involve many of the particles given in Table 20.1.

$$_4Be^9 + {_2He^4} \rightarrow {_6C^{12}} + {_0n^1}.$$
$$_{27}Co^{55} \rightarrow {_{26}Fe^{55}} + {_1e^0}.$$
$$_{26}Fe^{54} + {_1H^2} \rightarrow {_{27}Co^{55}} + {_0n^1}.$$
$$_{88}Ra^{226} \rightarrow {_{86}Rn^{222}} + {_2He^4}.$$
$$_6C^{14} \rightarrow {_7N^{14}} + {_{-1}e^0}.$$

Observe that, in a nuclear reaction, the sum of the superscripts and of the subscripts of reactants and products are related as follows:

$$(14 + 1) = (11 + 4)$$
$$_7N^{14} + {_1H^1} \rightarrow {_6C^{11}} + {_2He^4}.$$
$$(7 + 1) = (6 + 2)$$

20.5. Energy of Nuclear Reactions. The number of protons and neutrons in the atoms of the elements is known. The mass of a proton is equal to 1.00754 atomic mass units, and the mass of a neutron to 1.00893 atomic mass units. Since a helium nucleus consists of two protons and two neutrons, its mass should be:

$$(2 \times 1.00754) + (2 \times 1.00893) = 4.03294 \text{ atomic mass units.}$$

Actually, the helium nucleus has been found to have a mass of 4.003 atomic mass units. The difference:

$$4.033 - 4.003 = 0.030 \text{ atomic mass units}$$

represents the mass converted to energy in the process of formation of a helium nucleus. This process is probably occurring on the sun at the present time. The difference in nuclear weight is called the *mass defect* of the atom.

The amount of energy liberated may be calculated by means of the Einstein equation (Chapter 12). Then, for helium:

$$E = (2.2 \times 10^{13})(0.030) = 6.6 \times 10^{11} \text{ calories.}$$

That is, in the process of formation of 4.003 grams of helium from protons and neutrons, 6.6×10^{11} calories are liberated.

Problems

Part I

20.1. How much of a one gram mass of a radioactive substance would remain after 30 days if its half-life period is 5 days? *Ans.* 0.0156 g.

20.2. Explain the nuclear changes necessary to bring about the following transmutations: (a) ${}_{13}Al^{27} \rightarrow {}_{15}P^{30}$, (b) ${}_{3}Li^{7} \rightarrow {}_{4}Be^{6}$, and (c) ${}_{92}U^{238} \rightarrow {}_{90}Th^{234}$.

20.3. A radioactive isotope ${}_{11}A^{24}$ loses a β-particle, yielding a stable isotope B. What is the element B? *Ans.* Mg.

20.4. When ${}_{83}Bi^{209}$ is bombarded with α-particles, the bismuth nucleus captures one particle with the accompanying expulsion of two neutrons. What is the new element which is formed? *Ans.* At.

20.5. Calculate the mass defect for sodium. *Ans.* 0.194.

20.6. Calculate the energy in calories released when one gram-atom of sodium is formed from neutrons and protons. *Ans.* 4.3×10^{12} cal.

20.7. Complete the following nuclear reactions:

a. ${}_{7}N^{14} + {}_{2}He^{4} \rightarrow {}_{8}O^{17} +$ ——. *Ans.* ${}_{1}H^{1}$.

b. ${}_{1}H^{3} \rightarrow {}_{2}He^{3} +$ ——. *Ans.* ${}_{-1}e^{0}$.

c. ${}_{11}Na^{23} + {}_{2}He^{4} \rightarrow {}_{12}Mg^{26} +$ ——. *Ans.* ${}_{1}H^{1}$.

d. $2\ {}_{1}H^{2} \rightarrow {}_{2}He^{3} +$ ——. *Ans.* ${}_{0}n^{1}$.

Part II

20.8. The half-life period of ${}_{15}P^{30}$ is about 3 minutes. How much of a 16-microgram sample would remain after 15 minutes? *Ans.* 0.50 microgram.

20.9. The half-life period of radioactive carbon, C^{14}, is about 4700 years.

How many years would be required to reduce 32 micrograms of the isotope to one microgram? *Ans.* 23,500 years.

20.10. When U^{238} is bombarded with neutrons, each nucleus captures one neutron. What isotope of uranium is formed? *Ans.* U^{239}.

20.11. When N^{14} is bombarded with neutrons, the nitrogen nucleus captures a neutron with the formation of radioactive carbon, C^{14}. What particle is expelled from the nitrogen nucleus in order to bring about this transmutation? *Ans.* A proton.

20.12. Calculate the energy in calories released when one atom of beryllium is formed from neutrons and protons. *Ans.* 2.3×10^{-12} cal.

20.13. Diffusion methods are employed in the separation of U^{235} and U^{238} (Sec. 7.11). The fluorides, UF_6, of the two isotopes exist in the gaseous state above 56° C. at 760 mm. of Hg. What are the relative rates of diffusion of the two fluorides? *Ans.* 1.0043 : 1.0000.

20.14. What is the percentage of radium in a pitchblende which yielded 25 mg. of $RaCl_2$ from 15 tons of ore? *Ans.* $(1.4 \times 10^{-7})\%$.

20.15. The market price of radium is about $50,000 per gram. What would be the cost of 50 micrograms of $RaCl_2$ based on radium content? *Ans.* $1.90.

20.16. Calculate the mass defect for C^{12}. *Ans.* 0.088.

20.17. Calculate the energy in calories released when 1.00 gram-atom of C^{12} is formed from neutrons and protons? *Ans.* 1.94×10^{12} cal.

20.18. What is the mass number of the particle formed when an atom of naturally radioactive U^{238} emits an alpha particle? *Ans.* 90.

20.19. Complete the following nuclear reactions:

a. ${}_{15}P^{30} \rightarrow {}_{14}Si^{30} + $ ——. *Ans.* ${}_{1}e^{0}$.

b. ${}_{14}Si^{27} \rightarrow {}_{13}Al^{27} + $ ——. *Ans.* ${}_{1}e^{0}$.

c. ${}_{20}Ca^{43} + {}_{2}He^{4} \rightarrow {}_{21}Sc^{46} + $ ——. *Ans.* ${}_{1}H^{1}$.

d. ${}_{48}Cd^{113} + $ —— $\rightarrow {}_{48}Cd^{114}$. *Ans.* ${}_{0}n^{1}$.

20.20. Chlorine has two principal stable isotopes of atomic mass 34.965 and 36.964. The atomic weight of chlorine is 35.453. What is the per cent natural abundance of the above isotopes of chlorine, assuming other isotopes to be negligible in amount? *Ans.* 76%, 24%.

I. Natural Trigonometric Functions

angle	sine	tan	cot	cos	
0°	.0000	.0000	∞	1.0000	90°
1	.0175	.0175	57.29	.9999	89
2	.0349	.0349	28.64	.9994	88
3	.0523	.0524	19.08	.9986	87
4	.0698	.0699	14.30	.9976	86
5	.0872	.0875	11.43	.9962	85
6	.1045	.1051	9.514	.9945	84
7	.1219	.1228	8.144	.9926	83
8	.1392	.1405	7.115	.9903	82
9	.1564	.1584	6.314	.9877	81
10	.1737	.1763	5.671	.9848	80
11	.1908	.1944	5.145	.9816	79
12	.2079	.2126	4.705	.9782	78
13	.2250	.2309	4.332	.9744	77
14	.2419	.2493	4.011	.9703	76
15	.2588	.2680	3.732	.9659	75
16	.2756	.2868	3.487	.9613	74
17	.2924	.3057	3.271	.9563	73
18	.3090	.3249	3.078	.9511	72
19	.3256	.3443	2.904	.9455	71
20	.3420	.3640	2.748	.9397	70
21	.3584	.3839	2.605	.9336	69
22	.3746	.4040	2.475	.9272	68
23	.3907	.4245	2.356	.9205	67
24	.4067	.4452	2.246	.9136	66
25	.4226	.4663	2.145	.9063	65
26	.4384	.4877	2.050	.8988	64
27	.4540	.5095	1.963	.8910	63
28	.4695	.5317	1.881	.8830	62
29	.4848	.5543	1.804	.8746	61
30	.5000	.5774	1.732	.8660	60
31	.5150	.6009	1.664	.8572	59
32	.5299	.6249	1.600	.8481	58
33	.5446	.6494	1.540	.8387	57
34	.5592	.6745	1.483	.8290	56
35	.5736	.7002	1.428	.8192	55
36	.5878	.7265	1.376	.8090	54
37	.6018	.7536	1.327	.7986	53
38	.6157	.7813	1.280	.7880	52
39	.6293	.8098	1.235	.7772	51
40	.6428	.8391	1.192	.7660	50
41	.6561	.8693	1.150	.7547	49
42	.6691	.9004	1.111	.7431	48
43	.6820	.9325	1.072	.7314	47
44	.6947	.9657	1.036	.7193	46
45	.7071	1.0000	1.000	.7071	45
	cos	cot	tan	sine	angle

II. The Vapor Pressure of Water in mm. of Hg at Temperatures of from 0° C to 100° C

Temp. °C	Pressure	Temp. °C	Pressure	Temp. °C	Pressure
0	4.6	15	12.8	30	31.8
1	4.9	16	13.6	31	33.7
2	5.3	17	14.5	32	35.7
3	5.7	18	15.5	33	37.7
4	6.1	19	16.5	34	39.9
5	6.5	20	17.5	35	42.2
6	7.0	21	18.7	36	44.6
7	7.5	22	19.8	37	47.1
8	8.0	23	21.1	38	49.7
9	8.6	24	22.4	39	52.4
10	9.2	25	23.8	40	55.3
11	9.9	26	25.2	50	92.5
12	10.5	27	26.7	60	149.4
13	11.2	28	28.3	80	355.1
14	12.0	29	30.0	100	760.0

III. International Atomic Weights (Based on Carbon-12)

(Values in brackets represent the most stable known isotopes.)

Element	Symbol	Atomic Number	Atomic Weight
Actinium	Ac	89	[227]
Aluminum	Al	13	26.9815
Americium	Am	95	[243]
Antimony	Sb	51	121.75
Argon	Ar	18	39.948
Arsenic	As	33	74.9216
Astatine	At	85	[210]
Barium	Ba	56	137.34
Berkelium	Bk	97	[249]
Beryllium	Be	4	9.0122
Bismuth	Bi	83	208.980
Boron	B	5	10.811
Bromine	Br	35	79.909
Cadmium	Cd	48	112.40
Calcium	Ca	20	40.08
Californium	Cf	98	[251]
Carbon	C	6	12.01115
Cerium	Ce	58	140.12
Cesium	Cs	55	132.905
Chlorine	Cl	17	35.453
Chromium	Cr	24	51.996
Cobalt	Co	27	58.9332
Copper	Cu	29	63.54
Curium	Cm	96	[247]
Dysprosium	Dy	66	162.50
Einsteinium	Es	99	[254]
Erbium	Er	68	167.26
Europium	Eu	63	151.96
Fermium	Fm	100	[253]
Fluorine	F	9	18.9984
Francium	Fr	87	[223]
Gadolinium	Gd	64	157.25
Gallium	Ga	31	69.72
Germanium	Ge	32	72.59
Gold	Au	79	196.967
Hafnium	Hf	72	178.49
Helium	He	2	4.0026
Holmium	Ho	67	164.930
Hydrogen	H	1	1.00797
Indium	In	49	114.82
Iodine	I	53	126.9044
Iridium	Ir	77	192.2
Iron	Fe	26	55.847
Krypton	Kr	36	83.80
Lanthanum	La	57	138.91
Lawrencium	Lw	103	[257]
Lead	Pb	82	207.19
Lithium	Li	3	6.939
Lutetium	Lu	71	174.97
Magnesium	Mg	12	24.312
Manganese	Mn	25	54.9380
Mendelevium	Md	101	[256]
Mercury	Hg	80	200.59
Molybdenum	Mo	42	95.94
Neodymium	Nd	60	144.24
Neon	Ne	10	20.183
Neptunium	Np	93	[237]
Nickel	Ni	28	58.71
Niobium	Nb	41	92.906
Nitrogen	N	7	14.0067
Nobelium	No	102	[254]
Osmium	Os	76	190.2
Oxygen	O	8	15.9994
Palladium	Pd	46	106.4
Phosphorus	P	15	30.9738
Platinum	Pt	78	195.09
Plutonium	Pu	94	[242]
Polonium	Po	84	[210]
Potassium	K	19	39.102
Praseodymium	Pr	59	140.907
Promethium	Pm	61	[147]
Protactinium	Pa	91	[231]
Radium	Ra	88	[226]
Radon	Rn	86	[222]
Rhenium	Re	75	186.2
Rhodium	Rh	45	102.905
Rubidium	Rb	37	85.47
Ruthenium	Ru	44	101.07
Samarium	Sm	62	150.35
Scandium	Sc	21	44.956
Selenium	Se	34	78.96
Silicon	Si	14	28.086
Silver	Ag	47	107.870
Sodium	Na	11	22.9898
Strontium	Sr	38	87.62
Sulfur	S	16	32.064
Tantalum	Ta	73	180.948
Technetium	Tc	43	[99]
Tellurium	Te	52	127.60
Terbium	Tb	65	158.924
Thallium	Tl	81	204.37
Thorium	Th	90	232.038
Thulium	Tm	69	168.934
Tin	Sn	50	118.69
Titanium	Ti	22	47.90
Tungsten	W	74	183.85
Uranium	U	92	238.03
Vanadium	V	23	50.942
Xenon	Xe	54	131.30
Ytterbium	Yb	70	173.04
Yttrium	Y	39	88.905
Zinc	Zn	30	65.37
Zirconium	Zr	40	91.22

IV. Four-place Logarithm Table

N	L0	1	2	3	4	5	6	7	8	9	Proportional parts								
											1	2	3	4	5	6	7	8	9
10	0000	0043	0086	0128	0170	0212	0253	0294	0334	0374	4	8	12	17	21	25	29	33	37
11	0414	0453	0492	0531	0569	0607	0645	0682	0719	0755	4	8	11	15	19	23	26	20	34
12	0792	0828	0864	0899	0934	0969	1004	1038	1072	1106	3	7	10	14	17	21	24	28	31
13	1139	1173	1206	1239	1271	1303	1335	1367	1399	1430	3	6	10	13	16	19	23	26	29
14	1461	1492	1523	1553	1584	1614	1644	1673	1703	1732	3	6	9	12	15	18	21	24	27
15	1761	1790	1818	1847	1875	1903	1931	1959	1987	2014	3	6	8	11	14	17	20	22	25
16	2041	2068	2095	2122	2148	2175	2201	2227	2253	2279	3	5	8	11	13	16	18	21	24
17	2304	2330	2355	2380	2405	2430	2455	2480	2504	2529	2	5	7	10	12	15	17	20	22
18	2533	2577	2601	2625	2648	2672	2695	2718	2742	2765	2	5	7	9	12	14	16	19	21
19	2788	2810	2833	2856	2878	2900	2923	2945	2967	2989	2	4	7	9	11	13	16	18	20
20	3010	3032	3054	3075	3096	3118	3139	3160	3181	3201	2	4	6	8	11	13	15	17	19
21	3222	3243	3263	3284	3304	3324	3345	3365	3385	3404	2	4	6	8	10	12	14	16	18
22	3424	3444	3464	3483	3502	3522	3541	3560	3579	3598	2	4	6	8	10	12	14	15	17
23	3617	3636	3655	3674	3692	3711	3729	3747	3766	3784	2	4	6	7	9	11	13	15	17
24	3802	3820	3838	3856	3874	3892	3909	3927	3945	3962	2	4	5	7	9	11	12	14	16
25	3979	3997	4014	4031	4048	4065	4082	4099	4116	4133	2	3	5	7	9	10	12	14	15
26	4150	4166	4183	4200	4216	4232	4249	4265	4281	4298	2	3	5	7	8	10	11	13	15
27	4314	4330	4346	4362	4378	4393	4409	4425	4440	4456	2	3	5	6	8	9	11	13	14
28	4472	4487	4502	4518	4533	4548	4564	4579	4594	4609	2	3	5	6	8	9	11	12	14
29	4624	4639	4654	4669	4683	4698	4713	4728	4742	4757	1	3	4	6	7	9	10	12	13
30	4771	4786	4800	4814	4829	4843	4857	4871	4886	4900	1	3	4	6	7	9	10	11	13
31	4914	4928	4942	4955	4969	4983	4997	5011	5024	5038	1	3	4	6	7	8	10	11	12
32	5051	5065	5079	5092	5105	5119	5132	5145	5159	5172	1	3	4	5	7	8	9	11	12
33	5185	5198	5211	5224	5237	5250	5263	5276	5289	5302	1	3	4	5	6	8	9	10	12
34	5315	5328	5340	5353	5366	5378	5391	5403	5416	5428	1	3	4	5	6	8	9	10	11
35	5441	5453	5465	5478	5490	5502	5514	5527	5539	5551	1	2	4	5	6	7	9	10	11
36	5563	5575	5587	5599	5611	5623	5635	5647	5658	5670	1	2	4	5	6	7	8	10	11
37	5682	5694	5705	5717	5729	5740	5752	5763	5775	5786	1	2	3	5	6	7	8	9	10
38	5798	5809	5821	5832	5843	5855	5866	5877	5888	5899	1	2	3	5	6	7	8	9	10
39	5911	5922	5933	5944	5955	5966	5977	5988	5999	6010	1	2	3	4	5	7	8	9	10
40	6021	6031	6042	6053	6064	6075	6085	6096	6107	6117	1	2	3	4	5	6	8	9	10
41	6128	6138	6149	6160	6170	6180	6191	6201	6212	6222	1	2	3	4	5	6	7	8	9
42	6232	6243	6253	6263	6274	6284	6294	6304	6314	6325	1	2	3	4	5	6	7	8	9
43	6335	6345	6355	6365	6375	6385	6395	6405	6415	6425	1	2	3	4	5	6	7	8	9
44	6435	6444	6454	6464	6474	6484	6493	6503	6513	6522	1	2	3	4	5	6	7	8	9
45	6532	6542	6551	6561	6571	6580	6590	6599	6609	6618	1	2	3	4	5	6	7	8	9
46	6628	6637	6646	6656	6665	6675	6684	6693	6702	6712	1	2	3	4	5	6	7	7	8
47	6721	6730	6739	6749	6758	6767	6776	6785	6794	6803	1	2	3	4	5	5	6	7	8
48	6812	6821	6830	6839	6848	6857	6866	6875	6884	6893	1	2	3	4	4	5	6	7	8
49	6902	6911	6920	6928	6937	6946	6955	6964	6972	6981	1	2	3	4	4	5	6	7	8
50	6990	6998	7007	7016	7024	7033	7042	7050	7059	7067	1	2	3	3	4	5	6	7	8
51	7076	7084	7093	7101	7110	7118	7126	7135	7143	7152	1	2	3	3	4	5	6	7	8
52	7160	7168	7177	7185	7193	7202	7210	7218	7226	7235	1	2	3	3	4	5	6	7	7
53	7243	7251	7259	7267	7275	7284	7292	7300	7308	7316	1	2	2	3	4	5	6	6	7
54	7324	7332	7340	7348	7356	7364	7372	7380	7388	7396	1	2	2	3	4	5	6	6	7
N	L0	1	2	3	4	5	6	7	8	9	1	2	3	4	5	6	7	8	9

N	L0	1	2	3	4	5	6	7	8	9	Proportional parts								
											1	2	3	4	5	6	7	8	9
55	7404	7412	7419	7427	7435	7443	7451	7459	7466	7474	1	2	2	3	4	5	5	6	7
56	7482	7490	7497	7505	7513	7520	7528	7536	7543	7551	1	2	2	3	4	5	5	6	7
57	7559	7566	7574	7582	7589	7597	7604	7612	7619	7627	1	2	2	3	4	5	5	6	7
58	7634	7642	7649	7657	7664	7672	7679	7686	7694	7701	1	1	2	3	4	4	5	6	7
59	7709	7716	7723	7731	7738	7745	7752	7760	7767	7774	1	1	2	3	4	4	5	6	7
60	7782	7789	7796	7803	7810	7818	7825	7832	7839	7846	1	1	2	3	4	4	5	6	6
61	7853	7860	7868	7875	7882	7889	7896	7903	7910	7917	1	1	2	3	4	4	5	6	6
62	7924	7931	7938	7945	7952	7959	7966	7973	7980	7987	1	1	2	3	3	4	5	6	6
63	7993	8000	8007	8014	8021	8028	8035	8041	8048	8055	1	1	2	3	3	4	5	5	6
64	8062	8069	8075	8082	8089	8096	8102	8109	8116	8122	1	1	2	3	3	4	5	5	6
65	8129	8136	8142	8149	8156	8162	8169	8176	8182	8189	1	1	2	3	3	4	5	5	6
66	8195	8202	8209	8215	8222	8228	8235	8241	8248	8254	1	1	2	3	3	4	5	5	6
67	8261	8267	8274	8280	8287	8293	8299	8306	8312	8319	1	1	2	3	3	4	5	5	6
68	8325	8331	8338	8344	8351	8357	8363	8370	8376	8382	1	1	2	3	3	4	4	5	6
69	8388	8395	8401	8407	8414	8420	8426	8432	8439	8445	1	1	2	2	3	4	4	5	6
70	8451	8457	8463	8470	8476	8482	8488	8494	8500	8506	1	1	2	2	3	4	4	5	6
71	8513	8519	8525	8531	8537	8543	8549	8555	8561	8567	1	1	2	2	3	4	4	5	5
72	8573	8579	8585	8591	8597	8603	8609	8615	8621	8627	1	1	2	2	3	4	4	5	5
73	8633	8639	8645	8651	8657	8663	8669	8675	8681	8686	1	1	2	2	3	4	4	5	5
74	8692	8698	8704	8710	8716	8722	8727	8733	8739	8745	1	1	2	2	3	4	4	5	5
75	8751	8756	8762	8768	8774	8779	8785	8791	8797	8802	1	1	2	2	3	4	4	5	5
76	8808	8814	8820	8825	8831	8837	8842	8848	8854	8859	1	1	2	2	3	4	4	5	5
77	8865	8871	8876	8882	8887	8893	8899	8904	8910	8915	1	1	2	2	3	3	4	4	5
78	8921	8927	8932	8938	8943	8949	8954	8960	8965	8971	1	1	2	2	3	3	4	4	5
79	8976	8982	8987	8993	8998	9004	9009	9015	9020	9025	1	1	2	2	3	3	4	4	5
80	9031	9036	9042	9047	9053	9058	9063	9069	9074	9079	1	1	2	2	3	3	4	4	5
81	9085	9090	9096	9101	9106	9112	9117	9122	9128	9133	1	1	2	2	3	3	4	4	5
82	9138	9143	9149	9154	9159	9165	9170	9175	9180	9186	1	1	2	2	3	3	4	4	5
83	9191	9196	9201	9206	9212	9217	9222	9227	9232	9238	1	1	2	2	3	3	4	4	5
84	9243	9248	9253	9258	9263	9269	9274	9279	9284	9289	1	1	2	2	3	3	4	4	5
85	9294	9299	9304	9309	9315	9320	9325	9330	9335	9340	1	1	2	2	3	3	4	4	5
86	9345	9350	9355	9360	9365	9370	9375	9380	9385	9390	1	1	2	2	3	3	4	4	5
87	9395	9400	9405	9410	9415	9420	9425	9430	9435	9440	0	1	1	2	2	3	3	4	4
88	9445	9450	9455	9460	9465	9469	9474	9479	9484	9489	0	1	1	2	2	3	3	4	4
89	9494	9499	9504	9509	9513	9518	9523	9528	9533	9538	0	1	1	2	2	3	3	4	4
90	9542	9547	9552	9557	9562	9566	9571	9576	9581	9586	0	1	1	2	2	3	3	4	4
91	9590	9595	9600	9605	9609	9614	9619	9624	9628	9633	0	1	1	2	2	3	3	4	4
92	9638	9643	9647	9652	9657	9661	9666	9671	9675	9680	0	1	1	2	2	3	3	4	4
93	9685	9689	9694	9699	9703	9708	9713	9717	9722	9727	0	1	1	2	2	3	3	4	4
94	9731	9736	9741	9745	9750	9754	9759	9763	9768	9773	0	1	1	2	2	3	3	4	4
95	9777	9782	9786	9791	9795	9800	9805	9809	9814	9818	0	1	1	2	2	3	3	4	4
96	9823	9827	9832	9836	9841	9845	9850	9854	9859	9863	0	1	1	2	2	3	3	4	4
97	9868	9872	9877	9881	9886	9890	9894	9899	9903	9908	0	1	1	2	2	3	3	4	4
98	9912	9917	9921	9926	9930	9934	9939	9943	9948	9952	0	1	1	2	2	3	3	4	4
99	9956	9961	9965	9969	9974	9978	9983	9987	9991	9996	0	1	1	2	2	3	3	3	4
N	L0	1	2	3	4	5	6	7	8	9	1	2	3	4	5	6	7	8	9

Index